122 Advances in Polymer Science

Biopolymers II

Guest Editors: N. A. Peppas and R. S. Langer

With contributions by
K. S. Anseth, C. L. Bell, C. N. Bowman,
J. Klier, J. Kopeček, A. G. Mikos, S. M. Newman,
N. A. Peppas, D. Putman, B. Rangarajan,
A. B. Scranton, R. C. Thomson, M. C. Wake,
I. V. Yannas, M. J. Yaszemski

With 64 Figures and 10 Tables

 Springer

Guest Editors:

Prof. Nicholas A. Peppas
Purdue University, School of Chemical Engineering
West Lafayette, IN 47907-1283/USA

Prof. Robert S. Langer
Department of Chemical Engineering,
Institute of Technology,
25 Ames Street, Cambridge, MA 02139/USA

ISBN 978-3-662-14877-8   ISBN 978-3-540-49102-6 (eBook)
DOI 10.1007/978-3-540-49102-6

© Springer-Verlag Berlin Heidelberg 1995
Originally published by Springer-Verlag Berlin Heidelberg New York in 1995
Softcover reprint of the hardcover 1st edition 1995
Library of Congress Catalog Card Number 61-642

Typesetting: Macmillan India Ltd., Bangalore-25
SPIN: 10477712   02/3020 - 5 4 3 2 1 0 - Printed on acid-free paper

Editors

Preface

Significant developments have occurred in recent years in the fields of biopolymers and biomaterials. New synthetic materials have been synthesized and tested for a variety of biomedical and related applications from linings for artifical hearts to artifical pancreas devices and from intraocular lenses to drug delivery systems. Of particular interest in the future is the development of intelligent polymers or materials with special functional groups that can be used either for specialty medical applications or as templates or scaffolds for tissue regeneration.

In this second volume, following volume no. 107, we have collected review articles by leading authorities in the field of biomaterials who address the structure, properties and medical uses of a number of new polymers. All reviews have a strong emphasis on the polymer aspects of the biomedical development and offer ample evidence of how the structure can influence the medical behavior of these systems.

We hope that these reviews will become helpful, if not standard, references in the field and will contribute to our understanding of biopolymers.

December 1994

Nicholas A. Peppas
Robert S. Langer

Table of Contents

Table of Contents

Biomedical Applications of Polyelectrolytes

A. B. Scranton[1], B. Rangarajan[1], J. Klier[2]
[1] Michigan State University, Department of Chemical Engineering,
East Lansing, MI 48824, USA
[2] The Dow Chemical Company, Central Research, Advanced Polymeric
Systems Laboratory, Building #1712, Midland, MI 48640

Polyelectrolytes are used in a variety of biomedical systems, including dental adhesives and restorations; controlled release devices; polymeric drugs, prodrugs, and adjuvants; and biocompatible materials. This article provides a review of biomedical applications of polyelectrolytes with emphasis on recent developments. For completeness, an overview of the methods for polyelectrolyte synthesis is provided along with a description of the unique properties of polyelectrolyte solutions, gels, and complexes which make them useful in biomedical applications. The discussion of dental materials focuses on the recent developments in glass-ionomer cements and novel organic polyelectrolyte adhesives since these materials are replacing the traditional zinc carboxylates. The section on controlled release applications includes a brief overview of recent developments in the mature areas of coatings, matrices and binders; and provides more in-depth discussions of the advanced responsive, bioadhesive, and liposomal systems that have emerged in recent years. Finally, descriptions of the recent work in polyelectrolytes as biocompatible materials as well as drugs or prodrugs are provided.

---

[1] Authour to whom correspondence should be sent

1 Introduction

Polyelectrolytes are polymers which contain relatively high concentrations of ionizable groups along the backbone chain. Polyelectrolytes are distinguished from a related class of polymers, ionomers, by the density of the ionizable groups. Ionomers contain a relatively low concentration of ionizable groups (less than a few mole % of repeating units) while polyelectrolytes contain ionizable groups at levels ranging anywhere from a few mole % to 100 of the repeating units. Polyelectrolytes may be anionic, cationic, or amphophilic, and may be synthetic or naturally occurring. Examples of common polyelectrolytes include polymeric acids such as poly(acrylic acid) and poly(methacrylic acid), polymeric bases such as poly(vinyl amine) and poly(4-vinyl pyridine), and many naturally occurring proteins, polysaccharides, and nucleic acids.

The unique properties exhibited by polyelectrolytes have lead to their application in biomedical systems. Many biomedical applications of polyelectrolytes ultimately arise from their propensity to bind with oppositely charged surfaces and to associate to form complexes with oppositely charged polymers. For example, cationic polyelectrolytes have long been studied for their application in silicosis therapy and immunochemistry due to their ability to bind with negatively charged surfaces, and cationic polyelectyrolytes such as poly(vinyl pyridine N-oxide) are potent inhibitors of silica hemolysis of red blood cells. Similarly, complexes composed of two oppositely charged polyelectrolytes have been extensively employed as enteric coatings and controlled release devices, and many polyelectrolytes and their complexes have exhibited an antithrombogenic character.

This contribution will provide a review of polylectrolytes as biomaterials, with emphasis on recent developments. The first section will provide an overview of methods of synthesizing polyelectrolytes in the structures that are most commonly employed for biomedical applications: linear polymers, crosslinked networks, and polymer grafts. In the remaining sections, the salient features of polyelectrolyte thermodynamics and the applications of polyelectrolytes for dental adhesives and restoratives, controlled release devices, polymeric drugs, prodrugs, or adjuvants, and biocompatibilizers will be discussed. These topics have been reviewed in the past, therefore previous reviews are cited and only the recent developments are considered here.

2 Synthesis of Polyelectrolytes

Polyelectrolytes are synthesized by incorporating ionogenic functional groups into a polymer chain or by attaching oligomeric grafts which extend from the backbone chain. Moreover, the ionogenic groups may be introduced into the polymer chain at the time of synthesis by (co)polymerization of ionogenic monomers or may be added by chemical modification or functionalization of existing polymers. The first approach is generally more versatile and more commonly employed since the structure and properties of the final polyelectrolyte may be largely controlled through the composition of the reaction mixture or the reaction conditions. However, the method of choice for a particular application depends largely upon the type of ionogenic functional group to be added and the nature of the parent polymeric chain. For example, the post-functionalization approach is often employed for the synthesis of graft polyelectrolytes or for modification of natural polymers such as lignin. In this paper, an overview of the methods for synthesis of polyelectrolytes will be presented, including examples of the variety of synthetic schemes that have been developed. However, because the focus of this review is biomedical applications of polyelectrolytes the discussion presented here will be representative rather than exhaustive. Methods based upon polymerizations of ionogenic monomers will be considered first, followed by methods based upon chemical modification of existing polymers.

2.1 Polyelectrolyte Synthesis by Polymerization of Ionogenic Monomers

Polyelectrolytes are most commonly synthesized by free-radical chain polymerizations of ionogenic monomers containing a carbon double bond. The well-established mechanism for free radical polymerizations is described in many general polymer texts [1–3]. Polymerization begins when an initiating species produces active radical centers, typically by homolytic dissociation of a weak bond or by a redox reaction. Once formed, the active radical centers rapidly propagate through the carbon double bonds of many monomer units to form polymer chains. When two active radical centers meet, they may react with one another by combination or disproportionation to terminate the polymerization process. This free radical reaction scheme is very versatile and may be employed to produce polyelectrolytes in a variety of structures. Crosslinked polymeric structures may be readily produced by including small quantities (typically less than 1 mole %) of a divinyl crosslinking agent which may participate in the propagation of two radical chains and may therefore produce a crosslink between chains.

Many ionogenic monomers containing a polymerizable carbon double bond have been reported in the literature, and therefore a wide variety of anionic, cationic, and amphophilic polyelectrolytes may be synthesized using free radical polymerizations. Examples of anionic ionogenic monomers which have been used to synthesize anionic polyelectrolytes include acrylic acid [4–10], methacrylic acid [6–8, 11, 12], sodium styrenesulfonate [7, 13, 14], p-styrene carboxylic acid [7], vinyl sulfonic acid [7], vinyl phosphonic acid [15, 16], acrylamido *N*-glyconic acid [17], sulfoethyl methacrylate [13, 14], 3-acrylamido-3-methyl-butanoic acid [18, 19], and 2-acrylamido-2-methyl-propanesulfonic acid [18, 19]. These monomers react through the unsaturated carbon double bond producing a polymer chain containing ionizable pendent moieties. Chemical structures of some representative anionic polyelectrolytes are shown in Table 1. In most cases, the free radical polymerization of the ionogenic monomers may be carried out using either the acid form of the monomer or the sodium salt. One notable exception is sodium styrene sulfonate which may not be polymerized in its sulfonic acid form due to a spontaneous self-catalyzed cationic polymeriz-ation [14, 20].

Acrylic and methacrylic acid are by far the most common anionic ionogenic monomers due to their wealth of commercial applications [4–6, 9, 10, 21]. The utility of these monomers for biomedical applications arises in part from the ease with which they may be copolymerized with a host of comonomers [6, 22–33], including multifunctional crosslinking agents [4, 5, 9–11, 22] (to form polymer networks), and biologically active vinyl monomers [31, 32] such as *N*-2-methacryloyloxyethyl-5-fluorouracil. These copolymerizations have been extensively studied and may be readily performed using well-established procedures described in the references. In addition, these polymerizations may be carried out in bulk [12, 33], solution [4–6, 22–27], suspension [5, 6, 8, 34], or emulsion [25, 35], thereby allowing the geometry of the final product to range from a macroscopic polymeric device to a microparticle.

In some biomedical applications it is desired to synthesize grafts of a poly-electrolyte which extend from the surface of a substrate (typically a different polymeric material). For example, as discussed in more detail later in this review, polyelectrolyte grafts are often used to improve the biocompatibility of implants or other surfaces that are in direct contact with blood during service [36–39]. Free radical polymerizations may be used to graft polyelectrolytes onto a host of substrates using the radiation grafting method. In this technique, active radical centers are produced directly upon the substrate by irradiating its surface with an electron beam, ultraviolet light, or ionizing radiation. When the substrate is placed in contact with a solution containing the monomer, the active radicals on the surface initiate a polymerization reaction which results in the formation of a polymeric graft which is covalently attached to the surface. Due to its simplicity and versatility, the radiation grafting technique has been extensively studied and has been used to produce polyelectrolyte grafts onto a variety of substrates. For example, a number of investigators have studied radiation grafting technique for producing grafts of acrylic or methacrylic acid onto polyethylene [39–42], polyethyleneterephthalate, polyurethanes, silicone rub-bers, and Teflon [36, 37].

Free radical polymerizations may also be used to synthesize a host of cationic polyelectrolytes. Diallyl quaternary ammonium salts such as dimethyl-diallylammonium chloride, diallyldiethylammonium chloride, and diallylmethyl *b*-propionamido chloride are an interesting class of monomers which will

Table 1. Representative anionic polyelectrolytes

Name	Formula
poly(acrylic acid)	$\left[CH_2CH\right]_n$ with COOH
poly(methacrylic acid)	CH_3; $\left[CH_2C\right]_n$ with COOH
poly(acrylic acid-*co*-maleic acid)	$\left[CH_2CH-CH-CH\right]_n$ with COOH COOH COOH
poly(vinylsulfonic acid)	$\left[CH_2CH\right]_n$ with SO_3H
poly(p-styrenesulfonic acid)	$\left[CH_2CH\right]_n$ with phenyl ring and SO_3H
poly(p-styrene carboxylic acid)	$\left[CH_2-CH\right]_n$ with phenyl ring and COOH
poly(metaphosphoric acid)	$\left[O-P\right]_n$ with O and OH
poly(4-methacryloyloxyethyl trimellitate)	CH_3; $\left[CH_2C\right]_n$; $C=O$; OCH_2CH_2OCO — benzene ring with COOH, COOH

undergo free radical cyclopolymerization to produce cationic polyelectrolytes [7, 43–45]. Other monomers which may be readily polymerized to produce polyelectrolytes contain both a vinyl bond and a quaternary ammonium salt (or a primary, secondary or tertiary amine which may be protonated or quaternized). Examples of such monomers include vinylpyridine [7, 13, 14, 33, 46], dimethylaminoethyl methacrylate [7, 47], dimethylaminomethyl acrylamide [43], and vinylbenzene trimethylammonium chloride [13, 43]. For other examples see references 7, 43, 48–50. Finally, cationic polyelectrolytes may be produced by polymerizing vinyl monomers containing sulfonium or phosphonium groups [7]. Representative structures of some cationic polyelectrolytes are shown in Table 2.

Amphophilic polyelectrolytes are polymers which contain both anionic and cationic species. In general, the positive and negative charges may be located on the same pendent group, or on different pendent groups that are dispersed regularly or randomly on the backbone chain [51–54]. Several authors have reported methods of synthesizing amphophilic polyelectrolytes using free radical polymerizations. Salamone and collaborators studied vinyl imidazolium sulphobetaines [7, 55], (for example 1-vinyl-3-(3-sulphopropyl) imidazolium) which may be readily polymerized to yield amphophilic polyelectrolytes which contain both charges on the same pendent group (separated by two methylene units). A second class of monomers which has been extensively studied for the synthesis of amphophilic polymers are vinyl-containing aminimides [56–59]. These interesting monomers may be used to form amphophilic polymers in which the opposite charges are not only located on the same pendent group but are on adjacent nitrogen atoms. As described in a recent review [58], more than 40 distinct vinyl-containing aminimide monomers have been synthesized and polymerized. Finally, new monomers for synthesizing amphophilic polymers have been reported in recent years, including ammonium alkoxydicyano-ethenolates [60], and pyridium-carboxylates [61].

In comparison to free radical chain polymerizations, relatively few polyelectrolytes are generally synthesized using step-growth polymerization methods. However polyelectrolytes may be synthesized using step or condensation polymerizations, and some representative examples are described below. Poly-aminimides, whose synthesis by free radical polymerization was described above, may also be formed by step-growth polymerizations [58, 59]. In contrast to the radically polymerized analogs which contain the aminimide moiety in a pendent chain, the polymers prepared by step-growth reactions typically contain the aminimide functionality in the backbone chain. For example, bis-dimethylhydrazides and dihalides may undergo the Menschutkin reaction to produce polyelectrolytes with aminimide functionalities in the backbone polymer chain [59]. A second class of step-growth polyelectrolytes are the 'ionenes' produced by the polyalkylation reaction between α, ω-dibromoalkanes and N,N,N',N'-tetra-α, ω-alkanediamines [62–64]. Again, these polyelectrolytes contain the positively charged quaternary nitrogen atoms in the backbone polymer chain.

Table 2. Representative cationic or amphophilic polyelectrolytes

Name	Formula
poly(ethyleneimine)	$-\!\!\left[-CH_2CHN-\right]_n$ $\quad\quad\quad\quad H$
poly(vinylamine)	$-\!\!\left[-CH_2CH-\right]_n$ $\quad\quad\quad\quad NH_2$
poly(4-vinylpyridine)	$-\!\!\left[-CH_2CH-\right]_n$
poly(4-vinyl-N-dodecylpyridinium) chloride	$-\!\!\left[-CH_2CH-\right]_n$ $\quad\quad\quad\quad N^+$ $\quad\quad\quad\quad C_{12}H_{25}$
poly(vinyl imidazolium sulphobetaine)	$\left[-CH_2CH_2-\right]_n$ $\quad N$ $\quad+\!\!/\quad-H$ $\quad N$ $\quad (CH_3)_2$ $\quad SO_3^-$
poly(alkylmethyldialkylammonium bromide)	$\left[\quad\quad\right]_x$ $\quad N^+\quad Br^-$ $\quad CH_3\quad C_nH_{2n+1}$

2.2 Polyelectrolyte Synthesis by Polymer Modification

Polyelectrolytes may be synthesized by a variety of post-functionalization techniques in which ionogenic groups are introduced into the structure of an existing nonionic polymer. For an excellent review of polymer functionalization reactions the reader is referred to the recent book by Akelah and Moet [65]. In this paper, representative examples of the major polymer modification techniques for polyelectrolyte synthesis will be presented.

Sulfonation is a versatile technique for producing anionic polyelectrolytes containing sulfonic acid moieties [1, 14, 65, 66] since it can be applied to essentially any polymer that contains a carbon-hydrogen or a nitrogen-hydrogen bond [66]. Sulfonation may be carried out using concentrated sulfuric acid [67, 68], gaseous sulfur trioxide [66, 69–71], or aqueous complexes of sulfur trioxide [14] as the sulfonating agent. Polymers which are commonly sulfonated using this technique include polystyrene [1, 14, 66–68], polyethylene [66, 69–71], and polyurethanes [72, 73]. A related electrophilic substitution technique which is commonly used to functionalize polystyrene is chloromethylation followed by amination [1, 3, 50, 74]. The polymer is first reacted with a chloromethyl ether in order to attach chloromethyl groups to the aromatic ring. Quaternary ammonium salts are then formed by reaction of the chloromethyl groups with a tertiary amine to produce a cationic polyelectrolyte.

A number of investigators have reported polymer-modification techniques for synthesizing polyelectrolytes by addition or substitution reactions. These methods are based upon the identification of a polymer functional group which is susceptible to an addition or substitution reaction with an ionogenic reactant. For example, Bansleben and Jachimowicz [75] synthesized cationic polyelectrolytes by an addition reaction onto polymeric dienes. In their reaction scheme polyamines were first produced by the addition of primary or secondary amines to the backbone unsaturation of the diene polymer, then the pendent amine groups were quaternized to produce the cationic polyelectrolyte. The reaction of a secondary amine with a carbon double bond was also exploited by several other authors to produce amphophilic polyelectrolytes [76–78]. In this case, the addition of acrylic acid to a solution of polyiminoethylene or polyiminohexamethylene resulted in both a protonation and a Michael addition onto the polymer to produce the amphophilic polyelectrolyte. In other work, Dragan and collaborators [79, 80] synthesized the cationic polyelectrolyte poly(N,N-dialkylaminoalkylacylamide) by the reaction of the pendent nitrile groups of poly(acrylonitrile) with N,N-dialkylaminoalkylamines in the presence of water. Finally, Gieselman and Reynolds [81] used a two-step mechanism to derivatize poly(p-phenyleneterephthalamide) to form the anionic polyelectrolyte poly[(p-phenyleneterephthalamido)propanesulfonate].

A final technique commonly used to produce polyelectrolytes by functionalization of a nonionic polymer is the hydrolysis or oxidation of a labile bond or functionality. For example, the anhydride functionality is very easily hydrolyzed to yield carboxylic acid moieties on adjacent carbon atoms. Therefore, anionic polyelectrolytes containing pendent carboxylic acid moieties may be prepared by first copolymerizing anhydride-containing vinyl monomer such as maleic anhydride [1–3, 7] or 4-methacryloyloxyethyl trimellitate anhydride [82] with nonionic monomers, then hydrolyzing the resulting anhydride-containing polymers. An example of an oxidative scheme used to produce a polyelectrolyte was recently reported by Hassan and collaborators [83] who synthesized diketoalginic acid by the oxidation of sodium alginate.

3 Polyelectrolyte Properties

It is the unique properties exhibited by polyelectrolytes that have led to their use in a variety of biomedical applications. Therefore, any discussion of polyelectrolytes as biomaterials should provide some insight into the properties of polyelectrolyte systems. In this section, an overview of polyelectrolyte properties will be presented, including polyelectrolyte solutions, gels, and complexes. The purpose of this section is not to provide an exhaustive review of polyelectrolyte thermodynamics but to provide background information for the ensuing discussion of biomedical applications of polyelectrolytes.

The presence of the polymer-bound ionic moieties has a tremendous effect on the polyelectrolyte properties. For this reason, the solution properties of polyelectrolytes depend strongly on the degree of dissociation of the ionizable groups and may differ significantly from those of neutral polymers. The polymer-bound ionic groups of a polyelectrolyte may undergo electrostatic interactions with one another or with mobile charges in the solution. Therefore, solution properties such as viscosity and light scattering exhibit a strong dependence on ionic strength and polyelectrolyte concentration. For example, in salt-free aqueous solutions, polyelectrolytes generally exhibit expanded coil sizes and correspondingly higher viscosities than nonionic polymers due to the effects of the mutual electrostatic repulsions among the polymer-bound charges. The addition of a salt to the solution results in a decrease in the size of the polymer coil due to screening effects. In fact, at high ionic strengths, the viscosity of polyelectrolyte solutions will approach that of nonionic polymers.

Properties such as solution viscosity and angular dependence of light scattering depend upon the size of the polymer coil [3]. Therefore, the reduced viscosity (specific viscosity divided by the polymer concentration) of polyelectrolytes can be lowered dramatically by increasing the electrolyte content of the solvent. Similarly, aqueous solutions of a weak acid polyelectrolyte such as poly(acrylic acid) will exhibit a relatively low viscosity when the polymer is in its neutral form (at low pH) and a much higher viscosity in its ionized form (at intermediate pH values). As illustrated in Fig. 1, in contrast to nonionic polymers, the reduced viscosity of a polyelectrolyte may exhibit a sharp increase as the polymer concentration is decreased [3, 7]. This effect can be eliminated by the addition of a simple electrolyte salt. It is interesting to note that the amphophilic polyelectrolytes may exhibit "antipolyelectrolyte" behavior in aqueous solutions [55, 60]. In these systems, the addition of a salt may lead to increased chain expansion due to screening of the associative interaction between oppositely-charged moieties within the same chain.

The relatively high charge density along the backbone chain of a polyelectrolyte leads to some interesting properties. Titration data show that counterions (including protons in polyacids) bind to polyelectrolytes more readily than they bind to monomeric electrolytes. This effect arises from the additivity of electrostatic interactions between polymer-fixed ions and counterions. At high

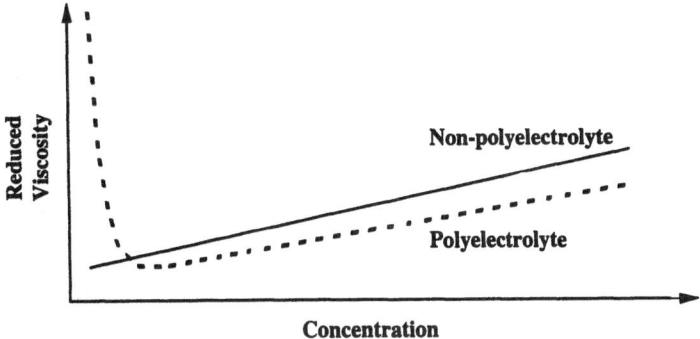

Fig. 1. Representative plots of the reduced viscosity vs concentration for polyelectrolytes and non-polyelectrolytes

degrees of ionization, even normally free counterions such as sodium or potassium will undergo nonspecific "atmospheric" binding to polyelectrolytes. Consequently, the activity coefficients of polyelectrolyte counterions are always less than unity, even at high dilutions where activity coefficients of simple salts approach unity. As expected, addition of simple electrolytes into a polyelectrolyte-containing system reduces this effect due to screening of the polymer-bound charged groups.

Significantly, counterion activity coefficients generally are not strongly affected by polyelectrolyte concentration. In contrast, simple salts exhibit a monotonic increase in activity upon dilution. This effect may be attributed to the fact that the charge density in the vicinity of the polyelectrolyte coil is insensitive to dilution, while dilution of a simple salt solution results in uniform separation of all of the charges and hence weaker ionic interactions.

In summary, polyelectrolyte solutions exhibit lower counterion activity coefficients, lower osmotic coefficients, higher extents of counterion binding and lower conductivity than corresponding solutions of low molecular weight electrolytes. Moreover, the charged nature of polyelectrolytes gives rise to properties distinct from those of nonionic polymers. They generally exhibit higher water solubility, expanded hydrated dimensions and a higher sensitivity to ionic strength and pH than corresponding nonionic polymers. Accordingly, thermodynamic descriptions of ionized polyelectrolytes differ significantly from descriptions of neutral polymers. The colligative, configurational and dynamic properties of polyelectrolytes in high dielectric constant media can be semiquantitatively described using one of several thermodynamic models outlined below.

Several models of polyelectrolyte thermodynamics have been proposed and have been recently reviewed [84]. Historically, two general approaches [84, 85] have been used to model polyelectrolyte thermodynamics, spherical and chain models. In the former approach the coiled polyions are treated as spherical domains with charge density distributed continuously within the sphere

[86–89]. However, these studies predicted larger polyion sizes and smaller electrostatic potentials than experimentally measured [85]. These discrepancies primarily arise from the assumption of smeared charge distributions in the polyion domain, whereas the potential around the polyion actually varies strongly with distance from the polyion.

In the 1960s, more realistic models were developed by incorporating the chain-like nature of the polymers. In this approach, rod-like polyelectrolytes are typically spaced in a parallel array with charges smeared uniformly on the rod surface, and the Poisson-Boltzmann equation is solved for the distribution of counterions [85, 90–92]. These models overestimate the counterion activity and chain expansion due to their failure to account for the reinforcement of electrostatic fields by coiled chains, nonuniform local dielectric constants, and site binding of counterions. The site-binding explanation, in particular, has considerable experimental support [85]. To explain the low experimental activity coefficients, several investigators postulated that some of the polyelectrolyte counterions bind reversibly to the polyion and therefore do not contribute to the free-ion concentration [93–95]. The polyelectrolyte was treated as an infinite linear array of discrete charges and an electrostatic free energy was formulated in terms of partition functions [85]. In these models, an adjustable parameter (lambda) proportional to the linear charge density of the hypothetical cylinder was used. Acceptable colligative properties were obtained only if a variable lambda was employed.

Most recent models also include the ideas of counterion condensation [89, 96, 97]. The specific assumptions in the counterion condensation theory [98, 99] are: i) the polyelectrolyte chain is replaced by an infinite line charge; ii) inter-polymer interactions are neglected, regardless of the polymer or salt concentrations, or in other words that all electrostatic interactions are intramolecular or between polymer and its counterions; iii) dielectric constant is that of the pure bulk solvent. It was shown that in the region near the polyion where there is no counterion screening, the Boltzmann factor of the counterions diverges above a critical degree of ionization. Therefore, above this degree of ionization, Manning's theory suggests that many counterions will condense on the polyion to lower the charge density parameter. More precise analyses based on a full solution of the Poisson-Boltzmann equation have been suggested by Stigter [100] and Russel [101]. In a recent model for the high electrolyte concentration case, a continuous string of charged beads surrounded by a salt solution shell was considered. Within the shell, the nonlinear Poisson-Boltzmann equation was solved, and outside the shell it was linearized. This approach gave limited agreement of activity coefficients with experimental values only at moderate Debye lengths [102].

The analysis described above is useful for modelling colligative properties but does not address polyelectrolyte conformations. Polyelectrolyte conformations in dilute solution have been calculated using the worm-like chain model [103, 104]. Here, the polymer conformation is characterized by a persistence length (a measure of the local chain stiffness) [96]. One consequence of the ionization of a polymer chain is an increase in the persistence length of the polymer. Two contributions to the persistence length are generally identified: the bare persistence length (the charge-independent contribution due to monomer structure and non-electrostatic interactions), and the electrostatic persistence length. Approximations to the electrostatic contributions were suggested by Odijk [105, 106] using linear counterion condensation theory and Fixman and Skolnick [107] under the assumption of large ratios of persistence length to Debye radius. The electrostatic persistence length can also be calculated using the nonlinear theory [101, 108]. At low ionic strengths, however, the electrostatic persistence lengths calculated by the linear theory are about twice as large as those calculated with the non-linear theory [101, 108, 109]. Furthermore, in these systems the excluded volume is accurately estimated only if the contour length is much larger than the persistence length. Also, for very small excluded-volumes (such as high electrolyte concentration) or very large persistence lengths (very stiff polymer chains), the excluded-volume effect is known only approximately.

Analysis of polyelectrolytes in the semi-dilute regime is even more complicated as a result of inter-molecular interactions. It has been established, via dynamic light-scattering and time-dependent electric birefringence measurements, that the behavior of polyelectrolytes is qualitatively different in dilute and semi-dilute regimes. The qualitative behavior of osmotic pressure has been described by a power-law relationship, but no theory approaching quantitative description is available.

The thermodynamics of weakly charged polyelectrolyte solutions have been the subjects of several recent investigations [110–112]. In these systems, the properties are intermediate between nonionic polymers and fully charged polyelectrolytes and both electrostatic and nonionic interactions can be important in determining the polymer characteristics.

In polyelectrolyte networks, ion repulsion and counterion osmotic effects promote swelling while network crosslinking restricts swelling. As in the case of linear polyelectrolytes, networks of polyelectrolytes generally swell more than their nonionic polymer counterparts. The swelling can be reduced by addition of salts or by inhibition of dissociation. The above-mentioned type of model [93] (or the Debye-Huckel type of model) has also been employed with qualitative success to the description of swelling of polyelectrolyte gels by Hasa and coworkers [113, 114], more recently by Konak and Bansil [115] and most recently by Yin and Prud'homme [116]. In the later model, counterion condensation, electrostatic persistence length, and osmotic effects are rigorously incorporated.

Polyelectrolyte complexes are formed by the ionic association of two oppositely charged polyelectrolytes [60, 117–119]. Due to the long-chain structure of the polymers, once one pair of repeating units has formed an ionic bond, many other units may associate without a significant loss of translational degree of freedom. Therefore the complexation process is cooperative, enhancing the stability of the polymeric complex. The formation of polyelectrolyte complexes can have a strong effect on polymer solubility, rheology, turbidity and conductivity of the polymer solution. In addition, complex formation can also dramatically alter the electrical conductivity, mechanical properties, and permeability of the system. The stability of the polyelectrolyte-complex is dependent on a variety of factors such as charge density, ionic strength, the nature and type of solvent, pH and temperature. Polyelectrolytes of high charge density may form relatively insoluble complexes. Several reviews of the properties and potential applications of polyelectrolyte complexes are available [117–119].

4 Polyelectrolyte-Based Dental Materials

Polymers used as dental materials must meet several stringent requirements. Dental restorative materials must be nontoxic, have aesthetic appearance, and good adhesive and mechanical properties. In addition, these materials must exhibit long term stability in the presence of water, enzymes, and various oral fluids, and withstand thermal and load cycles. Finally, a desirable dental restorative material should be convenient to work with at the time of application.

The properties exhibited by polyelectrolytes make them nearly-ideal candidates for dental material formulations. Dental polyelectrolytes are generally considered to be nontoxic and are able to adsorb chemically to the hydrophilic surface of tooth material through ionic interactions. Ionic cross-linking of the polyelectrolyte with multivalent cations (Zn^{2+}, Mg^{2+}, Al^{3+}, Ca^{2+}) results in the formation of a rigid and insoluble cement matrix. The stability and strength of the cement is attributed to the fact that, if a bond is broken, it can be reformed as long as the other bonds are maintained. Even today, polyelectrolytes are the only materials which are known with certainty to form a bond, which is stable with time, to tooth material [120]. In addition to long-term stability, many polyelectrolytes are translucent and possess cariostatic properties [121].

Polyelectrolyte-based dental cements or restorative materials include zinc polycarboxylates, glass ionomers, a variety of organic polyelectrolyte adhesives; as well as alginate-based impression materials. Dental cements are primarily used as luting (cementing) agents for restorations or orthodontic bands, as thermal insulators under metallic restorations, and as sealants for root canals, pits and fissures. They are also sometimes used as temporary or permanent (anterior) restorations. For further introduction to dental materials the reader is referred to standard texts [122, 123].

Zinc polycarboxylate, the first polyelectrolyte dental material, was developed and used as early as 1968 [124]. These materials are formed by the reaction of a zinc oxide powder with an aqueous solution of poly(acrylic acid). The zinc ions cross-link the polyacid chains and form a cement. A few years after the development of zinc polycarboxylate cements, Wilson and Kent introduced the first glass-ionomer cement (GIC) [125]. Glass-ionomer cements are formed by the reaction of an aqueous solution of poly(acrylic acid) with a glass powder. The acid degrades the glass causing the release of calcium and aluminum ions, which cross-links the polyacid chains. The cement sets around the unreacted glass particles to form a reaction-bonded composite. A perspective on the historical development and the expected future of GICs was recently published by Wilson [120]. In addition several other reviews of polyelectrolyte dental cements are available [126–128].

A variety of organic adhesives which are capable of forming strong bonds between a polymeric (acrylate) restoration and the hydrophilic tooth material have recently been developed. A number of these monomers, which possess a pendent ionizable group, are polymerized in the mouth to form an adhesive layer. Alginates, which are used as impression materials, are formed by the reaction of the sodium salt of anhydro-*beta-d*-mannuronic acid with calcium sulfate. Calcium ions crosslink the linear polymer to form a gel. This reaction is carried out inside the mouth, and the gel formed retains the shape of the oral cavity.

The early polyelectrolyte dental restorative materials were based on a chemical curing scheme. The reaction started when the powder and liquid were mixed before application. During a limited working time (while the material hardened) the cement was shaped and placed in the mouth. Setting was completed in the oral cavity, a hostile environment in which the material is subject to the influence of water and other oral fluids. Limited working times and prolonged setting times were characteristic of these early materials. (The terms working time and setting time are somewhat ambiguous in the literature; however working time essentially refers to the time (immediately after mixing) in which the material is plastic enough that it can be adapted to the margins of the cavity preparation, while the setting occurs primarily in the mouth and is said to be complete when the material reaches a threshold hardness.) Recently developed light-cured GICs are able to overcome these disadvantages, since they provide extended working time and rapid sets. This review will focus on GICs and on the new organic polyelectrolyte adhesives since both classes of materials are gaining popularity. Both the traditional and modern light-cured polyelectrolyte dental materials will be discussed with emphasis on research done over the past ten years. Wherever possible the reader will be referred to comprehensive reviews available in the literature.

4.1 Glass-Ionomer Cements – Compositions and Reactions

Glass-ionomer cements (GICs) are widely used as restorations for anterior teeth, as filling material for cavities and as luting cements and adhesive-liners for the attachment of metal or synthetic restorations. Glass-ionomer cements exhibit a minimal exotherm on setting, offer good appearance since they are translucent, and are cariostatic. The flexural strength and compressive strength of these cements actually increase with time [121, 129, 130]. However these materials have some disadvantages which have limited their application. For example, GICs are water sensitive in the early stages of the setting process and once set they are brittle. Even after setting, their mechanical properties limit their use to low-stress bearing anterior restorations.

Glass-ionomer cements are formed by the reaction of an approximately 50 wt% aqueous solution of a poly-acid with a finely-divided (20–50 μm) glass powder. The glass is degraded by the acid, and the Ca^{2+} and Al^{3+} ions which are released ionically cross-link the polymer chains and result in the formation of a hard cement [128–134]. The poly-acid most commonly used is poly(acrylic acid) or a copolymer of acrylic and itaconic or maleic acid. The molecular weight of the polymer is typically of the order of 25000–50000. The effects of molecular weight on the properties of GICs have been studied by Wilson and others [135–137]. Small amounts of (+)-tartaric acid are also incorporated into the acid solution and serve as a reaction control agent. Various experimental techniques such as pH and conductivity [138], infra-red spectroscopy [132, 134, 139], C-13 NMR spectroscopy [140], and electron microprobe analysis [141] have been used to study the setting of GICs. Such studies have led to the current ideas of GIC setting and are described by Wilson and McLean in their book on glass-ionomer cements [133].

Initial mixing of the polyacid and the glass powder causes the acid to attack the glass powder sequentially liberating Ca^{2+} ions followed by Al^{3+} ions. Leaching of the cations from the glass leaves behind an ion-depleted layer of silica gel around unreacted glass particles. The material sets when sufficient neutralization has taken place – usually within 5–7 min (with working times of 1–2 min and setting times of ~ 5 min [121]) after mixing and continues to harden slowly as further neutralization takes place. It is claimed that the type of mixing (manual vs mechanical) has an effect on the performance of the cement, although the evidence is conflicting [121, 129, 142].

Although glass-ionomer dental cements harden primarily due to the ionic crosslinking mechanism described above, details of the reaction mechanism are still under investigation. Recent studies by Wasson and Nicholson [143–145] suggest that, instead of sequential leaching of Ca^{2+} and Al^{3+} ions, complete dissolution of a fraction of the glass occurs, and that a silicate network might be responsible for part of the cementation. This view is based on the fact that ionomer glasses and acetic acid form cements after a period of a week and that the strength of these cements increased over a period of six months. Since there is no poly-acid linkage, the strength of the cement formed in their study is attributed to the silicate network. However, since the time scale for the formation of a silicate network is orders of magnitude larger than that required for the formation of the poly-salt network, the major portion of the strength is due to the conventional poly-salt matrix.

During the early stages of the setting reaction, both water uptake and water loss can occur. The matrix is sensitive to water attack and can degrade. It is therefore desirable that the material set as soon as possible once placed in the mouth. Light-cured varnishes are often used to seal newly placed restorations and prevent or at least minimize water loss and water uptake. However, these varnishes are not completely waterproof [146, 147].

Setting may be accelerated by the addition of reaction control agents or by photoinitiation. Small amounts (< 5 %) of (+)-tartaric acid have been shown to alter the setting behavior of GICs, resulting in prolonged working times and sharper sets. Chemical and spectroscopic (C-13 NMR) studies [140, 148] have shown that ions leached from the glass react preferentially with (+)-tartaric acid to form tartrate and hence delay the formation of the polysalt. A similar observation is made with the (−)-enantiomorph [149] but not with the *meso* form. FTIR studies [139] showed that (+)-tartaric acid not only reacted more readily with the glass than the with poly(acrylic acid), thereby delaying setting, but also enhanced the rate of formation of aluminum polyacrylate within the cement. This enhanced rate of matrix formation results in a sharper set. The *meso* form of tartaric acid also rapidly formed calcium tartarate but did not result in an enhancement in the formation of aluminum polyacrylate. Therefore the *meso* form resulted in a delayed set rather than a sharp set. The effect of tartaric acid and other additives such as NaF and NaCl [150], as well as a variety of hydroxyacids [151] on the rheology of GICs has been studied. At low concentrations, tartaric acid accelerated the setting process, but at high concentrations it retarded the setting process by complex formation. The exact role of tartaric acid is not yet fully understood.

Setting characteristics of GICs may be manipulated by changing the reactivity of the glass powder, the powder to liquid ratio, the amount and type of additive, or the reaction temperature [152, 153]. However, the last cannot be increased much due to pulpal damage. The reactivity of the glass powder may be affected by a host of variables, including the glass composition, heat-treatment, and acid-washing (to remove reactive cations). These issues are discussed by Wilson and McLean [153]. Heat treatment of the glass can affect the reactivity and even change the phase of the glass. The effect of heat treatment on the performance of silicate cements [131, 133, 154–156] and non-silicate cements [157, 158] has been investigated. The powder to liquid ratio also has a profound effect on the kinetics of cement formation as well as on the properties of the cement [121, 138, 159]. Reducing the powder to liquid ratio reduces the surface area available to acid attack, and hence reduces the kinetics of the reaction.

The composition of the glass phase may influence the properties of the glass ionomer cement. The most common type of glass system used is a fluoroaluminosilicate glass, which typically contains about 25–35 wt% SiO_2, 14–20 wt% Al_2O_3, 13–35 wt% CaF_2, 4–6 wt% AlF_3, 10–25 wt% $AlPO_4$, 5–20 wt% Na_3AlF_6 [131, 154]. In these systems CaF_2 or $AlPO_4$ is found to enhance the strength of cements [154]. The fluoride present in these glasses plays an important role in the cariostatic properties of the dental cement, changes the refractive index of the glass, and makes the cement translucent. Fluorine also causes amorphous phase separation of the glass to give very reactive droplets and a less reactive matrix (filler), leading to the formation of a reaction-bonded composite. Fluorine is also responsible for disrupting the glass network and increasing the susceptibility of the glass to acid degradation [131]. Changing the composition of the glass phase not only affects the strength but also affects the reactivity, working times, setting times and optical clarity [131]. Compounds such as strontium, barium, lanthanum, zinc oxide or zirconium oxide are also added to impart radio-opacity to commercial glass-ionomer cements [128]. Recently there has been some research into non-silicate glasses. Neve et al. [160, 161] have described the preparation and properties of aluminoborate glasses. A variety of ternary systems comprising of Al_2O_3, B_2O_3, and an oxide of Mg, Ca, Sr, Zn and Ba have been studied by Wilson and Combe [157], who found that reactivity decreased as the alumina content of the glass was increased.

Although poly(acrylic acid) and copolymers of acrylic and itaconic acid, or acrylic and maleic acid are the most commonly used poly-acids for glass ionomer cements, other polymers have been proposed. Poly(sulfonic acid) cannot be used since it is hydrolytically unstable [162]. Polyphosphonates and phosphates have been shown to enhance adhesion to hydroxyapatite and tooth structure [163, 164]. Aqueous solutions of poly(vinyl-phosphonic acid) have been reacted with metal oxides and aluminosilicate glasses to form dental cements [15, 165, 166]. Poly(phosphonic acids) are stronger acids than poly(carboxylic acids), and therefore setting times are much lower than those for poly(acrylic acid) systems [15]. Working and setting times for these systems are still too short for practical use. Tartaric acid cannot be used as a reaction control agent since it is a weaker acid than phosphonic acid. However for this system, a diphosphonic acid (amino *tris* (methylene phosphonic acid)) was able to extend working and setting time. Other polymers that have been investigated include poly(hydroxyethylmethacrylate (HEMA) phosphate) [167]. de Groot et al. [167] found that the addition of 5 wt % of poly(HEMA phosphate) to the polycarboxylate liquid of a zinc cement increased adhesion, compressive strength and tensile strength of the cement.

A major development in glass-ionomer cements is the advent of the light-cured glass-ionomer cement [120, 128, 168]. These cements offer the advantages of increased working flexibility, and result in the formation of bonds which have high strength and are stable not only with time, but also with temperature cycles [169, 170]. These cement formulations consist of a concentrated solution of the polyelectrolyte in a HEMA/water mixture along with trace amounts of a photo-initiator capable of polymerizing HEMA. When this liquid is mixed with the glass powder the rate of the ionic-reaction is retarded due to the reduced water content, thereby providing extended working times. When placed in the mouth and illuminated by initiating light, the cement rapidly sets as the HEMA is polymerized. In addition to extended working times and rapid or "snap" sets, light-cured GICs are also resistant to early water contamination and do not require a protective varnish. Modifications to this formulation include the addition of hydroxyacrylates, bisphenol A glycidyl methacrylate (BisGMA), and a small amount of modified poly(acrylic acid) which is capable of forming cross-links and copolymerizing with HEMA. As expected, the bond strength of light-cured GICs increases with greater irradiation time [171] and correspond-ingly higher cure.

Attempts have been made to increase the strength of GICs by reinforcing them with metal fibers or flakes. Silver and gold powder have been mixed or fused with the glass powder part of the cement. Such approaches have resulted in improved wear resistance. The properties and clinical use of metal-reinforced glass-ionomer cements have recently been reviewed by Wasson (1993) and hence will not be repeated here [172].

4.2 Adhesion

One of the main reasons polyelectrolytes are used is their ability to adhere to tooth material. Adhesion of the cement is presumed to be due to chemical (ionic) bonding of the carboxylic acid group with the calcium cation in the tooth by ion-bridging or chelation [173, 174]. Potentiometric titration of aqueous solutions of polyacrylic acids and polyelectrolyte dental cements have shown that these materials behave as weak acids without Ca^{2+} and as strong acids in the presence of the counterion. However Misra [175] suggests that adsorption of compounds with OH or COOH groups onto hydroxyapatite is mediated by hydrogen bonding rather than ionic bonding and that the H-bonding is affected by interactions between the OH/COOH group and surface associated water molecules. The true nature of this bond has not yet been completely or satisfactorily explained.

Several attempts have been made to increase the strength of the bone-cement or tooth-cement interface. Loosening at the interface results in not only reduced mechanical performance but also provides sites for bacterial growth and infections. Several approaches have been proposed to solve the problem of lack of bonding at the interface and are summarized by Bowen [176]. Such techniques include: use of an adhesive layer between the tooth and resin, micromechanical locking of the adhesive with fibers in dentin, copolymerization of the acrylic resin to modified apatite, and even creating a flexible interface between the resin and dentin. The two major types of adhesives are the organic polyelectrolyte adhesives and glass-ionomers.

4.2.1 Organic-Polyelectrolyte Adhesives

Acrylates such as poly(methyl methacrylate) (PMMA), bisphenol A glycidyl methacrylate (BisGMA), tri(ethyleneglycol) dimethacrylate (TEGDMA), etc. are used extensively as dental restorative materials and are frequently referred to as dental composites. Due to their hydrophobicity, these polymers do not adhere well to the hydrophilic tooth material. Monomers containing both hydrophilic and hydrophobic groups, which are capable of bonding with the tooth as well as reacting with the acrylate [177], are used as adhesives. This results in better attachment of the restoration, as well as reduction in marginal leakage. For example, methacrylate-based monomers, developed by Nakabayashi et al. [178, 179], have been reported to improve the adhesive strength of resins to tooth structure by promoting interpenetration and entanglement into dentinal surfaces (micromechanical locking). These monomers include: 4-(methoxyethyl) trimellitic anhydride or 4-META [180], 2-(methacryloxy) ethyl phenyl hydrogen phosphate or phenyl-P [181] and phenyl-P derivatives [182]. The adhesive formulations typically include approximately 5 wt % of 4-META or phenyl-P in MMA, with partially oxidized tri-n-butylborane (TBBO) as a chemical initiator. Bonding agents containing 4-META have been successfully used to bond denture base resins to plastic denture teeth [183]. Cements based on 4-META are adhesive to both bone and metal [82]. Hydroxyapatite, a bone-compatible filler, has been incorporated into a 4-META/MMA cement [184]. These adhesive resins also have the ability to bond to treated precious-metal alloys [185–188] and possibly untreated precious-metal alloys [185, 186, 188–190]. A status report on these new adhesive resins was recently presented by Swift [191].

Prior to the use of the adhesives, the dentin surface is etched with a conditioning solution which is usually an aqueous acidic solution of citric acid-ferric chloride, phosphoric acid, or even polyacrylic acid [192]. These solutions tend to demineralize the dentin and expose the collagen (organic fibers in the dentin) [193], and thereby result in higher bond strengths. The effects of various pre-conditioning treatments on the bond strength of 4-META to dentin have been discussed [194].

The adhesive strength of phenyl-P resins, ranges from 3-13 MPa [195, 196], depending on the type of surface treatment. Phenyl-P improved bond strength by promoting monomer diffusion and impregnation into demineralized dentin. Solubility parameters and fractional polarities of the adhesive, as well as the type of preconditioning, are all factors which affect the bond strength [197, 198].

The dentin-adhesive interface has been studied using a Raman microprobe technique [199], which shows the formation of resin-reinforced dentin and the penetration of resin into dentin substrate to a depth of 5–6 microns. Further study of the interface showed that only small molecules such as MMA, 4-MET (hydrolyzed 4-META) or oligomers infiltrated the dentin, and that all of the resin in the dentin originated from the monomer solution [200]. SEM and TEM studies of the ultrastructure of the resin-dentin interdiffusion zone showed a 2 micron zone with closely packed collagen fibrils running parallel to the interface [201].

The nature and type of initiation scheme plays an important role in the performance of the adhesive [194, 202–204]. Stresses due to polymerization shrinkage lead to the creation of a gap between the adhesive and tooth material. In the case of bulk chemical initiation, shrinkage stresses tend to create gaps at all interfaces, drawing material inward isotropically. With a photoinitiation scheme, polymerization begins at the free surface and pulls the material away from the dentin towards the free surface [194]. Thus the gap is created at the dentin adhesive interface. With interfacial initiation, polymerization tends to occur at the dentin interface first and actually draws in material towards the dentin. Such a system could have significant advantages since the gap created would be at the free surface. The concept of interfacial initiation in 4-META/ MMA adhesive bonding systems has been investigated [194, 202–204]. Imai et al. [194] used ferric ions or t-butyl peroxymaleic acid adsorbed onto dentin as an initiator and found that bond strengths significantly improved with the use of interfacial initiation. In a recent study the effect of the cure mode (chemical cure, light cure, or dual cure) was investigated [205], but no significant differences were observed in the performance of the various adhesives.

Another type of polyelectrolyte system that has been studied is the addition product of pyromellitic dianhydride (PMDA) and HEMA, which is referred to as PMDM. The adhesion of these compounds to conditioned dentin is enhanced by small amounts of N-containing surface active compounds [204, 206–208]. Monomer (PMGDM) synthesized from PMDM and glycerol dimethacrylate has been used as a dental adhesive [209]. PMGDM could be applied in thicker layers (13–24 microns) than PMDM (6 microns). It was noticed that PMDM broke preferentially at the adhesive/dentin layer, while the layers of PMGDM (~ 20 microns thick) always broke cohesively. In addition PMGDM was reported to wet the surface better and penetrate the dentinal tubules completely, and therefore yield higher strengths. SEM studies have shown that PMDM does not completely fill the dentinal tubules and that cohesive-fracture occurs in the tubular plugs [210].

Other types of dentin bonding agents which have been investigated include phosphate ester containing monomers [211], N-methacryloyl-5-aminosalicylic acid (MASA) [212], aqueous mixtures of o-methacryloyl-tyrosinamide (MTYA), glutaraldehyde and HEMA [213]. Solutions of MTYA or MASA diluted in HEMA promoted adhesion to dentin [214]. The studies discussed above have demonstrated the importance of the ability of the monomer to wet the dentin surface to attain good adhesion. The effect of conditioning dentin on the efficacy of the bonding has been reported by a number of researchers [215–217].

4.2.2 Glass-Ionomer Adhesives

Conventional glass-ionomer cements as well as recently developed light-cured GICs have also been studied as adhesives [133, 169, 170, 218]. GICs adsorb to the hydrophilic dentin surface and can develop good seals. While the mechanism of bonding of GIC with enamel is generally accepted to be due to ionic and polar forces [133], more complex interactions apparently occur with dentin. Recently developed light-cured GICs showed higher bond strengths, compressive strengths and diametral tensile strengths than conventional GICs [170]. Bond strengths were stable with time (over a period of 7 months) as well as with thermal cycling. Lin et al. [169] have studied bonding of GIC to dentin by a wide variety of techniques such as SEM, confocal microscopy, X-ray photoelectron spectroscopy, secondary-ion mass spectroscopy and bond strengths. Bond strengths of the GICs were highest for light-cure followed by dual cured cements and were lowest for chemically cured GICs [169, 170]. Light-cured GICs failed cohesively, where as chemically cured GICs failed in the smear layer. The light-cured cement was also more effective in penetrating into the dentin surface. XPS studies showed that the dentin attracted silicon, aluminum, fluorine, and zinc ions from the glass-cement. A larger amount of carboxyl groups at the failed-bond site for chemically cured GIC indicated a lower amount of ionic bonding at the failure site for light-cured cements, and suggested different bonding mechanisms. It is suggested that mechanical interlocking between the dentin and resin is better with the use of light-cured GICs. Adhesive strengths (or shear bond strengths) of the GIC to the tooth range from 5–20 MPa [219–222] and may be enhanced by proper pre-treatment of the dentin surface. This includes cleaning with aqueous acidic solutions of citric acid [223], phosphoric acid [224], or polyacrylic acid [171, 192, 219]. Glass-ionomer cements have also been used to repair cracks or erosions in existing GICs [222, 225–227]. Finally, the ability of GICs to bond to a precious metal alloy has been reported [185, 190, 228, 229].

4.3 Glass-Ionomer Cements – Performance

The performance of polyelectrolyte materials is determined by several factors, including the extent to which the dental material adheres to the tooth, cariostatic properties of the restoration, pulpal and tissue sensitivity in the vicinity of the restorative material, long-term stability of the dental material, and perhaps most importantly, the aesthetic appeal of the restorative material. Poor adhesion leads to the formation of gaps, which become sites for infection. Biodegradation of the cement can cause increased pulpal and oral-tissue sensitivity, as well as systemic responses. Several recent reviews on the performance of GICs [121, 173, 230–232] are available, so the subject is only briefly discussed here.

4.3.1 Marginal Leakage

Lack of adhesion of a dental restoration to tooth structure results in microleakage at tooth-restoration interface. This occurrence can result in discoloration at the margin of the restoration, or in the formation of caries. Occlusal forces on the restoration and differences between the coefficients of thermal expansion of the cement and tooth material can lead to leakage. In addition, oral fluids and moisture may affect the adhesion. Microleakage of composite resin restorations has been reviewed by Ben-Amar [233]. Microleakage is not as serious a problem with glass-ionomer cements as it is with resin-based restorative materials, due to reduced polymerization shrinkage [234].

A five year clinical study of the performance of GIC found that these cements had an exceptionally low rate of secondary caries [235]. An *in vivo* study of GICs showed minimal bacterial penetration along the GIC-tooth interface [236], however an *in vitro* study comparing microleakage between six different restorative materials found maximum leakage in the zinc polycarboxylate cements and substantial leakage in the GICs [237]. Other in vitro studies have also found microleakage with GICs and have found no significant differences between capsulated (pre-measured) and manually measured GICs [142, 238]. No microleakage was detected in an in vitro study using light-cured GICs [239, 240]. Light-cured GICs did not exhibit any microleakage even after undergoing in vitro thermal cycles for six months [241].

4.3.2 Tissue and Pulpal Responses

Glass ionomer cements are able to withstand wear and dissolution in the acidic and humid oral environment. The only major problem is disintegration of the surface layer of the cement during the initial phase of the setting process [242]. As mentioned previously, light-cured GICs are more resistant to initial dissolution than conventional GICs. Biodegradation of GICs has been reviewed by Oilo [230]. Degradation of the cement, either in the initial period of the setting reaction or during the course of use, can result in side effects such as irritation and increased pulpal sensitivity. These side-effects are minimal for GICs [235, 243, 244]. Local and systemic responses to both glass-ionomers and composites have been reviewed by Stanley [231]. Clinical reports about the use of GICs have been summarized by Bayne [232].

It was mentioned before that *in vitro* studies showed significant marginal leakage occurring with the use of GICs. However very few significant clinical problems associated with marginal leakage have been reported. The excellent response to GICs is due to the bactericidal and cariostatic properties of the cement [245–252]. A number of GICs were tested for and were found to possess cariostatic properties [245]. While the mechanism for antibacterial action is not completely understood, it is thought to be due to both the presence of fluoride as well as the low pH.

The excellent performance of GICs as dental materials has lead to their evaluation as biocompatible bone reconstruction materials. Initial results from *in vivo* and *in vitro* studies appear very promising [253–257].

4.3.3 Controlled-Release from Dental Materials

Polyelectrolytes have been widely used for the preparation of controlled/sustained release of drugs. It is therefore not surprising that attempts have been made to use dental implants for the slow-release of antibiotics and fluorides. It has been demonstrated that caries-like lesions are significantly reduced with the use of glass-ionomer cements or fluoride releasing dental materials [252].

Glass-ionomer cements are known for their slow release of fluoride [133, 252, 258–270]. The initial rate of F^- release is high, but after a few hours it stabilizes to a nearly constant value [258]. The influence of fluoride results in about a 3-mm thick zone of resistance to demineralization around the glass-ionomer restoration [251, 260, 261]. Mitra [263] has studied the *in vitro* fluoride release from a light-cured glass ionomer and compared the release with conventional self-cured glass ionomers. No significant differences were found in the rates of fluoride release between the two. The rate of fluoride release was independent of cure time, and long-term release of fluoride did not affect the strength of the material. Fluoride release was reported to be diffusion controlled, and significant amounts of fluoride were released after 12 months [258], 18 months [133], and even after two years (740 days) [263]. Novel studies have shown that GICs can absorb fluoride from fluoride-toothpastes, and subsequently slowly release the fluoride [264, 265].

The effect of variability in fluoride release between hand-mixed and capsulated systems was studied by Verbeeck et al. [266] who found that the mean value and variance of fluoride release were greater for the capsulated system than for the hand-mixed system. A two-process mechanism, consisting of a short-term elution (with a half life of nine hours) followed by diffusion controlled long-term release, for the release of F^- was suggested based on an empirical correlation of the data. The differences in the amounts of F^- released are attributed to the different mixing processes.

Wilson and Combe [271] discuss a novel GIC which releases strontium, aluminum and boron, instead of fluoride. These elements are reported to have anticariogenic properties, and their release, unlike with fluoride, is claimed not to result in discoloration.

5 Controlled Release Applications of Polyelectrolytes

Polyelectrolytes have several important applications in the area of controlled drug delivery [272–277]. For example, polyelectrolytes are commonly used as coatings in gastrointestinal pharmaceutical applications, as rate limiting membranes for sustained-delivery oral formulations, and as dissolution and binding agents in tablets. More recently, polyelectrolytes have found application as matrices for oral, buccal and nasal administration of drugs due to their bioadhesive properties. In the realm of injectable controlled delivery systems, polyelectrolytes have been investigated as polymeric drugs and polymer/drug complexes for site-specific delivery. In addition, polyelectrolytes and their complexes have been evaluated as depots for intraperotineal drug compositions to limit or control rates of drug delivery. Finally, recent exploratory activity has focused on the use of polyelectrolyte matrices, gels, coatings and membranes for responsive or self-regulated controlled delivery devices. These devices make use of the pH dependent swelling of polyelectrolytes to regulate solute release in response to specific physiological stimuli. In this section, each of these controlled release applications of polyelectrolytes will be considered, with emphasis on recent developments.

5.1 Polyelectrolytes in Controlled Release Coatings, Matrices and Binders

Enteric coatings were among the earliest biomedical applications of polyelectrolytes. In this relatively simple application, a drug-containing core is coated with a pH-sensitive polymer which protects the core from the degradative action of the low gastric pH. In the relatively basic environment of the small intestine, however, erosion of the coating is activated and the resulting polymer membrane controls drug release [273–277]. Common polymers for enteric coatings include methacrylic acid copolymers, cellulose acetate polymers [278], and poly(vinyl acetate phthalate) systems. The polymers are traditionally applied as solutions in organic solvents [279].

Although enteric coating technology is relatively mature, several developments during the past 10 years are noteworthy. Specifically, water-based colloidal dispersions of polymers such as cellulose acetate phthalate (CAP) have been cast onto tablets from latex-like dispersions, thereby avoiding the use of environmentally deleterious solvents [280–283]. Gumowski and coworkers demonstrated that coatings produced from aqueous dispersions of CAP give performance similar to that of coatings cast from organic solvents [284]. Water-based acrylic resin formulations have also been investigated for the formation of enteric coatings [285–288]. Coacervate techniques have been used to form aqueous-based controlled release coatings from copolymers of ionizable acrylatemethacrylate and sucrose [289]. The use of coacervates and complexes in controlled release systems is described in more detail later in this section.

Polyelectrolytes (most notably ionic cellulose derivatives and crosslinked polyacid powders) are also commonly used as matrices, binders and excipients in oral controlled release compositions. In these applications, the polyelectrolytes provide hydrophilicity and pH sensitivity to tablet dosage forms. Acidic polyelectrolytes dissociate and swell (or dissolve) at high pH values whereas basic polyelectrolytes (for instance, polyamines) become protonated and swell at low pH. In either case, swelling results in increased permeability [290], thereby allowing an incorporated drug to be released.

Cellulose disintegrants have been studied as insoluble matrices for sustained release tablets. Anionically charged carboxymethyl cellulose (sodium salt) was found to be inferior to methyl cellulose and poly(vinyl pyrollidone) as a binding agent for oxyphenbutazone tablets [291]. However sodium carboxymethyl cellulose has found application as a dispersing agent for ibuprofen microspheres

[292], as a component of zero-order release tablets of theophylline and ephedrine hydrochloride, and in sustained release theophylline tablets [293]. Ionizable polymers have also been used as matrices in adhesive controlled release formulations [294, 295]. For example, xanthan gum has been used as a binding agent for caffeine controlled release formulations [296].

Polyelectrolytes such as poly(acrylic acid) have been studied as prospective bases for controlled release ointments. For example, ointments for controlled release of antigens have been evaluated for novel immunization systems [297]. These types of systems may provide a safe and effective alternative method of immunization by effectively acting as a series of continuous minishots. Polyelectrolyte-based controlled release pharmaceuticals containing ionizable, water-soluble active agents have also been studied [298]. In one example, the coprecipitation of tetracycline with cationic or anionic polyelectrolytes gave materials with pH dependent dissolution rates [299]. The cationic polymers gave lower initial rates of absorption than the precipitated drug alone, while the anionic polymers gave the lowest rates and extent of absorption. In a second example, anionic, cationic and zwitterionic acrylic polymers retarded dissolution of coprecipitated ibuprofen [290]. The anionic materials were most effective followed by zwitterionic and cationic.

Crosslinked water-swollen polymers known as hydrogels have been widely studied matrices for controlled delivery devices [33, 300, 301]. Concentrated solutions of many therapeutic agents can be imbibed or dispersed into hydrogels and administered orally. For example crosslinked poly(acrylic acid) (Carbopol 934) has been extensively studied as a polymer matrix for swelling-controlled drug release formulations [302, 303] including zero order release devices. Poly(acrylic acid) has been incorporated into controlled release formulations for oral delivery of several drugs [304] including insulin [305] quinine sulfate [306], mepyramine maleate [307], sulfomethazole drugs [308], salicylate [309], acetylsalicylic acid [310], and furosemide [311]. Other applications of polyelectrolyte gels include acrylic resins for controlled release of pseudoephedrine hydrochloride [312], calcium alginate gels for proflavin pellets [313], cellulose gels [314], and polysaccharide scleroglucan gels [315] and poly(acrylic acid) hydrogels for release of theophylline and neophedrine hydrochloride [316].

Other hydrogels for drug delivery have been based upon polysaccharide xanthan gums. For example xanthan gum was crosslinked with epichlorohydrin and its swelling properties were correlated with crosslinking [317]. These hydrogels were loaded with theophylline, isosorbite dinitrate or methoxyprogesteron by diffusion, while neomycin was linked to the xanthan, and furazolidone was incorporated into the neomycin derived gel. Zero order release kinetics were observed for neomycin and furazolidone. Similarly, crosslinked chondroitin sulfate matrices were investigated for controlled release of indomethacin in the colon [318]. It was concluded that the indomethacin release from these systems was dependent upon the biodegradation action of the caecal content.

Responsive polyelectrolyte hydrogels that have been used for controlled release systems include gels which are hydrophobic in their neutral state

[319–321] and acrylic copolymer gels based on hydroxyethyl methacrylate [322]. For instance, hydrophobically modified poly(acrylic acid) has been synthesized by radical copolymerization of acrylic acid with varying amounts of butyl acrylate and crosslinker, giving pH dependent hydrogels [323]. In one example, as illustrated in Fig. 2, hydrophobic polyamine gels were loaded with caffeine, and the drug release was found to vary sharply with pH [320]. Neutral pH gave no caffeine release, while between pH 3 and 5 near zero order release kinetics were found. This effect arises from the ionization of the gel at low pH leading to polyelectrolyte-like character, enhanced swelling and thus solute release. Insulin release from hydrogels based upon two different cationic polymers, poly(2-hydroxyethyl methacrylate-*co*-2-diethylaminoethyl methacrylate) and poly(2-hydroxyethyl methacrylate-*co*-diethyl aminoethyl acrylate), was also studied [324]. A detailed model incorporating the diffusion and relaxation characteristics of the polymer was developed and solved numerically.

Ion exchange resins have also been used as matrices for controlled release. For example, ion-activated drug delivery was achieved by complexing an ionic drug with an ion exchange resin. [325, 326]. In these systems, the drug release rate is controlled by the rate of ion influx. In electrolyte media, ions diffuse into the system, react with the drug/resin complex and trigger drug release.

Several investigators have recently studied fundamental aspects of dissolution and film-forming properties of polyelectrolyte coatings, binders, and matrices. For example, Nakano and coworkers [327] studied carboxyethylmethyl cellulose enteric coatings and demonstrated that enteric-coated microcapsules are not all emptied from the stomach simultaneously, but follow a first order pattern. Isolated polymer films of Eudragit RL were studied by Navarro and coworkers [328] with regard to swelling, capacity, water vapor permeability and mechanical properties. The role of solute solubility in polymer films was identified as the main mechanism for steroid diffusion, providing an explanation

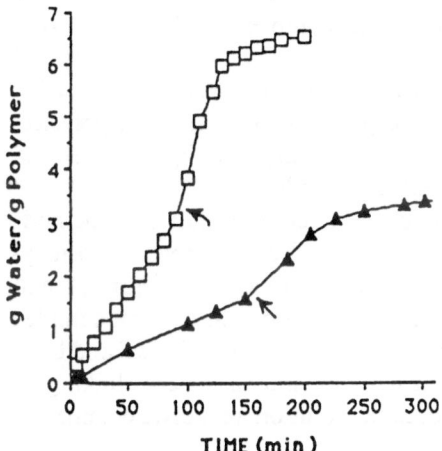

Fig. 2. Swelling kinetics of 8 caffeine-loaded methyl methacrylate/*N,N*-dimethylaminoethyl methacrylate copolymer hydrogels. (□) pH 3, (▲) pH 5. Siegel RA, Falamarzian M, Firestone BA, Moxley BC (1988) J Contr Rel 8: 179. Reproduced with permission

for the lower permeability of synthetic progestin than that of progesterone through methacrylic films [329]. Furthermore, the effects of plasticizers on compatibility, mechanical properties and adhesion strength of drug-free Eudragit E films were elucidated by Lin et al. [330].

The release kinetics of polyelectrolyte-containing controlled release compositions were modeled by Ozturk et al. [331]. According to this analysis the drug release rate depends on intrinsic solubilities as well as pK_a values of the drug and polymer. Explicit relationships between release rates and these factors were derived, resulting in successful predictions of experimental data.

Other studies have concentrated on the manufacturing processes of polyelectrolyte-containing pharmaceutical compositions. Fluidized-bed film formation processes for poly(methacrylic acid-co-ethyl acrylate) were studied by Fukumuri et al. [332], and the spraying of acrylic resin slurries was studied by Li and coworkers [312]. Recently available commercial enteric coating materials were investigated using dissolution studies, SEM, TMA and DSC analysis to find that they are all good film formers [279]. The rheological and enteric properties of organic solutions, ammonium salt aqueous solutions, and latex systems of a variety of enteric polymers were compared by Chang [333]. The relation between the force applied during formation of Carbomer 934-based pills and the *in vitro* release rate was also investigated [334].

Novel encapsulation techniques based on polyelectrolyte complexes have been developed and were reviewed by Philipp et al. [335] and Petrak [336]. One of the most common examples of encapsulation based upon polyelectrolyte complexes is coacervation, which involves the association of oppositely charged polymers to form an insoluble film or barrier. For example, calcium-gelled sodium alginate capsules reinforced by complex formation between alginate and poly-*l*-lysine have been investigated [337], along with haemophilized membranes prepared by the reaction of quaternary polyelectrolyte with anionic materials [338]. Novel controlled release theophylline granules coated with polyelectrolyte complex of sodium polyphosphatechitosan were also developed by Kawashima and coworkers [339, 340]. In this case, prolonged release allowed reduced dosage levels. Polyelectrolyte complex coatings were substituted for traditional ethylcellulose, acrylic resin and chitosan coatings and resulted in zero-order kinetics and significantly reduced release rates. Finally, Kikuchi and Kubota [341, 342] studied the permeability control of polyelectrolyte complex membranes based on glycol chitosan/poly(vinyl sulfate) and methyl glycol chitosan/carboxymethyl dextran complexes.

Several other investigators have reported microencapsulation methods based upon polyelectrolyte complexes [289, 343]. For example, oppositely-charged polyelectrolytes (Amberlite IR120-P (cationic) and Amberlite IR-400 (anionic)) were recently used along with acacia and albumin to form complex coacervates for controlled release microcapsule formations [343]. Tsai and Levy [344, 345] produced submicron microcapsules by interfacial crosslinking of aqueous poly(ethylene imine) and an organic solution of poly(2,6 dimethyl phenylene oxide). Salt in the microcapsules could be released under controlled conditions with release rates determined by osmotic pressure, membrane quaternization or modification of the capsule wall with surfactant. Non-stoichiometric polyelectrolyte complexes have also been prepared by reacting sodium polymethacrylate with poly(N ethyl vinyl pyridinium bromide) [335, 346].

Several techniques for microencapsulation involve placing acidic drugs into solution with an alkali solubilized enteric polymer. In these systems, the polymer becomes locally insoluble and adheres to the drug surface to form a seamless film. This technique was used for aspirin and indomethicin-containing formulations by Takahata et al. [347]. Similarly, human albumin and heparin were linked into a soluble conjugate and then crosslinked using glutaraldehyde in a water-in-oil emulsion to form microspheres. The swelling of the spheres was pH sensitive with low swelling at low pH and high swelling at high pH, and this effect was attributed to ionization of albumin amino acids [348].

In a related application, polyelectrolyte microgels based on crosslinked cationic poly(allyl amine) and anionic poly(methacrylic acid-co-epoxypropyl methacrylate) were studied by potentiometry, conductometry and turbidimetry [349]. In their neutralized (salt) form, the microgels fully complexed with linear polyelectrolytes (poly(acrylic acid), poly(acrylic acid-co-acrylamide), and poly(styrene sulfonate)) as if the gels were themselves linear. However, if an acid/base reaction occurs between the linear polymers and the gels, it appears that only the surfaces of the gels form complexes. Previous work has addressed the fundamental characteristics of these complexes [350, 351] and has shown preferential complexation of cationic polyelectrolytes with crosslinked carboxymethyl cellulose versus linear CMC [350]. The departure from the 1:1 stoichiometry with the non-neutralized microgels may be due to the collapsed nature of these networks which prevents penetration of water soluble polyelectrolyte.

Polyelectrolytes have also been used to regulate drug delivery rates in transdermal devices. For instance, a rate-controlled transdermal delivery systems based on an alginate polymer has been reported [352]. In this application, the polymer was used to reduce the rate of drug release in systems with high skin permeability. The mechanisms and release profiles of drugs from devices containing a monolithic drug/alginate matrix and a rate controlling alginate membrane were varied by using alginates with different crosslinking, metal contents or mannuronic and guluronic acid residues.

Polyelectrolyte complexes have also been studied as tablet matrices for controlled release applications. For example, the interpolymer complexes of chitosan with pectin and acacia were investigated as tablet matrices for release of chlorpromazine HCL [353]. The complex formed in situ by mixing chitosan with either pectin or acacia displayed the most efficient sustained release (compared to either pectin, acacia or a preformed complex). The results were attributed to the swelling and gel-forming capacity of the freshly formed complex in contrast to the preformed version.

5.2 Polyelectrolytes in Novel Responsive Delivery Systems

In recent years there has been considerable interest in the development of responsive drug delivery systems based upon physiologically responsive polymers [354–357]. This research is motivated by the fact that while current commercial drug delivery devices offer many advantages over conventional drug delivery methods, they have some limitations. For example, current controlled release devices may maintain drug concentration in a desired range, localize delivery, lower systemic drug levels, preserve medications rapidly destroyed by the body and improve patient compliance [354]. However, the constant drug levels characteristic of such devices are not always desired. Examples of drugs whose need varies with time include insulin for diabetes, antiarrhythmics for patients with heart disease, gastric acid inhibitors for ulcer control, nitrates for patients with angina pectoris, selective beta blockers, birth control, hormone replacement, immunization and chemotherapy drugs [354]. The basic objective of the research on responsive drug delivery is to develop systems which more closely mimic normal physiological processes by delivering the drug in response to changing physiological needs [354]. In a second class of novel responsive polyelectrolyte systems, the drug release rate is reversibly regulated by an external stimulus such as an electrical signal. In these systems, the charged moieties on the polyelectrolyte interact with the externally applied electric field to provide the desired drug release characteristics.

Much of the work on physiologically responsive controlled release has focused on the development of glucose-responsive devices for self-regulated release of insulin. In most of these exploratory systems, polyelectrolytes are used in the design of the glucose-sensitive trigger. For example, in one scheme, the enzyme glucose oxidase was immobilized in porous poly(dimethylaminoethyl-methacrylate-*co*-tetraethyleneglycol dimethacrylate) gels [358–363]. Upon diffusion of glucose into the gel, the enzyme catalyzes the conversion of glucose to gluconic acid, thereby lowering the pH and causing the tertiary amine moieties on the network to become protonated. The resulting swelling of the network allows the release of insulin. Insulin transport rates in these systems were found to be enhanced by a factor of 5.5 by the presence of glucose. In flowing systems the gluconic acid produced by the enzyme was depleted and no large changes were observed. Modeling efforts [362, 363] indicated that low glucose oxidase loading is needed in these systems to avoid oxygen depletion while sufficiently low amine concentrations are required to prevent buffering by the amine.

Two similar approaches were adopted by Ishihara et al. [364–368]. Membranes based on hydroxyethyl acrylate, dimethylaminoethyl methacrylate and trimethyl silyl styrene were solvent cast, and capsules containing insulin and glucose oxidase were prepared by interfacial precipitations. The authors reported dramatic changes in permeability in response to pH changes between 6.1 and 6.2. Moreover, addition of glucose induced an increase in the permeation rate of insulin, and upon removal of the glucose the permeability rates returned to their original levels. However, the conclusions were criticized [361] due to questions about the sensitivity of the insulin detection technique that was used [365].

Siegel and coworkers [290, 369, 370] proposed a mechanochemical insulin pump based upon a pH sensitive polyelectrolyte gel which expands in the presence of glucose. The proposed device, illustrated schematically in Fig. 3, contains three compartments, one containing the glucose sensitive gel, a second containing water, and a third containing the insulin formulation to be delivered. As shown in the figure, in the presence of glucose the polymer swells and pushes the diaphragm to the right. The increased pressure in the adjoining water chamber is also transmitted to the insulin chamber. Since the one-way valve in the water chamber is closed by the increased pressure while the valve in the

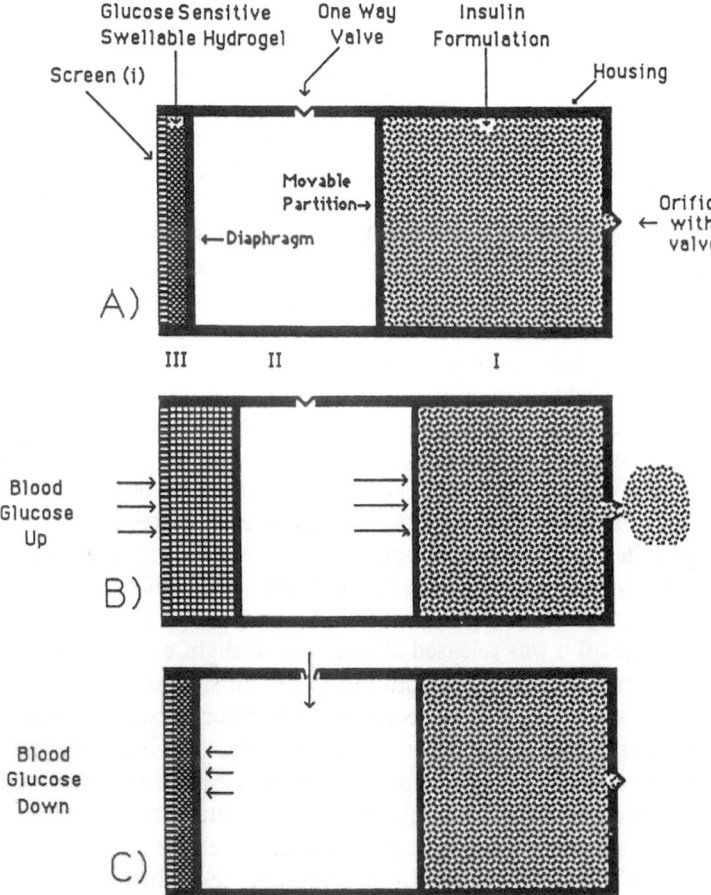

Fig. 3A–C. Schematic diagram of a proposed mechanochemical pump: A diagram of components; **B** response of pump to increase in glucose concentration; **C** response of pump to decrease in glucose concentration. Siegel RA (1990). In: Kost J (ed) Pulsed and self-regulated drug delivery. CRC Press, Boca Raton FL, p 129. Reproduced with permission

insulin chamber is opened, insulin is released until the pressures are again equalized (Fig. 3B). Finally, as illustrated in Fig. 3C, when the glucose level falls the polymer gel shrinks to its original size, water fills the center chamber and the system is ready for a second application. Kabra and Gehrke [371] also discuss hydrogel-driven osmotic pumps in which a drug and a hydrogel are encapsulated by a semipermeable membrane. The release of insoluble drugs through an orifice in the membrane was enhanced by swelling of the hydrogel.

A number of investigators have devised polyelectrolyte controlled release systems which respond to an externally applied electrical signal. Grodzinsky and coworkers [372, 373] used electrical fields to produce electro-osmotic flow in poly(methacrylic acid) membranes. In these systems, the flow through the membrane ceased when the current was turned off and reversed direction when the current was reversed. Valentine and Miller [374] studied the electrochemically-controlled cation binding and release using conductive polymer composites. Ion binding and transport studies were performed upon composites containing poly(N methyl pyrrole) as the cationic polyelectrolyte and either polystyrene sulfonate or Nafion as the anionic polyelectrolyte. These studies demonstrated that cations were incorporated by the composite under reduction and released by oxidation. These materials were suggested for transdermal drug delivery devices.

Osada and coworkers used an applied voltage to induce reversible contraction of polyelectrolyte gels [375–377]. Using materials based on poly(methacrylic acid), gelatin or collagen, it was demonstrated that pilocarpine could be released in response to an electric field. Upon application of the field, the gels collapsed and the pilocarpine was convected out [119]. Osada and coworkers also demonstrated the electrically-stimulated drug delivery via microparticles [378]. A similar approach was used by Uchida and coworkers who studied polyelectrolyte gels and electromagnetic composites for on-off switching release [379]. Methylene blue was used as a model drug, and matrices were based upon crosslinked copolymers of sodium acrylate or polymer composites of ferrite.

Kwon and coworkers described solid polyelectrolyte complex systems which dissolve rapidly in response to small electric currents. The solid doses were based on poly(ethyl oxazoline) and poly(methacrylic acid) with a repeating unit stoichiometry of 1:1. Insulin was released in response to slight electric currents due to electrically induced polymer dissolution [380]. In similar work Kwon and coworkers [381] studied release of edrophonium chloride and hydrocortisone from poly(2-acrylamido-2-methylpropane sulfonate-co-n-butyl methacrylate). An on/off mechanism of the edrophonium chloride release was observed and was attributed to ion exchange of solute and hydroxonium ion. The cationic solute release was assisted by electrostatic forces, whereas release of the neutral hydrocortisone solute was only affected by swelling and deswelling.

Novel responsive controlled release systems based upon polyelectrolyte-grafted membranes have also been reported. Iwata and Matsuda prepared novel environmentally sensitive membranes by grafting poly(acrylic acid) onto poly(vinylidene fluoride) membranes [382, 383]. Under basic conditions, the grafted chains were highly swollen and therefore blocked the membrane pores and restricted solute permeation. In contrast, at low pH the grafted chains were collapsed, allowing permeation of the solute. The insulin permeability of cellulose membranes grafted with poly(N acryloyl glycine) was also studied [384]. In this system, interactions between the uncharged polymer and the membrane were responsible for the slow release. Shatageva and Samsonov [385] studied insulin transport as a function of ionization of the membranes and the insulin. It was found that pH gradients increased flux, possibly due to gradients of ionization which affected swelling and thus transport rates. Similarly, Ito and coworkers [386] grafted poly(acrylic acid) onto cellulose membranes containing immobilized glucose oxidase. In this system the insulin permeability increased almost 2-fold upon the addition of glucose.

Systems using a polyelectrolyte-grafted porous membrane were also studied by Pefferkorn et al. [387] and Idol and Anderson [388]. Changes in solution pH altered the solvent hydraulic permeability of the membranes by adsorbing a polymer which undergoes a conformational transition at a threshold pH. In related work, Osada et al. [389, 390] showed pH sensitivity of water permeability (up to 1000 fold change) through poly(methacrylic acid) grafted membranes, while Ito et al. demonstrated pH-sensitive membranes of grafted PAA onto polycarbonate [391]. Finally, Okahata and coworkers used a similar approach to synthesize poly(4 vinyl pyridine) and PMAA grafted nylon capsules [392, 393].

Membranes with enhanced permeability at high pH have been derived from poly(glutamic acid) [394] and polyvinyl polypeptide graft copolymers [395–397]. The polypeptide-containing membranes exhibited pH-induced permeation changes due to conformational changes of the polypeptide. Membranes with enhanced low pH permeability were derived from aminolysis of poly(gamma methyl l glutamate) with ethylenediamine and were used to control lysine permeation in a pH sensitive manner [394].

5.3 Polyelectrolyte Bioadhesive Delivery Systems

Bioadhesive polymers may provide platforms for drug release by allowing a prolonged localized residence time, thus ensuring optimal contact between the drug depot and the release site. Bioadhesive materials may be useful for oral, nasal, buccal, rectal and vaginal applications. These materials may ultimately provide a means for peroral peptide delivery; however more work is needed for this application to be realized. Polyelectrolytes commonly used in bioadhesive formulations include sodium carboxymethylcellulose, poly(acrylic acid), gum tragacanth, poly(vinyl methyl ether-co-maleic anhydride), sodium alginate and karaya gum [398, 399].

The mucin to which bioadhesive polymers adhere is secreted by columnar epithelial cells and is a network comprising 0.5–2% of highly-hydrated flexible glycoprotein chains [400], which contain negatively charged moieties, and are crosslinked by disulfide bonds [401]. These network polymers consist of oligosaccharides linked to hydroxyamino acids along a polypeptide backbone. The molecules contain glycoprotein units, heavily glycosylated protein segments, and naked protein segments.

For bioadhesive applications, anionic polymers appear to provide the most effective balance between adhesiveness and toxicity, with carboxylic materials preferred over sulfonic polymers [400]. Poly(acrylic acid) microparticles have been identified as particularly effective bioadhesive materials [402]. Studies with poly(acrylic acid) microparticles have indicated that, while water-swollen particles exhibit good bioadhesion, dry polymer particles give no adhesion at all. In addition, adhesive strength increases as the degree of ionization of the polymer is increased [402]. Thus the expanded nature of the polymer network is important to mucoadhesion, probably via polymer interdiffusion and entanglement with mucin [403].

As recently discussed by Harris and Robinson [404], the use of bioadhesive polymers in peptide drug delivery has been an area of active research. As of 1989, 150 peptide or protein drugs were undergoing clinical trials [404, 405]. The most significant obstacles to their successful application have been associated with their difficulty in administration by injection and their high susceptibility to gastrointestinal degradation. Low drug availability arises from low tissue permeability, short residence times and enzymatic degradation. Thus, gastrointestinal routes are not viable with current delivery technology. Bioadhesive compositions may be used to solve these problems since they may enhance permeability, increase residence time, and protect drugs against enzymatic degradation [404].

The contributions of anionic polymer structural features to mucoadhesion were recently elucidated for a series of lightly-crosslinked acrylic acid and methyl methacrylate copolymers [401]. In addition, factors which effect mucoadhesion have been approximately evaluated in order to determine the relative bioavailability of the drug [400]. Through an analysis which included the effects of residence time and drug diffusivities but neglected enzymatic degradation, the authors concluded that bioadhesive compositions may have applicability in buccal and occular applications. However, successful nasal applications would require improvements in residence times and retardation of the mucosal clearance of the entire mucus layer. For gastrointestinal tract applications, extended residence times would help drug delivery, but in practice the polymer must attach to the surface of the mucosa (which itself is eroding) and the delivery systems must withstand the motility of the gastrointestinal tract (which is highly efficient at transiting ingested materials).

The bioadhesive characteristics of tablets for oral use made from modified starch, poly(acrylic acid), poly(ethylene glycol) and sodium carboxymethyl cellulose were recently investigated [406]. In this work, the force and energy adhesion were determined in vitro, and maximum adhesion times were evaluated in vivo in humans [406]. In the in vitro, studies, the poly(acrylic acid) gave the best performance, however in vivo bioadhesion was not strongly correletated with *in vitro* data. Poly(acrylic acid) was found to be irratating to the mucosa and poly(ethylene glycol) based materials showed good adhesion but poor mechanical strength. Modified maize starch containing approximately 5 % poly(acrylic acid) and poly(ethylene glycol) was found to be the most suitable in the in vivo tests.

A host of bioadhesive controlled release systems have been proposed in recent years. Among the most commonly studied applications of bioadhesive materials is the area of buccal controlled delivery [408]. The buccal delivery of small peptides from bioadhesive polymers was studied by Bodde and coworkers [409], and a wide range of compositions based on poly(butyl acrylate) and/or poly(acrylic acid) gave satisfactory performance. Bioadhesive poly(acrylic acid)-based formulations have also been used for oral applications [402, 410] for the sustained delivery of chlorothiazide [410] and for a thin bioadhesive patch for treatment of gingivitis and periodontal disease [411]. Other bioadhesive applications of polyelectrolytes include materials for ophthalmic vehicles [412, 413], and systems for oral [410, 414, 415–419], rectal [420, 421] vaginal [422] and nasal [423] drug delivery.

5.4 Liposome Controlled Release Systems

Polyelectrolytes have recently found application in the development of pH sensitive liposomal controlled release systems. This application arises from the fact that polyelectrolytes may be used both to stabilize liposomes, and to disrupt liposomes in a pH dependent manner. Although the use of liposomes in oral pharmaceutical compositions has been discussed [424], liposomes generally suffer from poor stability and are therefore prone to leakage of the entrapped active agents. To overcome this problem, several authors have stabilized the liposomes using polyelectrolytes. For example, Tirrell and coworkers have employed ionene [425], and poly(ethylene imine) [426] to stabilize liposomes. Similarly, Sato and coworkers have studied maleic acid copolymers [427], and Sumamoto and coworkers have studied liposomes [428] coated with polysaccharides. In related work, Kondo and coworkers have emphasized the use of carboxymethyl chitin to produce artificial red blood cells [429–435].

Seki and Tirrell [436] studied the pH-dependent complexation of poly(acrylic acid) derivatives with phospholipid vesicle membranes. These authors found that poly(acrylic acid), poly(methacrylic acid) and poly(ethacrylic acid) modify the properties of a phospholipid vesicle membrane. At or below a critical pH the polymers complex with the membrane, resulting in broadening of the melting transition. The value of the critical pH depends on the chemical structure and tacticity of the polymer and increases with polymer hydrophobicity from approximately 4.6 for poly(acrylic acid) to approximately 8 for poly(ethacrylic acid). Subsequent photophysical and calorimetric experiments [437] and kinetic studies [398] support the hypothesis that these transitions are caused by pH dependent adsorption of hydrophobic polymeric carboxylic acids on the phospholipid films. A strong molecular weight dependence was observed [438] with low molecular weight (12000) PEAA behaving differently than polymers of higher molecular weight (43000 and 164000). Finally, quasi-elastic light scattering and electrophoretic light scattering measurements [439] demonstrated that a reorganization of the vesicles does occur upon reduction in pH, and is accompanied by a substantial reduction in vesicle size [440], as shown in Fig. 4. It was suggested that as the pH was lowered, the collapsed PEAA chains provided hydrophobic sites for stabilization of hydrocarbon tails resulting in disruption of the structural organization of the phospholipids.

Several controlled release applications of the polymer-complexed phospholipid materials were investigated. Semisynthetic vesicular membranes of chemically immobilized poly(ethacrylic acid) on the surface of phospholipid bilayers gave membranes intrinsically sensitive to proton concentration and gave large permeability changes with small pH changes and rapid quantitative release by mild acidification [441, 442]. Other studies have considered photoregulated [443, 444] and glucose dependent disruption of the membranes [445] by photoinduced and glucose induced pH changes in the membrane structure.

Similarly comb-like copolymers of vinyl pyrollidone and vinyl alkyl amines were shown [446] to influence the permeability of negatively charged phospholipid liposomes containing encapsulated carboxyfluorescein. At a pH of approximately 7, the copolymers allowed permeability and solute release due to polymer/liposome complex formation and disruption of the phospholipid membrane.

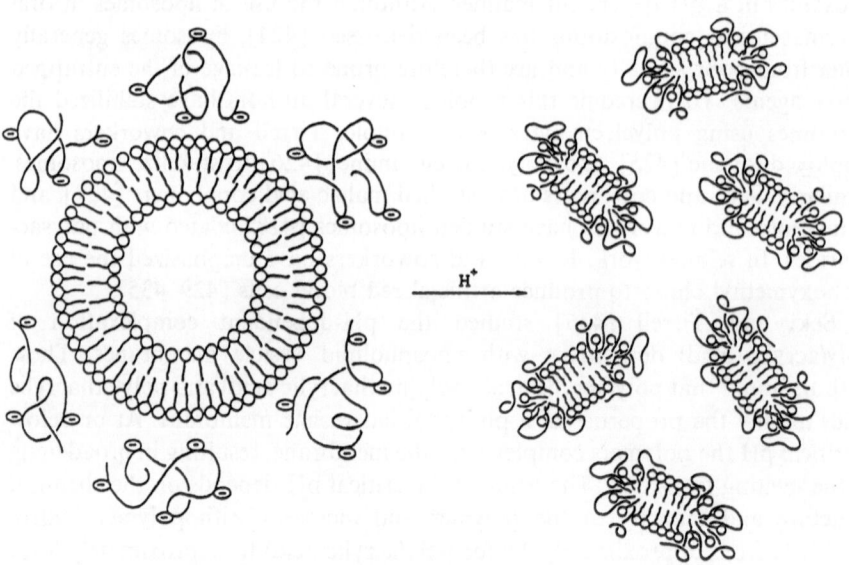

Fig. 4. pH-dependent reorganization of phospholipid vesicle membranes by poly(ethacrylic acid). Tirrell DA (1990). In: Kost J (ed) Pulsed and self-regulated drug delivery, CRC Press, Boca Raton FL, p 109. Reproduced with permission

Polymeric phospholipids based on dioctadecyldimethylammonium methacrylate were formed by photopolymerization to give polymer-encased vesicles which retained phase behavior. The polymerized vesicles were more stable than non-polymerized vesicles, and permeability experiments showed that vesicles polymerized above the phase transition temperature have lower permeability than the nonpolymerized ones [447–449]. Kono et al. [450, 451] employed a polypeptide based on lysine, 2 aminoisobutyric acid and leucine as the sensitive polymer. In the latter reference the polypeptide adhered to the vesicular lipid bilayer membrane at high pH by assuming an amphiphilic helical conformation, while at low pH the structure was disturbed resulting in release of the encapsulated substances.

6 Polymeric Drugs, Prodrugs and Adjuvants

In addition to their use as drug vehicles, polyelectrolytes have found utility as drugs, prodrugs and adjuvants. One of the most established uses of polyelectrolytes as active ingredients in pharmaceuticals is the application of ion exchange resins for lowering serum cholesterol ("Cholestyramine" drugs). Although this technology is not new, patent and literature activity in this area continues. Polyelectrolytes have also been studied as polymeric prodrugs. This area was recently reviewed in detail [452], and only a brief summary will be presented here.

In particular, water dispersible polymeric anticancer prodrugs containing the active group methacryloyloxyethyl 5 fluoroacil (MAFU) and ionic acrylic, methacrylic acid or maleic acid functional groups were recently reviewed [452]. Most investigations focused on controlling the hydrolysis rate of these prodrugs to give sustained release of the FU group. Dispersed systems gave slow hydrolysis rates, while the water soluble maleic acid products gave rapid FU release. Other derivatives giving good ester hydrolysis included MAFU copolymer with acrylic acid and vinyl imidazole.

The relationship between charge density and hydrolysis rates of polymeric prodrugs was studied by Mora and Pato [453]. The prodrugs in this investigation were poly(N vinyl pyrollidone-co-maleic acid) of different carboxyl contents, copolymerized with chymotrypsin cleavable pendant groups of phenylalanyl nitroanalide. Half and fully amidated poly(N vinyl pyrollidone-co-maleic acid) and polyanionic dextrans were also modified with different amounts of succinyl moieties containing peptidic side chains and used as model compounds. A correlation was found between the number of carboxyl groups and a decrease of the Michaelis constant of the hydrolysis.

Polyacrylate and polyhydroxylate polymers with pendent aspirin moieties were recently prepared for aspirin delivery [454]. The homogeneous and heterogeneous hydrolysis in acidic and alkali media of poly(acetylsalicylyloxy ethyl methacrylate-co-methacrylic acid) was studied and found to depend on the pH of the hydrolysis medium.

Several polyelectrolytes have been used as antiviral agents. Chemical binding or strong complexing of antigens or antigenic determinants with non-natural membrane active polyelectrolytes is thought to give effective immunostimulants with highly protective properties as discussed and reviewed by Kabanov [455]. This is believed to occur because polyelectrolytes are nonspecific polyclonal activators and they can partly substitute for T-helper functions. Here activity is thought to occur because adsorbed polyelectrolytes give globular complexes with cell membrane proteins, thereby increasing cell ion permeability and achieving the transmembrane transport of enzyme ATPase which triggers cell division. In general, polyelectrolytes above a certain critical length are effective in this manner. Specifically, the biological activity of poly(carboxylic acid) polymers was studied by Ottenbrite and coworkers [456], while acrylic acid-based pharmacologically active materials were studied by Bauduin et al. [457].

Anionic polyelectrolytes have been shown to enhance resistance to bacteria and fungi, enhance immune response, inhibit adjuvent arthritis and either depress or stimulate phagocytic activity of the reticuloendothelial system [458, 459]. Carboxylic acid polymers have shown interferon induction, antiviral activity, and tumor growth inhibition [460]. The effects include inhibition of sarcoma, leukemia, polyoma and vesicular stomatitis virus. In one application, the cytotoxicity of bleomycin toward cultured mammalian cells was synergistically enhanced by stirring in the presence of high molecular weight poly(acrylic acid) [461].

In addition, polyelectrolytes such as poly(acrylic acid-alt-maleic anhydride) (pyran) are reticuloendothelial stimulants, induce resistance to tumor growth [462, 463] and show interferon-inducing ability and antiviral activity. The pyran activity may be dependent upon molecular weight. Low molecular weight materials of narrow polydispersity activate macrophages; however antitumor activity does not appear to be molecular weight dependent [460].

Sulfonated and sulfated anionic polyelectrolytes have also received considerable attention as biologically active agents. Sulfated polysaccharides including ribofuranan, xylofuranan, mannopyranan, lentinan and courdlan sulfates were evaluated as potential anti-HIV agents. Courdlan sulfate showed the highest anti-HIV activity, low anticoagulant activity and low *in vivo* toxicity [464]. Novel sulfated polymers including poly(vinyl alcohol) derivatives and copolymers of acrylic acid with vinyl alcohol were identified as potent and selective inhibitors of HIV *in vitro* [465]. The antiviral and antitumor activity of poly(vinyl sulfonate) and divinyl ethermaleic anhydride copolymers through interferon inducing activity were also identified [466].

Other investigators have considered polyamines as polymeric drugs. Panarin and coworkers [467] studied the immunogenic and antibacterical activity of water-soluble cationic polyelectrolytes based on vinylamine aminoalkyl methacrylate quaternary ammonium salts. Macrophages were successfully activated in vivo to become cytotoxic to tumor cells *in vitro*, and cytotoxic or cytostatic activity toward Lewis and Ehrlic ascites was observed with no effects on normal cell populations. Linakis and coworkers showed that orally administered sodium polystyrene sulfonate lowers serum lithium concentrations when administered after an oral dose of lithium. This effect was dose related, and repeated polymer doses enhanced lithium elimination [468].

Protein drugs have been formulated with excipients intended to stabilize the protein in the milieu of the pharmaceutical product. It has long been known that a variety of low molecular weight compounds have the effect of preserving the activity of proteins and enzymes in solution. These include simple salts, buffer salts and polyhydroxylated compounds such as glycerol, mannitol, sucrose and polyethylene glycols. Certain biocompatible polymers have also been applied for this purpose such as polysaccharides and synthetic polymers such as polyvinyl pyrrolidone and even nonionic surfactants.

Methods were described for the incorporation of proteins in the form of noncovalent complexes with polycationic reagents, into sustained release systems where the polycation stabilizes the protein against inactivation while it resides in the delivery device, and retards release of the protein from the delivery device [469, 470]. A variety of polycations have been used, including simple polyamino acids such as polylysine or polyarginine, protamine and chitosan. The end result was the release of the active agent with retention of biological activity, with a high cumulative field and over a sustained period of time.

7 Polyelectrolytes as Biocompatible Materials

Polyelectrolytes have been widely investigated as components of biocompatible materials. Biomaterials come into contact with blood when used as components in invasive instruments, implant devices, extracorporeal devices in contact with blood flow, implanted parts of hard structural elements, implanted parts of organs, implanted soft tissue substitutes and drug delivery devices. Approaches to the development of blood compatible materials include surface modification to give blood compatibility, polyelectrolyte-based systems which adsorb and/or release heparin as well as polyelectrolytes which mimic the biological activity of heparin.

Exposing a foreign material to blood results in protein adsorption then platelet adsorption followed by white blood cell and red blood cell interactions which are mediated by the adsorbed protein. Eventually, healing may lead to covering with a passive layer of protein and overgrowth by cells. The surface composition of the material influences the composition and organization of the protein layer, while other factors such as flow, topography and molecular motion may also be important. At short exposure times, before lipid or protein adsorption has taken place, the polymer material surface has a significant effect on blood interactions. At long times, the interactions are generally mediated by adsorbed species. In addition, long term performance is influenced by mechanical properties of the implant, inflammatory processes, protein and cellular digestion of foreign bodies, and encapsulation in fibrous collagen scar tissue.

The surface properties contributing to blood compatibility were reviewed by Andrade et al. [471] and by Kim and Feijens [36]. These reviews identified surface charge as one important property governing the blood clotting response. In particular, negatively charged surfaces were suggested to give plasma protein and cell repulsion [472] due to the Hageman factor and contact activation [473, 474].

The most widely studied synthetic polymers for blood contact applications are polyether urethane ureas ("Biomer" (Ethicon)). These materials have been used in artificial hearts, as coatings for lead wires in pacemakers, have been used and are being considered for blood vessel prostheses. The success of these materials is believed to be due to preferential adsorption of albumin rather than globulin or fibrinogen which promote a clotting response. However, these materials are hydrophobic and questions of long-term effectiveness are unresolved. Particularly, these materials may shed emboli or may be susceptible to surface calcification. Thus, it may be desirable to have synthetic polymers which are hydrophilic and better resemble blood vessels [475].

The effect of carboxylate and/or sulfonate ion incorporation on the physical and blood contacting properties of polyurethanes was studied by Cooper and collaborators [476–478]. Specifically, propyl sulfonate and ethyl carboxylate groups were grafted onto polytetramethylene oxide-based polyurethanes. Carboxylate polymers had no statistically significant effect on canine *ex vivo* blood contact response, but propyl sulfonate incorporation significantly reduced platelet deposition for very short blood contact times.

Silver and coworkers [479] found that similar sulfonate functionalized polyurethanes interacted with fibrinogen or promoted antithrombin in a heparin-like manner. Evidence showed that the anticoagulation effect was more potent in plasma than in purified thrombin solutions. In other work, modification of polyether urethanes by heparin grafting were compared as antithrombogenic materials with anionically charged polyurethanes [480, 481], and the mechanisms of antithrombogenicity were investigated. The anionic material adsorbed albumin but did not cause conformational changes in adsorbed plasma protein. In addition, the material suppressed platelet adherence and deformation but did not deactivate blood clotting, and therefore were moderately antithrombogenic. In contrast, the heparin-bound materials did not selectively adsorb albumin, caused plasma protein denaturation, induced platelet deformation and adhesion, but deactivated the clotting system.

Santerre and collaborators [72, 73] also found that the plasma interaction of polyether urethanes was significantly altered by incorporation of sulfonate groups [73] which could be derivatized with amino acids [72]. The synergistic effect of pendant PEO and sulfonate groups grafted onto polyurethanes resulted in prolonged *ex vivo* occlusion times [482]. These results corresponded well with *in vitro* results, showing less adhesion, shape change and longer partial thromboplastin time and prothrombin time with the ionic material than with the unmodified material.

The blood compatibility of polyethylene terephthalate (PET) radiation grafted with acrylic acid was studied by Miller et al. [38]. Although PET is popular for autologous vein vascular prostheses, it is not effective in the femoropopliteal/distal region due to unfavorable mechanical and surface characteristics. The graft copolymers resulted in smaller rise in circumferential elastance, significantly slower pulse wave velocity and lower longitudinal impedance, and had more organized and thinner pseudointima than the non-grafted material. Thus the polyelectrolyte-grafted copolymer exhibited improved haemodynamic characteristics and surface properties which appeared to discourage fibrin and platelet adherence.

Chlorosulfonated styrene resins and carboxyaminoacid polymers were also found to possess thromboresistant properties by Josefonwicz and coworkers [483]. Studies included investigation of the effect of spacer length between amine and carboxylic groups as well as modification of styrene/isoprene/styrene blocks with chlorosulfonyl isocyanate giving sulfamate and carboxylic functionality [484].

Tumoko and coworkers [485] showed that polymers based on acrylamide, methyl propane sulfonic acid and butyl methacrylate in conjunction with poly(2 methacryloyloxyethyl phosphorylcholine-*co*-butyl methacrylate) are capable of suppressing platelet adherence. Similar results [486] were found on poly(gamma benzyl 1 glutamate-co-leucine) neutralized with sodium.

Significant research has been directed toward the use of polyelectrolyte complexes as blood compatible materials. Several investigators found that water-insoluble polyelectrolyte complexes can suppress blood coagulation [487–490]. Davison and coworkers reviewed and studied the biological properties of water-soluble polyelectrolyte complexes [491] between quaternized poly(vinyl imidazole) or poly(vinyl pyridine) and excess sulfonated dextran or poly(methacrylic acid). By forming complexes with a stoichiometric excess of anionic charge, a more compact conformation with anionic character was obtained.

Other investigations [459, 492, 493] showed that the complexes form blocked structures with complex cores and polyanion shells. When these non-stoichiometric polyelectrolyte complexes were subjected to platelet aggregation studies, the toxic effects of the polycations and, to a lesser extent, polyanions were reduced. *In vivo*, complexation retarded but did not eliminate interactions with negatively charged biological surfaces, but *in vivo* ionic bonds were not sufficiently strong to hold the systems together in biological fluids. Here the polyanions gave relatively mild immunoresponse [459, 492], polycations gave undesirable effects arising from ionic interactions with cell surfaces [494, 495].

When similar polyelectrolyte complexes were used as drug carriers [336, 496–498], steric barriers to antigenicity and increased circulation times were observed. Reductions in interactions between carriers and proteins and cells of the vasculature gave rise to increased avoidance of mononuclear phagocytic system.

The hemocompatibility of poly(amido-amine) polyelectrolyte complexes was recently studied by Xi, Zhang and coworkers [499, 500]. The poly(amido-amine) was based on piperazine and methylene bisacrylamide, and the polyelectrolyte complexes were obtained from the reaction of poly(amido-amine) with alginic acid, carboxymethyl cellulose or poly(methacrylic acid). Complexes of polyamido-amine and alginic acid with a 1:2 ratio gave the best hemocompatibility. Finally, the blood compatibility of polyelectrolyte complexes based on anionic and cationic cellulose derivatives were studied by Ito et al. [338]. *In vivo*, good blood compatibility of complexes formed from quaternary hydroxyethyl cellulose reacted with carboxymethyl cellulose and cellulose sulfate was observed.

The most widely used approach to formation of biocompatible surfaces has been heparinization. Both physical adsorption and covalent binding of heparin onto polymer surfaces have been attempted. The immobilization of heparin, heparin analogs (including heparin-like synthetic polyelectrolytes) and heparin/prostaglandin and heparin/fibrinolytic enzyme conjugates was reviewed by Jozefowicz [501]. Other approaches including albuminated surfaces, fibronectin coatings and immobilization of fibrinolytic enzymes were discussed by Merrill [502]. Among the most common approaches to surface modification is grafting of a polyelectrolyte onto the surface, followed by binding of antithrombogenic or fibrinolytic agents onto the graft copolymers [503].

Polyetherurethaneureas with tertiary amino groups were synthesized, quaternized and complexed with heparin [504–508] in order to slow release of heparin from side chains. The polymers contained either poly(tetramethyleneglycol) or an aminoether oligomer and were synthesized from dimethylaminomethyl 2-hydroxymethyl propanol and 1,6 dihydroxyhexane to give the pendant dimethylaminomethyl side group. Thrombus formation was reduced significantly (by in vitro blood clotting test) on the quaternized materials, and it was thought that high swelling contributes to high heparin release rates, giving the improved antithrombogenicity.

Benvenuti and co-workers [37, 503] studied synthesis strategies for tertiary amino polymers (poly(amidoamines)) which complex heparin from aqueous solutions. Calorimetric studies showed electrostatic interactions between the polyamine and heparin at physiological pH, and anticoagulation properties of heparin were retained even if it was linked to poly(amido amine) [509, 510]. These polymers were used to modify silastic, glass, PVC, Dacron and polyurethane. Chemical grafting was studied using different spacer arms, and the surfaces were heparinized by dipping into heparin solutions. Similarly, block copolymers of poly(ether amidoamine urethane) in which the poly(etherurethane) segments are compatible with heparinized poly(amidoamine) segments were also investigated. The antithrombogenic properties of these materials were better than the non-heparinized materials [511, 512].

Ikada and coworkers also studied the blood compatibility and protein denaturation properties of heparin covalently and ionically bound onto polymer surfaces [513]. Both types of bound heparin gave deactivation of the coagulation process. Clotting deactivation was attributed to a heparin/antithrombin III complex by covalently bound heparin which gave adsorbed protein denaturation and platelet deformation as compared with lack of these features with ionically bound heparin.

Problems of desorption and loss of activity encountered with natural heparin have led numerous workers to explore synthetic heparin-like polymers or heparinoids, as reviewed by Gebelein and Murphy [475, 514, 515]. The blood compatibility of 5% blended polyelectrolyte/poly(vinly alcohol) membranes was studied by Aleyamma and Sharma [516, 517]. The membranes were modified with synthetic heparinoid polyelectrolytes, and surface properties (platelet adhesion, water contact angle, protein adsorption) and bulk properties such as permeability and mechanical characteristics were evaluated. The blended membrane had a lower tendency to adhere platelets than standard cellulose membranes and were useful as dialysis grade materials.

Synthetic heparinoid polymers such as polyanionic polyesters containing sulfamate and carboxylate groups were synthesized by hydrolysis of $p(N$ chlorosulfonyl beta lactam) [475]. These materials possess anti-coagulant activity (although lower than heparin) presumably due to the similarity of functional groups of this material and those found in heparin.

Van der Does and coworkers [518, 519] reviewed the use of polyelectrolytes derived from *cis*- and *trans*-polyisoprene as antithrombogenic materials. These polyelectrolytes were synthesized by the reaction of polyisoprene with chlorosulfonyl isocyanate to give sulfonate and amide or sulfonate and caroboxylate functionality. The resulting polymers were used to coat poly(vinyl chloride), polystyrene and silicone rubber to increase hemocompatibility. The polyelectrolytes were linked ionically to polymer surfaces by adsorptive coupling agents, or crosslinked into hydrogels. The ionically linked materials gave substantial reductions in platelet adhesion but were not stable in blood plasma due to leakage. While the hydrogels did not deplete antithrombin II, reduced platelet adhesion indicated good biocompatibility. Furthermore, irradiation of the above polymer solutions in the presence of silicone rubber under non-gel conditions gave polymer surfaces with reduced platelet adhesion and evidence of surface grafting [518]. The N sulfate and carboxylate groups were found essential for anticoagulant activity.

Chitin and chitosan derivatives have also been studied as blood compatible materials both *in vivo* and *in vitro* [520]. Anticoagulant activity was greatest with O sulfated N acetyl chitosan, followed by N,O sulfated chitosan, heparin, and finally sulfated N acetyl chitosan. The lipolytic activity was greatest for N,O sulfated chitosan followed by heparin. The generally poor performance of chitosan was attributed to polyelectrolyte complexes with free amino groups present on the membrane surface. The O sulfate or acidic group at the 6 position in the hexosamine moiety was identified as the main active site for anticoagulant activity.

Finally, the hetero polysaccharide hexagluconylhexasaminoglycan sulfate was identified as an anticoagulant and antithrombotic material compared to heparin by Sederel and coworkers [521]. Derivatives of the polysaccharide dextran containing sulfate, sulfonate or carboxylic groups gave antithrombotic activity due to the presence of these charged groups [483, 522–525].

Acknowledgments. The authors appreciate the secretarial and other help provided by Elizabeth Curtiss, Brenda Becker, and Maxine Soper in the preparation of this article.

8 References

1. Odian G (1991) Principles of polymerization, 3rd edn. Wiley, New York
2. Rodrigues F (1982) Principles of polymer systems, 2nd edn. McGraw-Hill, New York
3. Billmeyer FW (1984) Textbook of polymer science, 3rd edn. Wiley, New York
4. Yin YL, Prud'homme RK, Stanley F (1992). In: Harland RS, Prud'homme RK (eds) Polyelectrolyte gels, ACS Symposium Series 480. American Chemical Society, Washington DC, p 91
5. Buchholz FL (1990). In: Brannon-Peppas L, Harland RS (eds) Absorbent polymer technology. Elsevier Science, New York, NY, p 23
6. Greenwald HL, Luskin LS (1980). In: Davidson RL (ed) Handbook of water-soluble gums and resins. McGraw-Hill, New York, p 17
7. McCormick CL (1991). In: Shalaby SW, McCormick CL, Butler GB (eds) Water-soluble polymers, ACS Symposium Series #467 American Chemical Society, Washington DC, p 2
8. Nemec JW, Bauer W (1988). In: Kroschwitz JI (ed) Encylcopedia of polymer science and engineering. Wiley, New York, p 211
9. Samsonov GV, Kuznetsova NP (1992) Advances in Polym Sci 104: 1
10. Kazanskii KS, Dubrovskii SA (1992) Advances in Polym Sci 104: 97
11. Peppas NA, Mikos AG (1986) in: Peppas NA (ed) Hydrogels in medicine and pharmacy, Vol I, CRC Press, Boca Raton, Florida, p 1
12. Malavasic T, Osredkar U, Anzur I, Vizovisek I (1986) J Macromol Sci Chem A23: 853
13. Brondsteo H, Kopecek J (1992) In: Harland RS, Prud'homme RD (eds) Polyelectrolyte gels, ACS symposium series 480. American Chemical Society, Washington DC, p 285
14. Glass JE, McDonald WF, Lundberg DL (1989) ACS Division of polymer chemistry, polymer Preprints 30: 387
15. Ellis J, Anstice M, Wilson AD (1991) Clinical materials 7: 341
16. Rath NC, Dimitrijevich S, Anbar M (1984) Chem-Biol Interations 48: 339
17. Muqbill R, Muller G, Fenyo JC, Selegny E (1979) J Polym Sci: Polym Lett 17: 369
18. Middleton JC, Cummins D, McCormick CL (1989) ACS Division of polymer chem. Polymer Preprints 30: 348
19. McCormick CL, Middleton JC, Cummins DF (1992) Macromolecules 25(4): 1201
20. Salamone JC, Ellis EJ, Bardoliwalla DF (1974) J Polym Sci Polym Symp 45: 51
21. Gross JR (1990) In: Brannon-Peppas L, Harland RS (eds) Absorbent polymer technology. Elsevier Science Publishing Company Inc. New York, NY, p 3
22. Schosseler F, Ilmain F, Candau SJ (1991) Macromolecules 24: 225
23. Wen S, Yin X, Stevenson WTK (1991) J Appl Polymer Science 42: 1399
24. Bajoras G, Makuska M, (1986) Polm J 18: 955
25. Yoshida M, Tamada M, Kumakura M, Katakai R (1991) Radiat Phys Chem 38: 7
26. Pradny M, Kminek I, Sevcik S (1986) Polym Bull 16: 195
27. Candau F, Zekhnini Z, Heatly F, Franta E (1986) Colloid & Polymer Sci 264: 677
28. Brannon-Peppas L (1990) In: Brannon-Peppas L, Harland RS (eds) Absorbent polymer technology. Elsevier Science Publishing Company Inc. New York, NY, p 45

29. Klier J, Scranton AB, Peppas NA (1990) Macromolecules 23: 4944
30. Kudo S, Konno M, Saito S (1992) Makromol Chem Rapid Commun 13: 545
31. Schoo HFM, Challa G (1992) Macromolecules 25: 1663
32. Akashi M, Takemoto K (1990) Advances in Polym Sci 97: 107
33. Peppas NA, Khare AR (1993) Advanced Drug Del Rev 11: 1
34. Trijasson P, Pith T, Lambla M (1990) Makromol Chem Macromol Symp 35/36: 141
35. Hunkeler D, Hamielec AE (1992) In: Harland RS, Prud'homme RK (eds) Polyelectrolyte gels, ACS Symposium Series 480 American Chemical Society, Washington, DC, p 24
36. Kim SW, Feijens J (1985) In: Williams D (ed) CRC Critical reviews in biocompatibility CRC press, Boca Raton FL, 1(3): 229
37. Benvenuti M, Dal Maso G (1989) In: Dawids S (ed) Polymers: their properties and blood compatibility p 259
38. Miller RM, Taylor DEM, Ringrose BS (1985) ESAO Proceedings 12: 466
39. Harada J, Chern RT, Stannett VT (1992) In: Harland RS, Prud'homme RK (eds) Polyelectrolyte gels, ACS Symposium Series 480 American Chemical Society, Washington, DC, p 80
40. Ishigaki I, Sugo T, Senoo K, Okada T, Okamoto J, Machi S (1982) J Appl Polym Sci 27: 1033
41. Ishigaki I, Sugo T, Takayama T, Okada T, Okamoto J, Machi S (1982) J Appl Polym Sci 27: 1043
42. Omichi H, Chundury D, Stannett VT (1986) J Appl Polym Sci 32: 4827
43. Butler GB, Zhang NZ (1991) In: Shalaby SW, McCormick CL, Butler GB (eds) Water-soluble polymers, ACS Symposium Series 467 American Chemical Society, Washington DC, p 25
44. Wandrey C, Jaeger W (1985) Acta Polymerica 36: 100
45. Yang YJ, Engberts Jan BFN (1991) J Org Chem 56: 4300
46. Bartels T, Arends J (1981) J Polym Sci Polym Chem 19: 128
47. Wen S, Yin X, Stevenson WTK (1991) Journal of Appl Polymer Science 43: 205
48. Goethals EJ (1980) Polymeric amines and ammonium salts, Pergamon Press, Oxford
49. Boussouira B, Ricard A (1987) J Macromol Sci Chem A24(2): 137
50. Korshak VV, Zubakova LB, Gandurina LV (1985) Env Protect Eng 11: 103
51. Salamone JC, Mahmud NA, Mahmud MU, Nagabhushanam T, Watterson AC (1982) Polymer 23: 843
52. Salamone JC, Ahmed I, Raheja MK, Elayaperumal P, Watterson AC, Olson AP (1988) In: Stahl GA, Shultz DN (eds) Water soluble polymers for petroleum recovery, Plenum, New York p 181
53. Salamone JC, Watterson AC, Olson AP, Tsai CC, Mahmud MU (1980) J Polym Sci Polym Chem Ed 18: 2983
54. Salamone JC, Raheja MK, Anwaruddin MU, Watterson AC, (1985) J Polym Sci Polym Lett Ed. 23: 12
55. Salamone JC, Volksen W, Olson AP, Israel SC (1978) Polymer 19: 1157
56. Taylor LD, Kolesinski HS (1986) J Polymer Science: Part C: Polymer Letters 24: 287
57. Salamone JC, Richard RE, Su CH, Watterson AC (1991) J Macromol Sci Chem A28(2): 225
58. Culbertson BM (1985) In: Mark HF, Kroschwitz J (eds) Encyclopedia of polymer science and engineering, John Wiley and Sons, Inc. New York, 11: 740
59. Salamone JC, Richard RE, Watterson AC (1986) J Polym Sci Polym Symp 74: 187
60. Pujol-Fortin M, Galin JC (1991) Macromolecules 24: 4523
61. Hsu YG, Hsu MJ, Chen KM (1991) Makromol Chem 192: 999
62. Schulz RC, Schmidt M, Schwarzenbach E, Zoeller J (1989) Makromol Chem Macromol Symp 26: 221
63. Dominguez L, Meyer RC, Wegner G (1987) Makromol Chem Rapid Commun 8: 151
64. Kataoka K, Ohki N, Tsuruta T (1979) Makromol Chem 180: 65
65. Akelah A, Moet A (1990) Functionalized polymers and their applications. Chapman and Hall, New York
66. Walles WE (1990) In: Koros WJ (ed) Barrier polymers and structures, ACS Symposium Series 423, American Chemical Society, Washington DC, p 266
67. Mays JW (1990) Polymer Comm 31: 170
68. Vink H (1981) Makromol Che 182: 279
69. Ihata J (1988) J Polym Sci Polym Chem 26: 167
70. Ihata J (1988) J Polym Sci Polym Chem 26:177
71. Ihata J (1988) J Polym Sci Polym Chem 26: 187
72. Santerre JP, ten Hove P, Brash JL (1992) J Biomed Mater Res 26: 1003

73. Santerre JP, ten Hove P, Vanderkamp NH, Brash JL (1992) J Biomed Mater Res 26: 39
74. Brydson JA (1982) Plastic materials, 4th edn., Butterworth, London
75. Bansleben DA, Jachimowicz F (1985) ACS Division of polymer chemistry, Polymer Preprints (ACS Div Polym Chem) 26: 106
76. Kobayshi S, Gros L, Muacevic G, Ringsdorf H (1983) Makromol Chem 184: 793
77. Perner T. Schulz RC (1987) Bitish Polymer J 19: 181
78. Fini A, Casolaro M, Nocentini M, Barbucci R, Laus M (1987) Makromol Chem 188: 1959
79. Dragan S, Grigoriu G, Petrariu I (1991) Polymer Bulletin 27: 17
80. Dragan S, Barboiu V, Petrariu I (1981) J Polym Sci Poym Chem 19: 2869
81. Gieselman MB, Reynolds JR (1992) Macromolecules 25: 4832
82. Ishihara K, Nakabayshi N (1989) J Biomed Mater Res 23: 1475
83. Hassan RM, Abd-Alla MA, El-Zohry MF (1993) Journal of Appl Polymer Science 47: 1649
84. Mandel M (1988) Encyclopedia of polymer science and technology, 11: 739
85. Armstrong RW, Strauss UP (1969) Enc Polym Sci and Tech 10: 781
86. Hermans JJ, Overbeek JTG (1948) Rec Trav Chim 67: 761
87. Kimball GE, Cutler M, Samelson H (1957) J Phys Chem 56: 47
88. Lifson S (1957) J Chem Phys 27: 100
89. Oosawa F, Imani N, Kajawa I (1954) J Polym Sci 13: 93
90. Alexandrowicz Z, Katchalsky A (1963) J Polym Sci A1: 3231
91. Katchalsky A, Alexandrowicz Z, Kedem O (1966) In: Conway BE, Barradas RG (eds) Chemical physics of ionic solutions, Wiley, New York, p 295
92. Kotin L, Nagasawa M (1961) J Am Chem Soc 83: 1026
93. Manning GS, Zimm BH, (1965) J Chem Phys 43: 4250
94. Gross LM, Strauss UP (1966) In: Conway BE, Barradas RG (eds) Chemical physics of ionic solutions, Wiley, New York p 361
95. Strauss UP, Ross PD (1959) J Am Chem Soc 81: 5295
96. Mandel M (1984) Die Angew Makromol Chem 123: 63
97. Manning GS (1972) Ann Rev Phys Chem 23: 117
98. Manning G (1969) J Chem Phys 51: 924
99. Manning G (1969) J Chem Phys 51: 934
100. Stigter D (1975) J Colloid Interface Sci 53: 296
101. Russel WB (1982) J Polym Sci Phys Ed 20: 1233
102. Nicolai T, Mandel M (1989) Macromolecules 22: 438
103. Benoit H, Doty P (1958) J Phys Chem 57: 958
104. Yamakawa G (1971) Modern theories of polymer solutions, Harper and Row, New York
105. Odijk T (1977) J Polym Sci Polym Phys Edn. 15: 477
106. Odijk T, Houwaart AC (1978) J Polym Sci Polym Phys Ed. 16: 627
107. Fixman M, Skolnick J (1978) Macromolecules 11: 863
108. Davis RM (1984) PhD Thesis Princeton University Princeton
109. Fixman M (1982) J Chem Phys 76: 6346
110. Borue VY, Erukhimovich IY (1988) Macromolecules 21: 3240
111. Khokhlov AR, Khachuturian KA (1982) Polymer 23: 1742
112. Khokhlov AR (1980) J Phys A: Math Nucl Gen 13: 979
113. Hasa J, Ilavsky M, Dusek K (1975) J Polym Sci Polym Phys Edn. 13: 253
114. Hasa J, Ilavsky M (1975) J Polym Sci Polym Phys Edn. 13: 263
115. Konak, Bansil M (1989) Polymer 30: 677
116. Prud'homme RK, Yin YL (1993) Proceedings of the ACS, polymeric materials science and engineering 69: 527
117. Tsuchida E, Abe K (1982) Adv Polym Sci 45: 1
118. Bekturov E, Bimendina LA (1980) Adv Polym Sci 41: 100
119. Osada Y (1987) Adv Polym Sci 82: 1
120. Wilson AD (1991) Clinical Materials 7: 275
121. Pearson GJ (1991) Clinical Materials 7: 325
122. Phillips RW (1982) Skinner's Science of Dental Materials, WB Saunder Co., Philadelphia, PA
123. Craig RG, O'Brien WJ, Power JM (1983) Dental Materials Properties and Manipulation, CV Mosby, St. Louis, MO
124. Smith DC (1968) Brit Dent J 125: 381
125. Wilson AD, Kent BE (1973) British Patent No. 1316129, Surgical Cement
126. Wilson AD (1978) Chem Soc Revs 7: 265

127. Wilson AD, Prosser HJ (1984) Brit Dent J 157: 449
128. Bowen RL, Marjenhoff WA (1992) Adv Dent Res 6: 44
129. Williams J, Billington RW (1989) J Oral Rehab 16: 475
130. Williams J, Billington RW (1991) J Oral Rehabil 18(2): 163
131. Wood D, Hill R (1991) Clinical Materials 7: 301
132. Crisp S, Wilson AD (1974) J Dent Res 53: 1408
133. Wilson AD, McLean JW (1988) Glass-Ionomer Cement, Quintessence Publishing Co. Chicago
134. Crisp S, Pringuer MA, Wardleworth D, Wilson AD (1974) J Dent Res 53: 1414
135. Wilson AD, Crisp S, Abel G (1977) J Dent 5: 117
136. Hill RG, Wilson AD, Warrens CP (1989) J Mat Sci 24: 363
137. Hill RG, Labok SA (1991) J Mat Sci 26: 67
138. Cook WD (1982) Biomaterials 3: 232
139. Nicholson JW, Brookman PJ, Lacy OM, Wilson AD (1988) J Dent Res 67(12): 1451
140. Prosser HJ, Richards CP, Wilson AD (1982) J Biomed Mater Res 16: 431
141. Barry TI, Clinton DJ, Wilson AD (1979) J Dent Res 58: 1072
142. Cooley RL, Train TE (1991) J Prosthet Dent 66: 773
143. Wasson EA, Nicholson JW (1991) Clinical Materials 7: 289
144. Wasson EA, Nicholson JW (1990) Brit Poly J 23: 179
145. Wasson EA, Nicholson JW (1993) J Dent Res 72(2): 481
146. Blagojevic B, Mount GJ (1989) Aust Dent J 33: 320
147. Haddad D, Mount GJ, Makinson OF (1992) Am J Dent 5: 286
148. Crisp S, Wilson AD (1976) J Dent Res 55: 1023
149. Crisp S, Lewis BG, Wilson AD (1979) J Dent 7: 304
150. Cook WD (1983) Biomaterials 4: 85
151. Hill RG, Wilson AD (1988) J Dent Res 67(12): 1446
152. Pearson GJ, Atkinson AS (1987) Dent Mater 3: 275
153. Brune D, Evje DM (1984) Scand J Dent Res 92: 156
154. Wilson AD, Crisp S, Prosser HJ, Lewis BG, Merson SA (1980) Ind Eng Chem Prod Res Dev 19: 263
155. Brune D (1982) Scand J Dent Res 90: 409
156. Wood D, Hill R (1991) Biomaterials 12: 164
157. Wilson MA, Combe EC (1991) Clinical Materials 7: 15
158. Neve AD, Piddock V, Combe EC (1993) Clinical Materials 12: 113
159. Billington R, Williams J, Pearson GJ (1990) Brit Dent J 169: 164
160. Neve AD, Piddock V, Combe EC (1992) Clin Mater 9: 13
161. Neve AD, Piddock V, Combe EC (1992) Clin Mater 9: 21
162. Hodd KA, Reader AL (1976) Br Polym J 8: 131
163. Anbar M, St. John GA (1971) J Dent Res 50: 778
164. Anbar M, Farley EP (1974) J Dent Res 53: 879
165. Ellis J, Wilson AD (1990) J Mater Sci Letters 9: 1058
166. Ellis J, Wilson AD (1991) Polym Int 24: 224
167. de Groot K, de Visser AC, Driessen AA, Wolke JGC (1980) J Dent Res (59)9: 1493
168. Wilson AD (1990) Int J Prosthod 3: 425
169. Lin A, McIntyre NS, Davidson RD (1992) J Dent Res 71(11): 1836
170. Mitra SB (1991) J Dent Res 70(1): 72
171. Hinoura K, Miyazaki M, Onose H (1991) J Dent Res 70(12): 1542
172. Wasson EA (1993) Clinical Materials 12: 181
173. Iioka A, Araki Y, Matsuda K, Ohno H (1989) Dental Materials Journal 8(2): 236
174. Wilson AD, Prosser HJ, Powis DR (1983) J. Dent. Res. 62:590
175. Misra DN (1988) Langmuir, 4: 953
176. Bowen RL (1985) Int Dent J 35: 159
177. Misra DN (1989) J Dent Res 68: 42
178. Nakabayashi N, Kojima K, Masuhara E (1982) J Biomed Mater Res 16: 265
179. Nakabayashi N (1984) CRC Crit Rev Biocomp 1: 25
180. Takeyama M, Kashibuchi N, Nakabayashi N, Masuhara E (1978) J Jpn Dent Appar Mater 19: 179
181. Yamaguchi J (1986) J Jpn Dent Mater 5: 144
182. Nakabayashi N, Kanda K (1988) Kobunshi Ronbunshu 45: 91
183. Suzuki S, Sakoh M, Shiba A (1990) J Biomed Mater Res 24: 1091

184. Ishihara K, Arai H, Nakabayashi N, Morita S, Furuya K (1992) J Biomed Mater Res 26: 937
185. Krabbendam CA, Ten Harkel HC, Duijsters PPE, Davidson CL (1987) J Dent 15: 77
186. Tanaka T, Atsuta M, Nakabayashi N, Masuhara E (1988) J Prosthet Dent 60: 271
187. Tanaka T, Hirano M, Kawahara M, Matsumura H, Atsuta M (1988) J Dent Res 67: 1376
188. Watanabe F, Powers JM, Lorey RE (1988) J Dent Res 67: 479
189. Dilts WE, Duncanson MG Jr., Miranda FJ, Brackett SE (1985) J Prosthet Dent 53: 505
190. Mojon P, Hawbolt EB, MacEntee MI, Ma PH (1992) J Dent Res 71(9): 1633
191. Swift EJ (1989) Am J Dent 2: 358
192. Hewlett ER, Caputo AA, Wrobel DC (1991) J Prosthet Dent 66: 767
193. Wang T, and Nakabayashi N (1991) J Dent Res 70(1): 59
194. Imai Y, Kadoma Y, Kojima K, Akimoto T, Ikakura K, Ohta T (1991) J Dent Res 70(7): 1088
195. Nakabayashi N, Yamashita S, Kojima K, Masuhara E (1981) Rep Inst Med Dent Eng 15: 37
196. Kiyomura M, Kanda K, Nakabayashi N (1987) J Jpn Dent Mater 6: 719
197. Asmussen E, Hansen EK, Peutzfeldt A (1991) J Dent Res 70(9): 1290
198. Asmussen E, Uno, (1993) J Dent Res 72(3): 558
199. Suzuki M, Kato H, Wakumoto S, Katagiri G, Ishitani A (1987) J Raman Spectrosc 18: 315
200. Suzuki M, Kato H, Wakumoto, S (1991) J Dent Res 70(7): 1092
201. Van Meerbeek B, Dhem A, Goret-Nicaise M, Braem M, Lambrechts P, Vanherle G (1993)
     J Dent Res 72(2): 495
202. Bowen RL, Cobb EN, Misra DN (1984) Ind Eng Chem Prod Res Dev 23: 78
203. Misra DN, Johnston AD (1987) J Biomed Mater Res 21: 1329
204. Bowen RL, Tung MS, Blosser RL, Asmussen E (1987) Int Dent J 37: 158
205. Zuellig-Singer R, Krejci I, Lutz F (1992) J Dent Res 71(11): 1842
206. Webb RE, Johnston AD (1991) J Dent Res 70(3): 211
207. Chen R-S, Bowen RL (1989) J Adhes Sci Technol 3: 49
208. Johnston AD, Asmussen E, Bowen RL (1989) J Dent Res 68: 1337
209. Venz S, Dickens B (1993) J Dent Res 72(3): 582
210. Eick JD, Robinson SJ, Cobb CM, Chappell RP, Spencer P (1992) Quint Int 23: 43
211. Hasegawa T, Manabe A, Itoh K, Wakumoto S, (1989) Dent Mater 5: 150
212. Tagami J, Hosoda H, Imai Y, Masuhara E (1987) Dent Mater J 6: 201
213. Hayakawa T, Endo H, Hara T, Fukai K, Horie K (1988) Dent Mater J 7: 19
214. Chigira H, Ito K, Wakumoto S (1988) Jpn J Conserv Dent 31: 667
215. Chigira H, Ito K, Wakumoto S (1991) Dent Mater 7: 103
216. Munksgaard EC, Irie M, and Asmussen E (1985) J Dent Res 64: 1409
217. Zidan O, Gomez-Marin O, and Tsuchiya T (1987) J Dent Res 66: 716
218. McLean JW, Powis DR, Prosser HJ, Wilson AD (1985) Brit Dent J 158: 410
219. Garcia-Godoy F (1992) Am J Dent 5: 283
220. McCaghren RA, Retief DH, Bradley EL, Denys FR (1990) J Dent Res 69: 40
221. Prati C, Montanari G, Biagini G, Fava F, Pashley DH (1992) Dent Mater 8: 21
222. Prati C, Mongiorgi R, Valdre G, Montanari G (1991) Clinical Mater 8: 137
223. Powis DR, Folleras T, Merson SA, Wilson AD (1982) J Dent Res 61: 1416
224. Garcia-Godoy F (1988) Am J Dent 1: 97
225. Brackett WW, Johnston WM (1989) J Prosthet Dent 62: 261
226. Pearson GJ, Bowen G, Jacobsen P, Atkinson AS (1989) Dent Mater 5: 10
227. Scherer W, Vaidyanathan J, Kaim JM, Hamburg M (1989) Oper Dent 14: 82
228. Button GL, Barnes RF, Moon PC (1985) J Prosthet Dent 53: 34
229. Tjan AHL, Peach KD, VanDenburgh SL, Zbaraschuk ER (1991) J Prosthet Dent 66: 602
230. Oilo G (1992) Adv Dent Res 6: 50
231. Stanley HR (1992) Adv Dent Res 6: 55
232. Bayne SC (1992) Adv Dent Res 6: 65
233. Ben-Amar A (1989) Am J Dent 2: 175
234. Lynch E, Tay WM (1989) J Irish Dent Assn 35: 66
235. Brackett WW, Metz JE (1992) J Prosthet Dent 67: 59
236. Heys RJ, Fitzgerald M (1991) J Dent Res 70(1): 55
237. White SN, Sorensen JA, Kang SK, Caputo AA (1992) J. Prosthet Dent 67: 156
238. Scherer W, Kaim J, Gottlieb-Schein E, Roffe-Bauer M (1989) Am J Dent 2: 355
239. Crim GA (1993) J Prosthet Dent 69: 561
240. Davis EL, Yu X, Joynt RB, Wieczkowski G Jr., Giordano L (1993) Am J Dent 6: 127
241. Crim GA (1993) Am J Dent 6: 192

242. Um CM, Oilo G (1992) Quint Int 23: 209
243. Osborne JW, Berry TG (1986) Dent Mater 2: 147
244. Osborne JW, Berry TG (1990) Am J Dent 3: 40
245. DeSchepper EJ, Wl..te RR, von der Lehr W (1989) Am J Dent 2: 51
246. Tobias RS, Browne RM, Wilson CA (1985) Int Endod J 18: 161
247. McComb D, Ericson D (1987) J Dent Res 66: 1025
248. Berg HJ, Brown LR, Farrell JE, Puente ES (1988) J Dent Res 67: 280 (Abstr 1336)
249. Lippman N, Scherer W, Kaim J (1988) J Dent Res 67: 263 (Abstr 1201)
250. Barkhordar RA, Kempler D, Pelzner RB, Stark M (1988) J Dent Res 67: 263 (Abstr 1202)
251. Hicks MJ, Flaitz CM, Silverstone LM (1986) Quintessence Int 17: 527
252. Hicks MJ, Flaitz CM (1992) Am J Dent 5: 329
253. Brook IM, Craig GT, Lamb DJ (1991) Clinical Mater 7: 295
254. Jonck LM, Grobbelaar CJ (1992) Clinical Mater 9: 85
255. Lindeque GP, Jonck LM (1993) Clinical Mater 14: 49
256. Jonck LM, Grobbelaar CJ, Starting H (1989) Clinical Mater 4: 85
257. Brook IM, Craig GT, Hatton PV, Jonck LM (1992) Biomaterials 13(10): 721
258. Swartz ML, Phillips RW, Clark HE (1984) J Dent Res 63: 158
259. Toumba KJ, Curzon MEJ (1993) Caries Research, 27(suppl 1): 43
260. Retief DH, Bradley EL, Denton JC, Switzer P (1984) Caries Res 18: 250
261. Forss H, Seppa L (1990) Scand. J Dent Res 98: 173
262. Wilson AD, Groffman DR, Kuhn AT (1985) Biomaterials 6: 431
263. Mitra SB (1991) J Dent Res 70(1): 75
264. Hatibovic-Kofman D, Koch G (1991) Swed Dent J 15: 253
265. Forsten L (1991) Scand. J Dent Res 99: 241
266. Verbeeck RMH, DeMoor RJG, Van Even DFJ, Martens LC (1993) J Dent Res 72(3): 577
267. Forsten L (1977) Scand J Dent Res 85: 503
268. Garcia-Godoy F, Chan DCN (1991) Am J Dent 4: 223
269. Meryon SD, Smith AJ (1984) Int Endodont J 17: 16
270. Cranfield M, Kuhn AT, Winter GB (1982) J Dent 10: 333
271. Wilson MA, Combe EC (1988) Clinical Materials 3: 273
272. Kulkarni PV, Rajur SB, Antich PP, Aminabhavi TM, Aralayuppi MI (1990) JMS Rev Macromol Chem Phys C30: 364
273. Tani N, Van Dress M, Anderson JM (1980) In: Lewis DH (ed) Controlled release of pesticides and pharmaceuticals, Plenum, New York, p 79
274. Chien YW (1990) Med Res Rev 10(4): 477
275. Chien YW (1983) Drug Dev Ind Pharm 9: 497
276. Bruck SD, Mueller EP (1988) Rev Ther Drug Carrier Syst 5(3): 171
277. Dumitriu S, Popa M, Dumitriu I (1990) J Bioact and Compatible Polymers 5: 89
278. Chang RK, Price JC (1988) J Biomaterials Applications 3: 80
279. Mortny KS, Kubert DA, Fawzi MB (1988) J Biomaterials Applications. 3
280. Davis M, Ichikawa I, Williams EJ, Banker GS (1986) Int J Pharm 28: 157
281. Davis MBG, Peck GE, Banker GS (1986) Drug Dev and Ind Pharm 12(10): 1419
282. McGinty JW (1989) Aqueous polymeric coatings for pharmaceutical dosage forms, Ed. Marcel Dekker, p 317
283. Lehman K, Klaus OR (1989) In: McGinity JW (ed) Chemistry and application properties of polymethacrylate coating systems in aqueous polymeric coating for pharmaceutical dosage forms. Marcel Dekker, New York, p 153
284. Gumowski F, Doelker E, Gurny R (1987) Pharm Tech 11(2): 26
285. Lehmann K (1984) Acta Pharm Fenn 93: 55
286. Lehmann K (1985) Proc Int Symp Controlled Release Bioactive Mat 12: 361
287. Osterwald HP (1985) Pharm Res 1: 14
288. Murthy KS, Enders NA, Mahjar M, Fawzi MB (1986) Pharm Tech 10: 36
289. Okor RS, Otimenjin S, Ijeh I (1991) J Contr Release 16(3): 349
290. Siegel RA (1990) In: Kost J (ed) Pulsed and self-regulated drug delivery, CRC Press, Boca Raton FL, p 129
291. Kislalioglu MS, Khan MA, Blount C, Goettsch RW, Bolton S (1991) J Pharm Sci 80(8): 799
292. Kawashima Y, Iwamoto T, Niwa T, Takeuchi H, Ito Y (1991) Int J Pharm 75(1): 25
293. Buraie AN (1991) Alexandria J Pharm Sci 5(1): 37
294. Morimoto Y, Kokubo T, Sugibyashi K (1992) J Controlled Rel 18: 113

295. Kokubo T, Sugibayashi K, Morimoto Y (1991) J Controlled Rel 17: 69
296. Talukdar MM, Plaizier-Vercammen JA (1991) In: Kellaway IM (ed) Proc Intern Symp Control Rel Bioact Mater 18: 247
297. Alekseev KV, Ginsburg OS, Solotarevich ME, Mustafin RI (1991) In: Bundulis J (ed) Novel Drug Formulation Syst Delivery Devices, Int Semin, Latvia State Pharm Co Riga, Latvia p 59
298. Dunn RL (1991) ACS Symp Ser 469: 11
299. Ghanem A, Meshali M, Hashem F (1980) Pharm Acta Helv 55(3): 61
300. Graham NB (1986) NATO ASI Ser E (Polym Biomater) 106: 170
301. Colombo P (1993) Adv Drug Delivery Rev 11(1-2): 37–57
302. Craig DQM, Tamburic S, Buckton G, Newton JM (1992) In: Kopecek J (ed) Proceed Intern Symp Control Rel Bioact Mater 19: 260
303. Peraz Marcos B, Iglesias R, Gomez Amoza JL, Martinez Pacheco R, Souto C, Concheiro AJ (1991) J Controlled Rel 17: 267
304. Brannon-Peppas L, Peppas NA (1989) J Controlled Release 8: 267
305. Greenly RL, Zia H, Garbow J, Rodgers RL (1991) ACS Symp Ser 469 Polym Drugs Drug Delivery Syst p 213
306. Choulis NH, Papadopoulus H (1975) J Pharm Sci 641: 1033
307. Salib N, El-Fattah S, El-Massik M (1983) Pharm Ind 45: 902
308. Agabeyoglu IT (1986) Drug Dev Ind Pharm 12: 569
309. Malley J, Rollet M, Taverdt JL, Vergnaud JL (1987) Drug Dev Ind Pharm 13: 67
310. Capan Y, Serol S, Calis ST, Hincal AA (1989) Pharm Ind 51: 443
311. Souto C, Concheiro A (1991) Int J Pharm 67: 113
312. Li SP, Feld KM, Kowaeshi CR (1991) Drug Cev Ind Pharm 17(12): 1655
313. Johnson M, Medlen J (1985) Eur Polym J 21(2): 147
314. Grignon J, Scallan AM (1980) J Appl Polym Sci 25: 2829
315. Alhaique F, Riccieri FM, Santucci E, Crescenzi V, Gamini A (1984) J Pharm Pharmacol 37: 310
316. Durrani MJ, Todd R, Andrews A, Whitaker RW, Greenberg E, Benner SC (1992)) In: Kopecek J (ed) Proceed Intern Symp Control Rel Bioact Mater 19: 811
317. Dumitriu S, Dumitriu M, Dumitriu D, Medvichi C (1992) In: Ottenbrite RM, Chiellini E (eds) Polymers in Medicine, Technomic, Lancaster, PA, p 115
318. Rubinstein A, Nakar D, Sintov A (1992) International Journal of Pharmaceutics 84(2): 141
319. Firestone BA, Siegel RA (1988) Polym Comm 29: 204
320. Siegel RA, Falmarzian M, Firestone BA, Moxley BC (1988) J Controlled Rel 8: 179
321. Siegel RA, Firestone BA (1988) Macromolecules 21: 3254
322. Kou JH, Amidon GL, Lee PI (1988) Pharm Res 5: 592
323. Pradny M, Kopecek J (1990) Makromol Chem 1919(8): 1887
324. Hariharan D, Peppas NA (1992) In: Kopecek J (ed) Proceed Intern Symp Control Rel Bioact Mater 19: 367
325. Raghunathan Y, Amsel L, Hinsvark O, Bryant W (1981) J Pharm Sci 70: 379
326. Amsel LP, Hinsvark ON, Turenberg K, Sheumaker JL (1984) Pharm Tech 8(4): 28
327. Nakano M, Itoh N, Juni K, Sekikano H, Arita T (1980) Int J Pharm 4: 291
328. Navarro A, Ballesteros MP, Vasquez F (1991) STP Pharma Sci 1(2): 151
329. Blanchon S, Couazzare G, Gieg-Falson F, Cohen G, Puisieux (1991) Int J Pharm 72(1): 1
330. Lin SY, Lee CJ, Lin YY (1991) Pharmaceutical Research 8(9): 1137
331. Ozturk SS, Palsson BO, Donohue B, Dressman JB (1988) Pharmaceutical Research 5(9): 550
332. Fukumari Y, Yamauka Y, Ichikawa H, Takeuchi Y, Fukada T, Osaka Y (1988) Chem Pharm Bull 36(12): 4927
333. Chang RK (1990) Pharm Technol 14(10): 62
334. Souto C, Perez-Marcos B, Gomez-Amoza JL, Martinez-Pacheco R, Concheiro A (1992) Congr Int Technol Pharm 6th 3: 349 Assoc Pharm Galenique Ind, France
335. Philipp B, Koetz J, Linow KJ, Dautzenberg H (1991) Polymer News 16(4): 106
336. Petrak K (1986) J. Bioactive and Compatible Polymers 1: 202
337. Imenson AP (July 1983) Gums and Stab Food Ind, Proc 2nd Int Conf 2: 189
338. Ito H, Shibata T, Miyamoto T, Noishiki Y, Inagaki H (1986) J Appl Polym Sci 31: 2491
339. Kawashima Y, Handa T, Kasai A, Takenaka H, Lin SY, Ando Y (1985) J Pharmaceutical Sciences 74(3): 264
340. Chang RK, Hsio CH, Robinson JR (1987) Pharm Tech 11: 56
341. Kikuchi Y, Kubota N (1988) Bull Chem Soc Jpn 61: 2943
342. Kikuchi Y, Kubota N, Mitsuishi H (1988) J Appl Polym Sci 35: 259

343. Burgess DJ, Marek TA, Kwork TKK, Singh ON (1991) Polym Prepr 32(1): 600
344. Tsai M, Levy M (1984) J. Polym Sci Polym Chem Ed. 22(10): 2523
345. Tsai MF, Levy M (1984) In: Naden D, Streat M (eds) Ion exch technol, Horwood, Chichester, UK p 533
346. Skorodinskaya AM, Kemenova VA, Efimov VS, Mustafaev MI, Kasaikin VA, Zezin AB, Kabanov VA (1984) Khim Farm Zh 18(3): 283
347. Takahata H, Koida Y, Kobayashi M, Samejima M (1990) Chem Pharm Bull 38 (9): 2556
348. Kwon GS, Bae YH, Kim SW, Cremers H, Feijen J (1991) J Colloid & Interface Science 143(2): 501
349. Kotz J, Paulke B, Philipp B, Denkinger P, Burchard W (1992) Acta Polymer 43(4): 193
350. Kotz J, Borrmeister B, Philipp B (1991) Acta Polymerica 42: 28
351. Kotz J, Hahn M, Philipp B (1993) Makromol Chem 194: 397
352. Oraceska B, Nixon JR, Solomon MC (1992) In Kopecek J (ed) Proc Intern Symp Control Rel Bioact Mater 19: 198
353. Meshali MM, Gabr KE (1993) International Journal of Pharmaceutics 89(3): 177
354. Kost J, Langer R (1991) Advanced Drug Delivery Reviews. 6: 19
355. Pitt CG (1986) Pharmacy International 88
356. Heller J (1993) Crit Rev Theo Drug Carrier Syst 10(3): 253–305
357. Ito Y (1992) In: Imanishi Y (ed) Synthesis of biocomposite materials, CRC Press Boca Raton, Fla p 137
358. Horbett TA, Kost J, Ratner BD (1983) Am Chem Soc Div Polym Chem 24: 34
359. Horbett TA, Kost J, Ratner BD (1984) In: Shalaby S, Hoffman AS, Horbett TA, Ratner BD (eds) Polymers as biomaterials, Plenum Press, New York, p 193
360. Kost J, Horbett TA, Ratner BD, Singh M (1985) J Biomed Mater Res 19: 1117
361. Albin G, Horbett TA, Ratner BD (1985) J Controlled Release 2: 153
362. Albin G, Horbett TA, Miller SR, Ricker NL (1987) J Controlled Release 6: 267
363. Albin G, Horbett TA, Ratner BD (1990) In: Kost J (ed) Pulsed and self regulated drug delivery, CRC Press, Boca Raton FL, p 159
364. Ishihara K, Kobayashi M, Shinohara I (1983) Makromol Chem Rapid Commun 4: 327
365. Ishihara K, Kobayashi M, Ishimaru N, Shinohara I (1984) Polym J 16: 625
366. Ishihara K Kobayashi M, Shonohara I (1984) Polym J 16: 647
367. Ishihara K, Matsui K (1986) J Polym Sci Polym Lett Edn. 24: 413
368. Ishihara K (1988) Proceed Intern Symp Controlled Rel Bioact Materials 15: 168
369. Demoor CP, Siegel RA (1989) Proceed Intern Symp Controlled Rel Bioact Mater 16: 157
370. Siegel RA, Firestone BA (1988) Proceed Intern Symp Controlled Rel Bioact Mater 15: 164
371. Kabra B, Gehrke SH (1989) Polym Prepr 30: 490
372. Grodzinsky AJ, Grimshaw PE (1990) In: Kost J (ed) Pulsed and self-regulated drug delivery, CRC Press, Boca Raton FL, p 47
373. Weiss AM, Grodzinsky AJ, Yarmish ML (1986) AIChE Symp Ser 82, 250: 85
374. Valentine JR, Miller LL (1982) Polym Prepr 29(2): 446
375. Osada Y, Gong JP, Ohnishi S, Sawahata K, Hori H (1991) J Macromol Sci Chem A28(11–12): 1189
376. Osada Y, Gong JP (1993) Prog Polym Sci 18: 187
377. Osada Y, Okuzaki H, Gong JP (1994) Trends in Polym Sci 2: 61
378. Sawahata K, Hara M, Yasunaga H, Osada Y (1990) J. Controlled Release 14: 253
379. Uchida K, Kaetsu I, Morita Y (1992) In: Kopecek J (ed) Proceed Intern Symp Control Rel Bioact Mater 19: 381
380. Kwon IC, Bae YH, Kim SW (1991) Nature 354: 291
381. Kwon IC, Bae YH, Okano T, Kim SW (1991) J Controlled Release 17: 149
382. Iwata H, Matsuda T (1988) J Membrane Sci 38: 185
383. Iwata H, Amemiyu H, Hata T, Matsuda T, Takano H, Akutsu T (1988) Proceed Intern Symp Controlled Rel Bioact Mater 15: 170
384. Barbucci R, Casularo M, Magnani A (1991) J Controlled Release 17: 79
385. Shateyeva LK, Samsonov GV (1979) J Appl Polym Sci 23: 2247
386. Ito Y, Casularo M, Kono M, Imanishi Y (1989) J Controlled Release 10: 195
387. Pefferkorn E, Schmitt A, Varoqui R (1982) Biopolymers 21: 1451
388. Idol WK, Anderson JL (1986) J Membr Sci 28: 269
389. Osada Y, Takeuchi Y (1981) J Polym Sci Polym Lett Ed. 19: 303
390. Osada Y, Hondo K, Ohta M (1986) J Membr Sci 27: 327
391. Ito Y, Kotera S, Inaba M, Konop K, Imanishi Y (1990) Polymer 31: 2157

392. Okahata Y, Ozaki K, Seki T (1984) J Chem Soc Chem Commun 519
393. Okahata Y, Noguchi H, Seki T (1987) Macromolecules 20: 15 (1987)
394. Kinoshita T,Yamashita I, Iwata T, Takizawa A, Tsujita Y (1983) J Macromol Sci Phys B 22: 1
395. Maeda M, Kimura M, Hareyama Y, Inoue S (1984) J Am Chem Soc106: 250
396. Chung DW, Higuchi S, Maeda M, Inoue S (1986) J Am Chem Soc 108: 5823
397. Higuchi S, Mozawa T, Maeda M, Inoue S (1986) Macromolecules 19: 2263
398. Devlin BP, Tirrell DA (1987) Polym Preprints 28(2)
399. Nagai T, Machida Y (1993) Adv Drug Delivery Rev 11(1-2): 179-191
400. Peppas NA, Buri PA (1984) J Controlled Rel 2: 257
401. Spence Leung SH, Robinson JR (1988) J Controlled Rel 5: 223
402. Park K, Chang HS, Robinson JR (1984) In: Anderson JM, Kim JW (ed) Recent advances in
     drug delivery systems, Plenum, New York NY, p 163
403. Park H, Robinson JR (1985) J Controlled Rel 2: 47
404. Harris D, Robinson JR (1990) Biomaterials 11: 652
405. Manning ML, Patel K, Burchardt RT (1989) Pharm Res 6: 903
406. Chitnis VS, Malshe VS, Lalla JK (1991) Drug Dev Ind Pharm 17(6): 879
407. Bottenberg P, Cleymaet R, De Muynck C, Remon J P, Coomans D, Michotte Y, Slop D (1991)
     J Pharm Pharmacol 43(7): 457
408. Bremecker KD, Strempel H, Klein G (1984) J Pharm Sci 73: 548
409. Bodde HE, DeVries ME, Junginger HE (1990) J Controlled Rel 13(2-3): 225
410. Longer MA, Chang HS, Robinson JR (1985) J Pharm Sci 74: 406
411. Rytting JH, Hefferren JJ, Qi H, Itoh T, Nishihata T, Boyce E (1991) In: Kellaway IW (ed)
     Proceed Program Intern Symp Control Rel Bioact Mater 18: 115
412. Park K, Robinson JR (1984) Int J Pharm 19: 107
413. Rozier A, Mazuel C, Grace J, Plazonet B (1989) Int J Pharm 57: 163
414. DeVries ME, Bodde HE, Bussler HJ, Junginger HE (1988) Biomed Mater Res 22: 1023
415. Gurny R, Ibrahim H, Aebi A, Buri P, Wilson C G, Washington N, Edman P, Camber O (1987)
     J Controlled Release 6: 367
416. Khisla R, Davis S (1987) J Parm Pharmacol 39: 47
417. Smart JD, Kellaway IW (1989) Int J Pharm 53: 79
418. Harris D, Fell JT, Sharma HL, Taylor DC (1990) J Controlled Rel 12: 45
419. Harris D, Fell JT, Taylor DC, Lynch J, Sharma HL (1990) J Controlled Rel 12: 55
420. Machida Y, Masuda H, Fujiyama N, Ito S, Iwata M, Najai T (1979) Chem Pharm Bull 27: 93
421. Morimoto K, Iwamoto T, Morisaka K (1987) J Pharm Dyn 10: 85
422. Morimoto K, Takeda T, Nakamoto Y, Morisaka K (1982) Int J Pharm 12: 107
423. Morimoto K, Morisaka K, Kamada A (1985) J Pharm Pharmacol 37: 134
424. Dong C, Rogers JA (1991) J Controlled Release 17: 217
425. Tirrell DA, Turek AB, Wilkinson DA, McIntosh TJ (1985) Macromolecules 18: 1512
426. Takigawa DY, Tirrell DA (1985) Macromolecules 18: 338
427. Sato T, Koyina K, Ihda T, Sunamoto J (1986) J Bioact and Comp Polym 1: 445
428. Sumamoto J, Iwamoto K, Takeda M, Yuzuriha T, Katayama K (1983) In: Chielling E, Giwit
     P (eds) Polymers in medicine, Plenum, NY p 157
429. Sumamoto J, Sakai K, Sato T, Kondo H (1987) Chem Lett p 1781
430. Sumamoto J, Iwamoto K (1985) CRC Crit Rev Ther Drug Carrier Syst 2: 117
431. Kato A, Kondo T (1987) Polym Sci Tech 35: 299
432. Kato A, Arakawa M, Kondo T (1985) Polym Mater Sci Eng 53: 654
433. Kato A, Arakawa M, Kondo T (1984) J Microencapsulation 1: 105
434. Kato A, Kondo T (1987) In: Gebelein CG (ed) Advances in biomedical polymers. Plenum
     Press, New York, p 285
435. Kato A, Arakawa M, Kondo T (1983) Biology 20: 593
436. Seki K, Tirrell DA (1984) Macromolecules 17: 1692
437. Borden KA, Eum KM, Langley KH, Tirrell DA (1987) Macromolecules 20: 454
438. Schroeder UKO, Tirrell DA (1989) Macromolecules 22: 765
439. Eum KM, Langley KH, Tirrell DA (1989) Macromolecules 22: 2755
440. Tirrell DA (1990) In: Kost J (ed) Pulsed and self-regulated drug delivery, CRC Press, Boca
     Raton FL, p 109
441. Maeda M, Kumano A, Tirrell DA (1991) Ann NY Acad Sci 618: 362
442. Maeda M, Kumano A, Tirrell DA (1988) J Am Chem Soc 110: 7455
443. Borden KA, Eum KM, Langley KH, Tan JS, Tirrell DA, Vogycheck CL (1988) Macro-
     molecules 21: 2649

444. Ferritto MS, Tirrell DA (1988) Macromolecules 21: 3117
445. Devlin BP, Tirrell DA (1986) Macromolecules 19: 2465
446. Maksimenko OO , Feldshtein MM, Panarin EF, Turchilin VP, Vasilev AE, Plate N A (1990) Vysokomol Soedin Ser A 32(12): 2362
447. Regen SL, Shin JS, Yamaguchi KJ (1984) J Am Chem Soc 106: 2446
448. Brady JE, Evans DF, Kachar B, Ninham B (1984) J Am Chem Soc 106: 4279
449. Alico KV, Ringsdorf H, Schlarb B, Leister KH (1984) Makromol Chem Rapid Commun 5: 345
450. Kono K, Kimura S, Imanishi Y (1990) Biochemistry 29: 3631
451. Kono K, Kimura S, Imanishi Y (1990) J Membr Sci 50: 85
452. Akashi M, Takemoto K (1990) Advances in Polymer Science. 97: 107
453. Mora M, Pato J (1992) J Controlled Rel 18: 153
454. Li F, Go ZX, Li G, Wang S, Feng X (1991) J. Bioact Compat Polym 6(2): 142
455. Kabanov VA (1986) Macromol Chem Macromol Symp 1: 101
456. Ottenbrite RM, Regelson W, Kaplan AM, Carchman R, Morahan P, Munson A (1978) In: Donaroma G, Vogl O (eds) Polymeric Drugs Acad p 263
457. Bauduin G, Bondon D, Martel J, Pietrasanta Y, Pucci B (1981) Makromol Chem 182: 773
458. Ottenbrite RM, Kuus K, Kaplan AM (1983) Polym Sci Technol (Polym Med) Plenum 23: 3
459. Ottenbrite RM (1985) In: Gleblein CG, Carreher GE (eds) Bioactive polymeric systems-an overview, Plenum Press, NY, pp 514
460. Mansen AE, White P, Klybken P (1981) Cancer Res 16: 329
461. Sarai K, Kuhda K, Hagatsu H, Migamae T, Kawazoe Y (1991) Anticancer Res 11: 957
462. Donaruma LG, Ottenbrite RM (1980) in: Vogl O (ed) Anionic Polymeric Drugs 1, Wiley NY
463. Carrale CE, Gebelein CG, Ottenbrite RM (1982) Eds ACS Symp Ser 186
464. Uryo T, Yogshida T, Ikoshima N, Hatanoada K, Kaneko Y, Mimura T, Nakashima H, Yamamoto N (1991) Polym Sci C Symp Proc Polym '91. 2: 989
465. Baba MP, Shols D, Clerck ED, Pauwels R, Nagy M, Gyorgyi J, Delenyi E, Lew M, Gorog S (1990) Antimicrob Agents Chemother 34(1): 134
466. Ottenbrite RM, Butler GB (1984) Anticancer and Interferon Agents. Dekker, New York, p 247
467. Panarin EF, Solovskii MV, Zaikina NA, Afinoyenov GE (1985) Makromol Chem Suppl 9: 25
468. Linakis JG, Eisenberg MS, Lacouture PG, Maher TJ, Lewander WJ, Driscoll JL, Woolf A (1992) Pharmacol Toxicol (Copenhagen) 70(1): 38
469. Gekko K, Timasheff SN (1981) Biochemistry 20: 4667
470. Gekko K, Morikawa T (1981) J Biochem 90: 39
471. Andrade JK, Nagaoka S, Cooper S, Okano T, Kim SW (1987) Trans Am Soc Artif Intern Organs 83: 75
472. Sawyer PN, Surface Charge and Thrombosis Ann NY Acad Sci 416: 561
473. Anderson JM, Koftke-Marchant K (1985) CRC Critical Review in Biocompatibility. p 111
474. Ratner BD, Johnston AB, Lenk TJ (1987) J Biomed Mater Res Applied Biomaterials. 21(A1): 59
475. Gebelein CG, Murphy D (1987) In: Gebelein CG (ed) Advances in biomedical polymers. Plenum Press, New York, pp 277
476. Grasel TG, Cooper SL (1989) J Biomed Mater Res 23(3): 311
477. Okkema AZ, Giroux TA, Grassel TG, Cooper SL (1987) Mater Res Soc Symp Proced Vol 110; (1988) Biomed Mater Devices, p 91
478. Okkema AZ, Cooper SL (1991) Biomaterials 12(7): 668
479. Silver J, Okkema AZ, Jozefowicz M, Cooper Sl (1990) Poplym Prepr 31(1): 222
480. Ito Y, Sisido M, Imanishi Y (1986) J Biomed Mater Res 20: 1157
481. Bamford CH, Middleton IP (1983) Eur Polym J 19: 1027
482. Han DK, Jeong YS, Kim YH, Min BG, Cho HI (1991) J Biomed Mater Res 25(5): 561
483. Fougnot C, Josefonwicz J, Samanna M, Bara L (1979) Ann Biomed Eng 7: 429
484. Fougnot C, Josefonwicz M, Rosenberg RD (1984) Biomaterials 5: 94
485. Tumoko U, Ishihara K, Nakabayashi N (1991) Konbunshi Ronbunshu 48(5): 289
486. Helmus MN, Gibbons DF, Jones RD (1984) J Biomed Mater Res 18: 165
487. Kikuchi Y, Okamura Y (1980) Makromol Chem Rapid Commun 1: 253
488. Katauka K, Sakurai Y, Okaike T, Tsuruta T (1980) Macromol Chem 181: 1363
489. Philipp B, Hong LT, Dawydoff W, Linow KJ, Arnold K, Ratzsch M (1980) Acta Polym. 31: 592
490. Philipp B, Hong LT, Linow KJ, Dawydoff W, Arnold K (1980) Acta Polym 31: 654
491. Davidson CJ, Smith KE, Hutchinson LEF, O'Mullane JE, Brookman L, Petrak K, Harding SE (1990) J Bioactive and Compatible Polymers 5(3): 267

492. Artussen P, OMullane JE (1989) Adv Drug Deliv Rev 3: 165
493. Ottenbrite RM, Kaplan AM (1985) In: Tirrell DA, Donaroma LG, Turek AB (eds) Macro-molecules as drugs and as carriers for biologically active materials, Ann NY Acad Sci 446: 160
494. Azori M, Pato J, Fehervari F, Tudos F (1986) Makromol Chem 187: 303
495. Ryser HJP, Shen WC (1980) Cancer. 45: 1207
496. Kabanov VA, Zezin AB (1984) Makromol Chem Suppl 6: 259
497. Tsuchida E, Abe K (1982) Adv Polym Sci 45: 1
498. Philipp B (1989) Prog. Polym Sci M 14(1): 91
499. Xi T, Zhang J, Tian W, Lei X, Song Q, Zheng P (1990) In: Feng H, Han Y, Huang L (eds) C MRS Int Symp Proc 3: 483
500. Xi T, Zhang J, Tian W, Lei X, Song Q, Zheng P (1991) Clinical Mater 8: 43
501. Jozefowicz M, Jozefowicz J (1985) ASAIO J 8: 218
502. Merrill EW (1987) Annals New York Acad. Sci 516: 196
503. Barbucci R, Benvenuti M, Ferruti P, Nocentini M (1987) In: Gebelein CG (ed) Advances in biomedical polymers. Plenum Press, New York, p 259
504. Shibutu R, Tanaka M, Sisido S, Imanishi Y (1986) J Biomed Mater Res 20: 971
505. Ito Y, Sisido M, Imanishi Y (1986) J Biomed Mater Res 20: 1139.
506. Sanada T, Ito Y, Sisido M, Imanishi Y (1986) J Biomed Mater Res 20: 1179
507. Ito Y, Sisido M, Imanishi Y (1984) Polym Prepar Jpn 33: 2163
508. Ito Y, Sisido M, Imanishi Y (1986) J Biomed Mater Res 20: 1017
509. Ferruti P, Barbuci R (1984) Adv in Polym Sci 58: 55
510. Ferruti P, Marcisio MA, Barbucci R (1985) Polymer 26: 1336
511. Azzuoli G, Barbucci R, Benvenuti M, Ferruti P, Nocentini M (1987) Biomater 8: 61
512. Levi M, Muttoni M, Martini A (1989) In: Dawids S (ed) Polymers: Their Properties and Blood Compatibility p 379
513. Ikada Y, Suzuki M, Tamada Y (1984) In: Shalaby SW, Hoffman AS, Ratner BD, Horbett TA (eds) Polymers as Biomaterials, Plenum Publ Co, p 135
514. Muramatsu N, Kondo T (1983) J Biomed Mater Res 17: 959
515. Sederel LC, van der Does L, van Duijl JF, Beugeling T, Bantjes A (1981) J Biomed Mater Res 15: 819
516. Aleyamma AJ, Sharma CP (1988) Polym Mater Sci Eng 59: 692
517. Aleyamma AJ, Sharma CP (1990) In: Gebelein CG (ed) Biomemetic Polym p 191
518. Van der Does L, Sederel LC, Beugeling T, Van Duijl JF, Bantjes A (1986) Dev Hematol Immunol 14: 215
519. van der Does L, van Duijl JF, Sederel LC, Bantjes A (1980) J Polym Sci Polym Lett Ed 18: 53
520. Hirano S, Noishiki Y, Kinugawa J, Higashijima H, Hayashi T (1987) In: Gebelein CG (Ed) Advances in Biomedical Polymers. Plenum Press, New York, p 285
521. Sederel LC, Kolar Z, Hummel A, van der Does L, Bantjes A, (1983) Biomater 4: 210
522. Margau M, Fougnot C, Josefonwicz J (1981) Proc Third MISAO 5: 504
523. Fougnot C, Josefonwicz M, Samaura M, Bara L (1979) Ann Biomed Eng 7: 441
524. Mauzac M, Aubert N, Josefonwicz J (1982) Biomaterials 3: 221
525. Mauzac M, Josefonwicz J (1984) Biomaterials 5: 301

Polymer Conjugates with Anticancer Activity

D. Putnam and J. Kopeček*

Departments of Pharmaceutics and Pharmaceutical Chemistry/CCCD, and of Bioengineering, University of Utah, Salt Lake City, Utah 84112, U.S.A.

Polymer conjugates may possess anticancer activity through a variety of mechanisms. The macromolecules themselves may have anticancer activities, or, more typically, inert biocompatible polymers serve as carriers for low molecular weight anticancer agents. Polymer conjugates may also be targeted to increase the concentration of conjugate in the vicinity of a specific subset of cells. This article reviews the recent literature that pertains to polymer conjugates with anticancer activity. The types of polymers chosen as drug carriers and the biodistribution of polymers in the body are discussed. Also, the synthesis, biological properties, and the means used to evaluate the anticancer activities of polymer conjugates are detailed.

The rationale for the design of targetable water soluble synthetic polymeric carriers of anticancer drug are explained using copolymers of N-(2-hydroxypropyl)methacrylamide as examples. Comparison of polymer conjugates with other drug delivery systems, i.e., liposomes, nanoparticles, microspheres and immunotoxins, is provided along with the prospects for the future of anticancer drug delivery.

* To whom correspondence should be addressed at Biomedical Polymers Research Building, Room 205, University of Utah, Salt Lake City, Utah 84112, U.S.A.

List of Symbols and Abbreviations

AA	amino acid
ADEPT	antibody directed enzyme prodrug therapy
ADR	adriamycin (doxorubicin)
APA	4-amino-4-deoxy-N^{10}-methylpteroic acid
ARS	arsanilic acid
AZT	azidothymidine
BOP reagent	benzotriazol-1-yloxytris(dimethylamino)phosphonium hexafluorophosphate
cis-DPP	cis-dichlorodiamminoplatinum (II)
CEA	carcinoembryonic antigen
CURL	compartment for uncoupling receptor and ligand
DMSO	dimethyl sulfoxide
DNM	daunomycin
EDTA	ethylene diamine tetraacetic acid
EEDQ	N-ethoxycarbonyl-2-ethoxy-1,2-dihydroquinoline
EPR	enhanced permeability and retention effect
Fab, (Fab)$_2$	antibody fragments
FITC	fluorescein isothioisocyanate
GalN	galactosamine
HAMA	human anti-mouse antibodies
hCG	human chorionic gonadotropin
HPMA	N-(2-hydroxypropyl)methacrylamide
IgG	immunoglobulin G
IMC	Institute of Macromolecular Chemistry
i.p.	intraperitoneal
i.v.	intravenous
k_{cat}	catalytic constant
K_M	Michaelis constant
LCST	lower critical solution temperature
LHRH	luiteinizing hormone releasing hormone
MA	methacryloyl
mAbs	monoclonal antibodies
Mce$_6$	meso-chlorin e$_6$ monoethylenediamine disodium salt
MTT	3-(4,5-dimethylthiazol-2-yl)-2,5-diphenyl tetrazolium bromide
MTX	methotrexate
mPEG	monomethoxypoly(ethylene glycol)
Mw	molecular weight
NAp	para-nitroanilide
OC125	anti-ovarian carcinoma monoclonal antibody
ONp	para-nitrophenol
P	polymer backbone

PDT	photodynamic therapy
PEG	poly(ethylene glycol)
PEG-ADA	PEG-adenosine deaminase
P-D	polymer-drug conjugate
⊃—P-D	targeted polymer-drug conjugate
Pi	enzyme active center subsite
RES	reticuloendothelial system
s.c.	subcutaneous
sFV	single-chain antigen binding proteins
Si	substrate subsite
SPDP	N-succinimidyl 3-(2-pyridyldithio)propionate
sulfo-MBS	m-maleimidobensoyl sulfosuccinimide ester
TEA	triethylamine
X	lysosomal enzyme
⊥	clathrin molecule
—●	cell surface receptor/antigen

1 Introduction

The development of chemotherapeutic agents for the treatment of cancer has resulted in a significant number of once fatal neoplasia to become classified as curable, or at the very least, treatable. However, despite the great advances in the synthesis of new anti-cancer compounds, these compounds tend to possess nonspecific toxicity restricting their efficiency in vivo. Furthermore, multidrug tumor cell resistance is fairly common.

To overcome these obstacles to effective chemotherapy, drug delivery systems have been created to allow lower drug doses to decrease the toxicity of the treatment, and to provide site-specific drug targeting to increase the concentration of drugs at the tumor site. It was in fact Paul Ehrlich in 1906 who coined the phrase "the magic bullet" when describing drugs which might be selectively directed to their site of action [1]. The level of the understanding of the basic scientific principles gradually increased with discoveries of lysosomes and lysosomotropism of macromolecules by De Duve and coworkers [2], conjugation of drugs to polymeric carriers by enzymatically cleavable bonds [3], and biorecognition of macromolecules at the cellular level [4] among many others. Finally, Ringsdorf [5] presented the first clear concept of the use of polymers as targetable drug carriers. He combined the concepts of site-specific drug release with site-specific recognition and predicted some important beneficial properties of polymer-drug conjugates as compared with their low molecular weight analogs.

Extension of these concepts has resulted in a milieu of drug delivery systems possessing anti-tumor activity [6], such as liposomes [7], nanoparticles [8], microspheres [9], immunoconjugates [10], and drug releasing implants [11]. The scope of this article is to review the literature and present the state-of-the-art in soluble polymer-drug conjugates. The rationale of design, tailor-made synthesis, physiological implications, biological properties, and efficacy of these diverse anti-tumor conjugates are discussed.

2 Macromolecules as Drug Carriers

Many macromolecules have been utilized as drug carriers in an attempt to decrease the toxicity and/or to increase the therapeutic index of a parent anticancer compound, and have been reviewed extensively [12–20].

Presently, clinically useful anticancer compounds are low molecular weight chemicals which possess a high pharamacokinetic volume of distribution, which suggests the ubiquitous presence of toxic compounds in the host. It is this total diffusion of drug throughout the body which results in the majority of toxic effects produced by chemotherapeutic regimens.

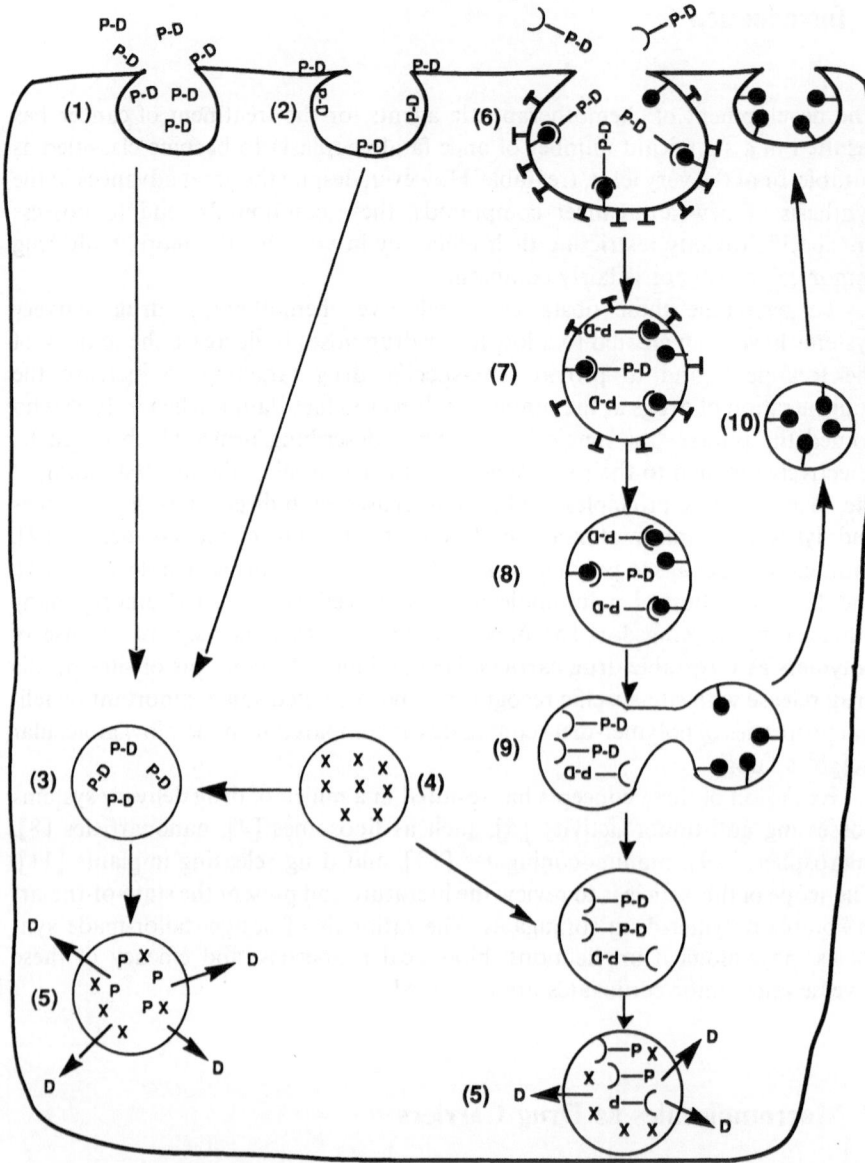

Fig. 1. Endocytic pathways used by cells to internalize soluble macromolecules [25]: fluid-phase pinocytosis (1), adsorptive pinocytosis (2), and receptor-mediated endocytosis (pinocytosis) (6). Each of these processes involves a formation of a sealed vesicle formed from the plasma membrane which encloses part of the extracellular medium. The internalization of a polymer-drug conjugate (P-D), and targeted polymer-drug conjugate (⊃ —P-D) is shown. Other abbreviations: — ● = cell surface receptor/antigen; ⊥ = clathrin molecule; X = lysosomal enzyme. Fluid-phase pinocytosis (1) and adsorptive pinocytosis (2) are nonspecific processes which direct the macromolecule into the lysosomal compartment of the cell. Once P-D is internalized, whether by (1) or (2), the resulting endosome (3) is ultimately fused with a primary lysosome (4) forming a secondary lysosome (5). In the latter compartment P-D is in contact with several types of lysosomal enzymes. The membrane of (5) is impermeable to macromolecules. Consequently, the structure of P-D may be designed in such

The entrance of low molecular weight drugs into the cell interior is limited primarily by diffusion. Macromolecules, on the other hand, tend to be limited in their distribution about the body, since their cellular uptake is limited to endocytosis. Macromolecules captured by this mechanism are channeled to the lysosomal compartment of the cell [19]. Endocytosis is a common term encompassing phagocytosis and pinocytosis. Phagocytosis describes the capture of vesicular material by specialized cells (macrophages and monocytes). Pinocytosis describes the capture of extracellular fluid, all solutes dissolved therein and any material adherent to the infolding surface [20]. It is a process common to most cell types. Depending on the structure of the macromolecule, three types of pinocytosis may occur: fluid-phase, adsorptive, and receptor-mediated pinocytosis. Fluid-phase pinocytosis occurs when no interaction of the macromolecule with cell surface takes place. Consequently, macromolecules are taken up slowly depending on their concentration in the extracellular fluid. Incorporation of hydrophobic moieties [21] or positive charges [22] into the macromolecular structure results in non-specific interactions with plasma membranes of different cells with a concomitant increase in the rate of macromolecular uptake. This process is known as adsorptive pinocytosis. Incorporation of moieties in the macromolecular structure which are complementary to cell surface receptors or antigens of a subset of cells renders the macromolecule biorecognizable [23, 24]. The macromolecule is internalized specifically by a select subset of cells and not only the rate of uptake, but also the body distribution is substantially altered. The latter mechanism, i.e., receptor-mediated pinocytosis, is the rationale for the design of targetable polymeric carriers (Fig. 1).

A drug delivery system based on the above mentioned rationale (Fig. 2) may be represented as follows [5]:

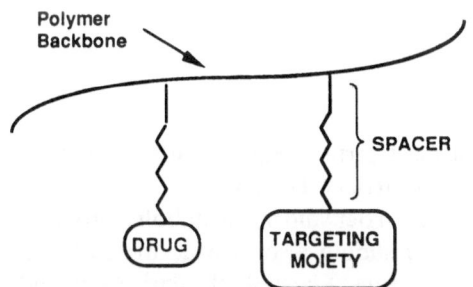

Fig. 2. A schematic design of a targetable polymeric conjugate. Drug and targeting moiety are bound to the soluble polymeric carrier via a spacer

a way to be recognized by one or several lysosomal enzymes resulting in the cleavage of the bond between the polymeric carrier and drug. During the internalization process the pH inside the P-D containing compartment gradually acidifies reaching pH 4.5–5.0. Receptor-mediated endocytosis of polymer-drug conjugates containing a targeting moiety recognizable by a cell surface receptor/antigen is initiated by the occupancy of a cell surface receptor. The receptors tend to reside within or near clathrin coated pits (6). Upon internalization, a coated vesicle, or primary endosome (7) is formed whereupon the clathrin coat is lost. The resulting endosome (8) eventually forms a CURL (compartment for uncoupling receptor and ligand) or sorting endosome (9). This organelle provides for the separation of the cell surface receptor from the targeted polymeric drug carrier. The cellular receptors are then recycled back to the cell surface (10) while the endosome that contains internalized macromolecules fuses with (4) to form a secondary lysosome (5)

The system consists of an inert carrier macromolecule, side-chains degradable in the lysosomal compartment to act as sites of drug attachment and release, and side-chains terminated in a biorecognizable (targeting) moiety. In other words, a targetable water-soluble drug delivery system has to be biorecognizable on two levels. First, on the cellular level at the plasma membrane to be internalized by a subset of cells by receptor-mediated pinocytosis. Second, it has to be recognized intracellularly in the lysosomal compartment on a molecular level by lysosomal enzymes. For example, the side-chains have to be recognized by the active site of lysosomal cysteine proteinases [26] to permit the release of the drug from the carrier. Due to the low permeability of the lysosomal membrane only the free drug can penetrate into the cytoplasm and start its biological action. Drugs active in the cell nucleus and bound to the macromolecular carrier via a nondegradable covalent bond are inactive. They are "buried" in the lysosomal compartment [27]. However, the second level of recognition (intracellular) is not required if the drug is bound via bonds which are either hydrolyzed or reduced in the prelysosomal or lysosomal compartments. For example, hydrazone derivatives of adriamycin were synthesized which are hydrolyzed at pH 4.5–5.5 [28]. In model experiments, Shen and Ryser [29] have shown that binding daunomycin to a polymeric carrier via cis-acotinyl spacers results in acid sensitive bonds. Disulfide bonds can also be used to conjugate drugs to carriers [30]. The reduction of these bonds starts at the cellular surface. Based on results of subcellular fractionation and kinetic analysis of the [^{125}I]iodotyramine-SS-poly(D-lysine) conjugate the Golgi apparatus was suggested as the most probable major site of cleavage [31]. To overcome the problems with the stability of these bonds in plasma, sterically hindered disulfide linkages have been developed [32].

2.1 Types of Macromolecules

Polymers chosen as drug carriers must meet certain requirements in order to maximize their potential as polymeric drug carriers. The polymers must be well characterized and easily synthesized. The carrier and all metabolic products should be nontoxic and nonantigenic, and should also provide drug attachment/release sites for the incorporation of drugs. Likewise, the carriers should display the ability to be directed to predetermined cell types, and, ideally, they should be biodegradable or eliminated from the organism after fulfilling their function [16, 20].

Macromolecules as drug carriers may be divided into degradable and nondegradable types based on their fate within the organism. Biodegradable polymeric drug carriers are traditionally derived from natural products {polysaccharides, poly(amino acids)} in the hope that the body's natural catabolic mechanisms will act to break down the macromolecular structure into small, easily eliminated fragments. However, the substitution of natural macromolecules with covalently linked drug molecules generally results in the hampering of the host's ability to enzymatically degrade the polymeric carrier effectively [33]. It is apparent that enzymes that cleave peptide or saccharide bonds have a considerably large active site to accomodate several "monomer" units. Substitution along the macromolecular backbone renders the formation of the enzyme-substrate complex energetically less favorable. Therefore, drug substitution of a polymeric carrier may result in the inability of a naturally occurring, normally biodegradable macromolecule, to be degraded into easily eliminated fragments. Degradation of drug modified macromolecules, such as poly(amino acids) [34] or polysaccharides [35] may yield fragments which cannot penetrate the lysosomal membrane. Moreover, they may possess biological activity. At the present state-of-the-art it is safe to treat any macromolecular drug carrier as nondegradable [12].

Synthetic macromolecules may be preferred as anticancer drug carriers since they can be tailor-made to have properties matching the biological situation. However, to minimize storage, their entire molecular weight distribution must be under the renal threshold. To prevent the nonspecific reuptake of the macromolecule after being released into the bloodstream following cell death, its structure must be designed in such a way as to be internalized by fluid-phase pinocytosis. The absence of nonspecific interactions with plasma membranes will minimize the probability of accumulation of the carrier in nontarget cells and thus increase the biocompatibility of the carrier. Examples of the use of macromolecules as drug carriers are shown in Table 1.

2.1.1 Macromolecules with Intrinsic Anticancer Activity

Some synthetic macromolecules have shown intrinsic anticancer activity. Their mechanism of action is probably either the direct action upon the tumor cell, or the stimulation of the host's immune system [59] (Table 2). We do not exclude the use of such macromolecules as drug carriers in the future. However, the combination of a biologically active carrier with a biologically active drug makes the evaluation of the structure-property relationships more difficult.

There are many natural and biological macromolecules that possess anticancer activity. Cytokines, topoisomerase inhibitors, monoclonal antibodies, thymic hormones, cell growth inhibitors, and enzymes have been used [68]. They have been recently reviewed [59, 69] and their detailed description is beyond the scope of this article. The main problems connected with the administration of such natural macromolecules is their short intravascular half-life, immunogenicity, and sometimes poor solubility. Their modification with synthetic macromolecules can dramatically increase their therapeutic potential as described below.

Table 1A. Synthetic polymers used as anticancer drug carriers

N-(2-hydroxypropyl)methacrylamide (HPMA) copolymers:

Polymeric Carrier	Drug	Type of Bond	Targeting Moiety	Type of Bond	Activity in vitro	Activity in vivo	Ref.
HPMA copolymer	Adriamycin	Amide bond between drug aminoribosyl and oligopeptide spacer carboxyl	Galactosamine	Amide bond between sugar amine and oligopeptide spacer carboxyl	N/T	Preclinical evaluation of targeted and non-targeted conjugates showed increased anticancer activity relative to free drug against numerous tumor models	18, 36, 264
HPMA copolymer	Adriamycin	Amide bond between drug aminoribosyl and oligopeptide spacer carboxyl	Melanocyte-stimulating hormone (MSH)	Amide bond between MSH amino group and oligopeptide spacer carboxyl	Targeted conjugates showed greater anticancer activity than non-targeted conjugates against mouse melanoma cells (M3S91)	Targeted conjugates showed greater anticancer activity than non-targeted conjugates in mice bearing established B16F10 mouse melanoma	37
HPMA copolymer	Melphalan	Amide bond between drug amine and variable oligopeptide spacer carboxyl	None	N/A	Conjugates showed concentration dependent toxicity against Walker sarcoma (LLC-WRC 256)	Conjugates showed ability to inhibit the development of Walker sarcoma (W256) and showed ability to cause regression of established Walker sarcoma	38
HPMA copolymer	Daunomycin	Amide bond between drug aminoribosyl and variable oligopeptide spacer carboxyl	Fucosylamine- and galactosamine	Amide bond between sugar amine and variable oligopeptide spacer carboxyl	Conjugates showed activity against L1210 leukemia, but less activity relative to free drug	Carbohydrate containing conjugates showed greater activity against L1210 leukemia compared to non-targeted conjugates depending upon experimental conditions	18, 39, 40
HPMA copolymer	Daunomycin	Amide bond between drug aminoribosyl and variable oligopeptide spacer carboxyl	anti-Thy 1.2 or anti-Ia^k antibodies	Amide bond between antibody lysine N$^\varepsilon$-amino groups and variable oligopeptide spacer carboxyl	Conjugates with degradable spacer showed twice the cytotoxicity of free drug	Targeted conjugates showed decreased toxicity and increased efficacy relative to free drug	24, 41

HPMA copolymer	Chlorin e_6	Amide bond between drug carboxyl and spacer amine	anti-Thy 1.2 antibody	A. Amide bond between antibody lysine N^ε-amino groups and glycyl spacer carboxyl B. alkylamine bond between aldehyde of oxidized antibody carbohydrate and spacer amine	Both targeted conjugates showed greater cytotoxicity than nontargeted, and B showed greater activity than A	N/T	23
HPMA copolymer	A. adriamycin and B. meso chlorin e_6 (combination chemotherapy and PDT)	A. amide bond between drug aminoribosyl and oligopeptide spacer carboxyl B. amide bond between drug amine and degradable or nondegradable oligopeptide spacer carboxyl	None	N/A	N/T	Polymer with degradable chlorin e_6 oligopeptide spacer showed greater tumor killing effect than polymer with nondegradable spacer. Combination therapy showed synergistic killing effect (Neuro 2A neuroblastoma)	42
HPMA copolymer	Chlorin e_6	Amide bond between drug carboxyl and spacer amine	Galactosamine	Amide bond between galactosamine amine and glycyl spacer carboxyl	Conjugate showed anticancer activity against human hepatocarcinoma (PLC/PRF/5)	N/T	43, 191
Poly(α-L-glutamic acid) (PGA): PGA	Adriamycin	Amide bond between drug aminoribosyl and either carrier carboxyl or variable oligopeptide spacer	None	N/A	Conjugates were not as active as free drug against L1210 and B16 cancer cell cultures	Conjugate with no oligopeptide spacer was completely inactive against L1210 leukemia, whereas conjugates with degradable spacers were active. Anticancer activity increased with increasing oligopeptide length and degradation rate	44, 45

Table 1. Continued

Polymeric Carrier	Drug	Type of Bond	Targeting Moiety	Type of Bond	Activity in vitro	Activity in vivo	Ref.
Poly(L-lysine): Poly(L-lysine)	5-fluorouridine	Alkylamine bond between aldehyde of oxidized drug ribose and carrier amine	Monoclonal antibody (SF25MAb) against human hepatoma	Alkylamine bond between aldehyde of oxidized oligosaccharide of antibody and amine of carrier lysine side chain	Conjugate showed activity against LS180 colon carcinoma	Conjugate was too toxic in vivo to be effective. Succinylation of lysine side chains decreased toxicity, but eliminated anticancer effect against LS180 colon carcinoma	46
Poly(L-lysine)	Methotrexate	Amide bond between drug carboxyl and polyglutamate spacer amine	None	N/A	Conjugate showed activity against methotrexate resistant H35 hepatoma cells	N/T	47
Block copolymers: Poly(ethylene glycol-co-aspartic acid)	Adriamycin	Amide bond between drug aminoribosyl and aspartic acid residue carboxyl in poly(aspartic acid) chain	None	N/A	Conjugate showed concentration dependent cytotoxicity against P388 mouse leukemia	Conjugate showed superior antitumor activity vs free drug against several murine and human tumors	48, 49

Table 1B. Natural and miscellaneous polymers used as anticancer drug carriers

Polymeric Carrier	Drug	Type of Bond	Targeting Moiety	Type of Bond	Activity in vitro	Activity in vivo	Ref.
Albumin: BSA	Mitomycin C	Amide bond between drug amine and glutaric acid spacer carboxyl	None	N/A	IC_{50} slightly less than free drug against S180 cells	75% and 100% 40 day survival rates in S180 bearing mice (control = 39%)	50
BSA	Doxorubicin	Alkylamine bond between drug aminoribosyl and glutaraldehyde spacer	None	N/A	Showed marked anticancer activity against doxorubicin resistant AH66DR cells	N/T	51
Chitins: Carboxymethyl chitin	Mitomycin C	Amide bond between drug amine and carrier carboxyl	None	N/A	N/T	Conjugate showed less anticancer activity than free drug against L1210 leukemia, but greater activity against B16 melanoma	52
N-Succinyl chitosan	Mitomycin C	Amide bond between drug amine and carrier carboxyl	None	N/A	N/T	Conjugate showed less anticancer activity than free drug against L1210 leukemia, but greater activity against B16 melanoma	52
Dextrans: Anionically charged dextran	Mitomycin C	Amide bond between drug amine and hexamethyl spacer carboxyl	Monoclonal antibody (A7) against human colon carcinoma and murine myeloma	Amide bond between antibody lysine amine and spacer carboxyl	Conjugate showed cytotoxicity against SW 1116 colorectal carcinoma, but less activity than free mitomycin C	N/T	53

Polymeric Carrier	Drug	Type of Bond	Targeting Moiety	Type of Bond	Activity in vitro	Activity in vivo	Ref.
Aminodextran	Daunomycin and doxorubicin	Amide bond between drug aminoribosyl and succinic acid spacer carboxyl	Monoclonal antibody (NP-4) against human colon carcinoma	Alkylamine bond between aldehyde of oxidized polysaccharide of antibody and amine of dextran	IC_{50} of conjugate greater than free drug against LoVo cells	Conjugate showed 81% inhibition of human colonic GW-39 tumor growth relative to saline control	54
Neoglycoprotein: Human serum albumin containing galactose residues	Daunorubicin	Amide bond between drug aminoribosyl and tetrapeptide spacer carboxyl	Galactose	Alkylamine bond between sugar aldehyde and lysine N^ϵ-amino groups of protein	N/T	Conjugate showed activity against human hepatoma xenografts, while free drug had no activity	55
Miscellaneous: SMANCS (Styrene-*co*-maleic acid/anhydride polymer bound to neocarzinostatin	Neocarzinostatin (an antitumor protein)	Amide bond between polymer carboxyl and protein amino	None	N/A	SMANCS showed anticancer activity against many tumor cell lines, and had lower IC_{50} values than five other anticancer agents tested	Liver tumors reduced more than 50% after 6 months in human subjects	15, 56, 57
DIVEMA (Hydrolyzed form of divinyl ether and maleic anhydride copolymer)	Cyclophosphamide	Ester bond between polymer derivative hydroxyl and polymer carboxyl	None	N/A	N/T	Anticancer activity of conjugate was comparable to free drug against L1210 leukemia bearing mice	58

Abbreviations used: N/A, not applicable; N/T, not tested; BSA, bovine serum albumin

Table 2. Polymers with intrinsic anticancer activity

Polymer	Anticancer Activity and Proposed Mechanism	Reference
Polyetheneimine (Mw not reported)	Active against Ehrlich ascites through direct antitumor activity	60
Polylysine (Mw = 50 000)	Active against the nonlymphocytic cell line K562 through direct antitumor activity	61
DEAE-dextran (diethylamino-ethyl-dextran (Mw = 2 000 000)	Active in vivo against NJA leukemia, JBI ascites, plasmocytoma in mice, and Yoshida ascites in rats	62
Poly(Arg-Gly-Asp) (Mw = 10 000)	Inhibits formation of lung metastasis and migration of B16-BL6 melanoma in mice	63
SCM-Chitin (sulfated and carboxymethylated chitin) (Mw = 10 000 to 60 000)	Inhibits cell attachment and migration of B16-BL6 melanoma cells through extracellular matrix	64
MVE-2 (Divinyl ether-maleic anhydride {DIVEMA} copolymer) (Mw < 18 000)	Activity demonstrated against numerous tumor models through stimulation of host immune system. L1210 bearing mice treated with MVE-2 and 5-aza-2-deoxycytidine led to a number of cures	65
Carbetimer (Maleic anhydride-ethylene copolymer) ("Intermediate" Mw reported)	Activity shown against HM5-Carb/S melanoma in vitro and in mice, presumably by inhibition of nucleoside uptake	66
Copovithane [1,3 bis (methylaminocarboxy)-2 methylene propane carbamate and N-vinylpyrrolidone copolymer] (Mw = 5 800)	Active against sarcoma 180, P388 leukemia, and carcinoma F0771 in mice, and Walker 1098 in rats. Inactive in vitro, and is suspected to act through stimulation of host immune system	67

2.1.1.1 Poly(ethylene glycol)-Protein Conjugates

Various enzyme or cytokine-poly(ethylene glycol) (PEG) conjugates were developed as new therapeutics taking advantage of their decreased immunogenicity, resistance to proteolytic degradation, and persistance in the bloodstream [70–75]. The attachment of PEG to modify the biorecognition of proteins [72], cytokines [73], or chimeric toxins [75] is a general procedure (Table 3). One conjugate, PEG-adenosine deaminase (PEG-ADA) has been recently approved by the FDA for the treatment of ADA-deficient severe combined immunodeficiency syndrome [71]. PEG-modified L-asparaginase is in Phase III clinical trials [71]. L-asparaginase, an enzyme which converts L-asparagine to L-aspartic acid, exerts its anticancer activity by depriving tumor cells, which cannot synthesize L-asparagine, of this amino acid [76]. As a protein of non human origin, it has a short intravascular half-life and is immunogenic. Its conjugation with synthetic macromolecules such as PEG improves its biological properties.

Usually, PEG-protein conjugates are prepared by the reaction between methoxypolyethylene glycol (mPEG) carrying a reactive electrophile group

Table 3. Poly(ethylene glycol) modified proteins

Protein	Polymer and molecular weight	Degree of amino group substitution and linkage method	Stage in Clinical Trials	Ref.
L-Asparaginase	PEG (5 000)	18%, cyanuric chloride	Phase III	77
Adenosine deaminase	PEG (5 000)	60%, cyanuric chloride	FDA Approved	78, 79
Superoxide dismutase	PEG (1 900, 4 000, 5 000)	95%, cyanuric chloride; 95%, carbonyldiimidazole; 40% phenylchloroformate	Phase III	80–82
Uricase	PEG (5 000)	43%, cyanuric chloride	Phase I/II	83
Interleukin-2	PEG (5 000)	10–20%, succinimidyl active ester	Phase II	84
Streptokinase	PEG (2 000, 4 000, 5 000)	% Substitution not specified, carbonyldiimidazole activated PEG	Preclinical evaluation	85
Hemoglobin	PEG (3 600)	25%, succinimidyl active ester	Preclinical evaluation	86
G-CSF	PEG (10 000)	100% cyanuric chloride	Preclinical evaluation	87

("activated PEG") with nucleophilic residues on the surface of a protein [88]. Use of PEG-based acylating reagents, which react selectively with amino groups of proteins under very mild conditions, often lead to good preservation of activity of the conjugates [74]. However, altered kinetics, broadened pH optima, and changed substrate specificity of enzymes may be observed [72]. These changes are not so pronounced toward a low molecular weight substrate. On the other hand, high molecular weight substrates have a decreased biorecognition by PEG-modified enzymes and are cleaved at a substantially lower rate [74].

The decreased biorecognition of PEG-modified proteins by proteolytic enzymes and by the immune system is consistent with the ability of PEG to exclude proteins from its surroundings [89, 90] and with the protein repulsion properties of PEG-modified surfaces [91–93]. Other types of hydrophilic polymers may be used to modify the structure and properties of proteins. For example, albumin was used to modify L-asparaginase [94] and N-(2-hydroxypropyl)methacrylamide copolymers to modify acetylcholinesterase [95]. However, to compete with PEG, semitelechelic polymers have to be synthesized [96] and used. Okano et al. [97, 98] synthesized semitelechelic oligomers of N-isopropylacrylamide using 3-mercaptopropionic acid as a chain transfer agent. Using collagen [97], bovine serum albumin or bovine plasma fibrinogen [98] as model biomolecules, they have demonstrated that protein-poly(N-isopropylacrylamide) conjugates are temperature responsive and possess a lower critical solution temperature (LCST). The LCST can be manipulated by the composition of the modifying macromolecule. By using oligo(N-isopropylacrylamide-co-butyl methacrylate) macromolecules, the LCST of protein-polymer conjugates can be lowered by increasing the butyl methacrylate content [98].

An interesting synthetic approach was introduced by Ito et al. [99]. They covalently bound an azoinitiator, 4,4'-azobis(4-cyanovaleric acid) to trypsin. In the presence of the monomers, 3-carbamoyl-1-(p-vinylbenzyl)pyridinium chloride, or a mixture of methacrylic acid and methyl methacrylate, trypsin-polymer conjugates were synthesized (Fig. 3) which were sensitive to external signals, i.e., redox sensitive and pH sensitive, respectively.

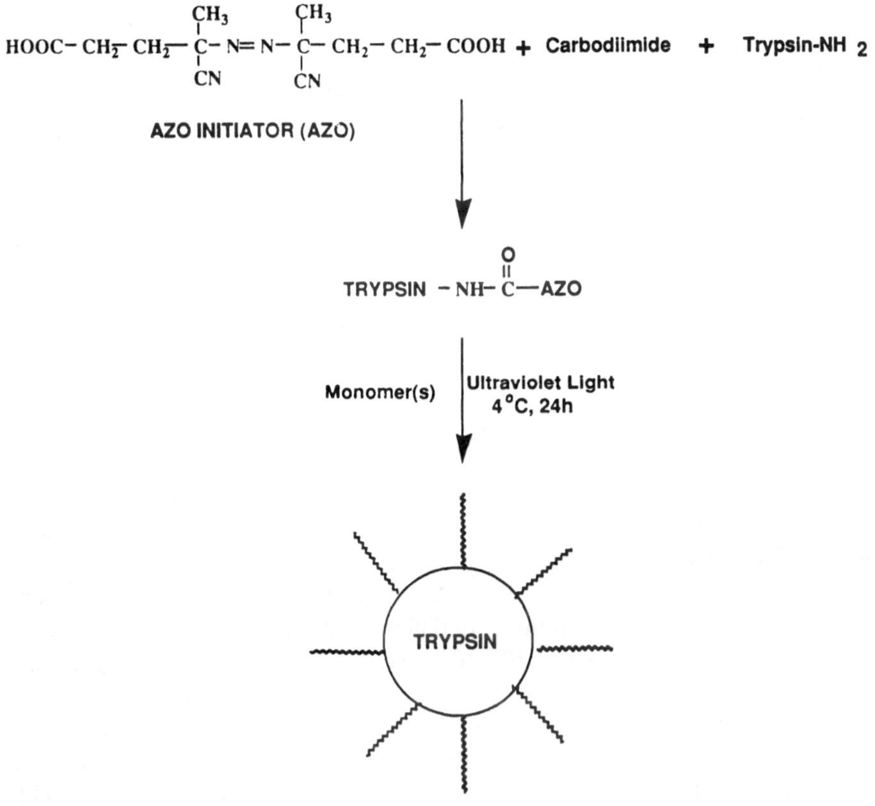

Fig. 3. Modification of trypsin molecules with covalently bound polymer chains renders the conjugate stimuli sensitive depending on the structure of polymeric chains (according to Ito et al. [99])

2.2 Biodistribution of Soluble Macromolecules

There are four main compartments a soluble macromolecule can enter: the central compartment (blood and lymphatic system), interstitium, intestinal lumen, and lysosomes [100, 101]. Minor compartments are primary urine, liquor, bile, etc. There is no experimental evidence that clearly indicates the penetration of synthetic macromolecules into the cytoplasm, i.e., into the intracellular compartment (inside the cell but outside the endosomes or lysosomes) [101]. The movements of soluble macromolecules between body compartments have been extensively reviewed [14, 20, 100–104] and will not be covered in detail here. We shall concentrate on the discussion of main factors influencing the movement of soluble macromolecules when administered into the bloodstream. Depending on the structure and molecular weight distribution, part of the polymeric molecules are excreted in the urine. Simultaneously, the macromolecules are cleared from the bloodstream by endocytosis. It is important to note that nonspecific capture of soluble macromolecules by the specialized cells of the reticuloendothelial system is generally much less (orders of magnitude) when compared to vesicular carriers of a comparable structure.

There are differences in the ease of extravasation of macromolecules from the bloodstream into different tissues [14, 104, 105]. Capillaries in the liver, spleen, and bone marrow have incomplete basal membranes and are lined with endothelial cells which are not continuously arranged. Capillaries in the muscle have a somewhat tighter arrangement, and there is an almost impermeable barrier which isolates the central nervous system from circulating blood. The rate of glomerular filtration of macromolecules depends on their hydrodynamic radius, the threshold being approx. 45 Å [106]. Structure of the macromolecule is of utmost importance, since shape, flexibility, and charge influence the penetration and possible readsorption in the tubular epithelia [100].

2.2.1 Extravasation in Tumor Tissue

Macromolecules having molecular weight above the renal threshold may cross the vasculature during inflammation and localize in tumor tissue [107]. Several biologically active molecules are involved in the modulation of vascular permeability in tumor tissue [15]. Vascular permeability factor is a protein (about 38 kD) produced in a number of different tumors. It possesses angiogenic and mitogenic activity in vivo [108, 109]. Many tumor cells produce plasminogen activator, which activates one or more steps in the kinin-generating cascade [15]. Resulting higher levels of bradykinin influence the permeability of tumor vasculature. It is now well documented that macromolecules having a long intravascular half-life do passively accumulate in solid tumors [13, 110, 111]. The combination of poor tissue drainage with increased tumor vascular permeability and lack of lymphatic capillaries results in a phenomenon termed by Maeda et al. as the [15, 107, 111] "enhanced permeability and retention effect" (EPR). This accumulation provides a convenient means by which to carry chemotherapeutic agents to the site of a tumor via soluble polymer conjugates.

2.2.2 Targeting Potential of Macromolecular Drug Carriers

The accumulation of polymer-drug conjugates in the tumor tissue described above is termed "passive targeting" (for review see [14]). Incorporation of a biorecognizable (targeting) moiety into the polymer carrier structure (Fig. 2) leads to active targeting [112]. Monoclonal and polyclonal antibodies, carbohydrates, peptide hormones, growth factors and other such molecules have been proposed as targeting moieties. It is important to distinguish between organ-/compartment (first order) targeting and tumor-specific (second order) targeting [13]. Whereas good results (on animals) were obtained with the localization of carbohydrate containing polymeric carriers [113], the targeting with antibodies has a number of problems [114] such as heterogeneity of target antigen expression, macromolecular transport barriers in solid tumors and immunogenicity.

To overcome the heterogeneity of tumor antigen expression the use of bispecific antibodies and their mixtures has been proposed [115, 116]. To improve transport, antibody fragments may be used [117]. At the same time this approach improves the immunogenicity of the conjugates as mentioned below. The majority of antibodies studied in clinical investigations are monoclonal antibodies (mAbs) of murine origin. With some exceptions, like one dose regimens, or treatment of B lymphomas, Hodgkin's disease and chronic lymphocytic leukemia, where the humoral immune response is suppressed [114], administration of such antibodies results in the formation of human anti-mouse antibodies (HAMA) [118]. To avoid the immunogenicity of antibodies, antibody fragments, F(ab)$_2$ and Fab, single-chain antigen binding proteins (sFv) [119], or chimeric antibodies are used in designing conjugates (for review see [120]). In chimeric antibodies variable, antigen binding sequences of murine antibodies are combined with those of the constant region of human IgGs [114]. However, anti-idioptype responses may still be observed [121]. Another approach is to use humanized antibodies in which only the murine hypervariable sequences encoding the complementary determining regions are inserted into a human immunoglobin gene construct [114, 122]. In spite of these difficulties, there is a window for the design of functioning conjugates. Trail et al. [123] have recently studied the properties of a conjugate of adriamycin with chimeric mAb binding to an antigen related to Lewis Y which is abundantly expressed at the surface of cells from many human carcinomas. Moreover, this antibody has a high tumor selectivity and internalizes after binding. This conjugate induced complete regressions and cures of xenografted human lung, breast, and colon carcinomas (growing subcutaneously in athymic mice). The authors presented an example of successful treatment when targeting an antigen which is expressed also in normal tissues [123]. We believe that the specific toxicity of such conjugates could be further increased when using a polymeric intermediate. Examples of the synthesis and properties of targetable conjugates will be described in Sect. 3.

2.2.3 Structural Factors Influencing the Biodistribution of Macromolecules

The biodistribution of soluble macromolecules depends on their structure, charge, molecular weight distribution, conformation and changes in conformation when crossing compartmental barriers, and biodegradability [12]. These factors are interrelated, since chemical structure influences conformation, charge, and biodegradability, etc. All these aspects were discussed in detail previously. The effect of molecular weight on the intravascular half-life and rate of elimination was studied for poly[N-(2-hydroxypropyl)methacrylamide] [124, 125], polysaccharides [126, 127] and many other polymers (for review of early work see [128]). The influence of molecular weight distribution on the elimination of poly(vinylpyrrolidone) was studied by Hespe et al. [129]. The molecular weight dependent deposition of macromolecules in tumor-bearing mice was described by Takakura et al. [130]. They have concluded that anionic charge and a higher molecular weight enhance tumor accumulation.

The rate of pinocytosis (cellular uptake) may be influenced by molecular weight [131], charge [22] and incorporation of hydrophobic comonomers [21] among others.

The behavior of macromolecules may and usually will change when a drug and/or a targeting moiety is attached to the polymer backbone. For example, when hydrophobic side-chains terminated in drug are attached to hydrophilic macromolecules, the solution properties of the copolymer may change [132–134]. Such copolymers (drug-polymer conjugates) may associate in water forming micelle-like structures, the hydrophobic side-chains being inside and the polymeric chains outside. The association number and compactness of the micelles depend [134] on the copolymer structure, copolymer concentration, and temperature. Micellar shells were shown to hinder the penetration of enzymes into the micellar core [132], thus reducing the rate of drug release. The increased size of macromolecular aggregates impairs their movement between body compartments. On the other hand, Kataoka et al. [49, 135, 136] took advantage of micelle forming properties of adriamycin conjugated to poly(ethylene glycol)-poly(aspartic acid) block copolymers (Fig. 4). Hiding of the hydrophobic drug in the core of the micelle permits the administration of high concentrations of this otherwise toxic compound. Whereas it was clearly demonstrated [49, 136] that these conjugates are biologically active, the mechanism of drug release in the lysosomal compartment remains to be solved.

The attachment of targeting (biorecognizable) moieties introduces serious changes in the properties of polymer conjugates when compared to the properties of the carrier macromolecule. If antibodies are used, a dramatic increase in molecular weight is observed. The main consequence, however, is the biorecognition of the conjugate at an organ or subset of tumor cells resulting in a

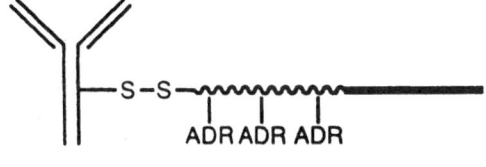

 Poly(aspartic acid)

━━━━━━━━━━ Poly(ethylene glycol)

IgG

ADR   Adriamycin

Fig. 4. A block copolymer of poly(ethylene glycol) and poly(aspartic acid) containing adriamycin and IgG (according to Yokoyama et al. [135])

different biodistribution when compared to the polymer-drug conjugate without a targeting moiety.

3 Synthesis of Conjugates

There are several possibilities how to attach biologically active compounds (drugs, targeting moieties) to macromolecular carriers. Covalent bonds are preferred with the exception of noncovalent bonds possessing a very high binding constant, e.g., the avidin-biotin system. For example, Trouet and Jolles [55] found that complexes of anthracycline antibiotics with DNA are not stable in the bloodstream (in vitro affinity constant $10^6 \, M^{-1}$). Consequently, the design of their conjugates was changed to include covalent bonds. Serum albumin was used as a carrier and anthracycline antibiotics were attached via oligopeptide spacers stable in the bloodstream and degradable by lysosomal enzymes [137]. Another example of a delivery system using noncovalent bonds for the attachment of drug to polymeric carriers are conjugates of cis-dichlorodiammineplatinum (II) (cis-DPP) [138]. Macromolecules which contain carboxyls, amides, and alcohols form low affinity bonds with platinum(II) and are therefore potentially suitable to serve as cis-DPP carriers [139].

Examples of polymeric carriers are presented in Table 1 and typical methods of covalent conjugation are shown below. Drugs may be bound to macromolecular carriers via, e.g., ester, amide, urethane, hydrazone, thioether, and disulfide bonds. The bond between the drug and polymeric carrier may be either hydrolyzable at the acid pH of the prelysosomal or lysosomal compartments (e.g., aliphatic ester, and hydrazo bonds or bound via *cis*-acotinyl spacers), reduced in prelysosomal compartments (disulfide bonds), or cleaved by enzymatically catalyzed hydrolysis in the lysosomal compartment (specific amide or ester bonds). Designs have also been proposed where a drug is released after the main chain is degraded. However, the drawbacks of this approach are discussed in Sect. 2.1.

3.1 Attachment of Drugs

Generally there are two possible ways of synthesis of macromolecular-drug conjugates (Fig. 5) [140]:

a) synthesis of a polymerizable drug derivative followed by copolymerization with suitable monomer(s);
b) polymeranalogous reactions, i.e., attaching the drug to a preformed polymeric carrier.

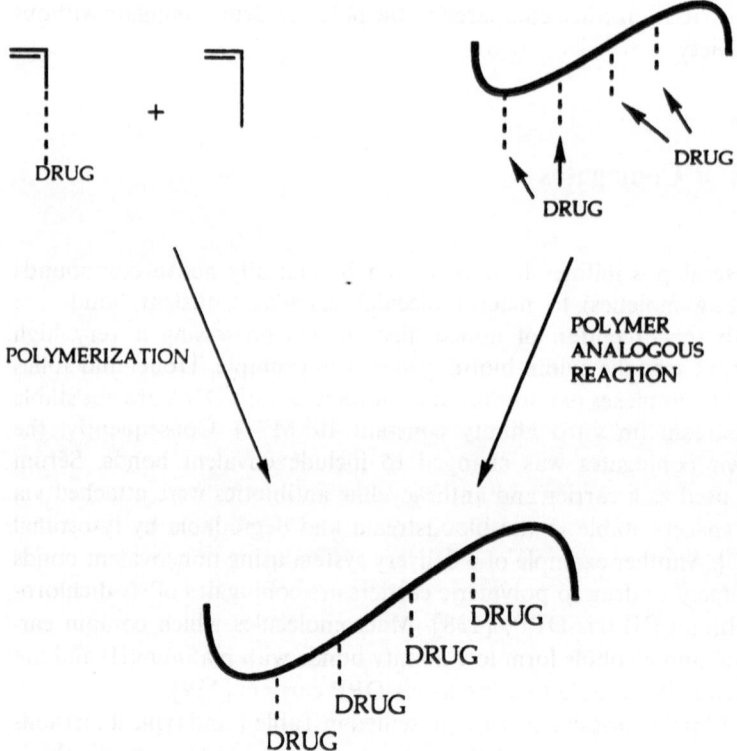

Fig. 5. Synthetic approaches to polymer-drug conjugates

It is important to note that many modifications of this general scheme occur, e.g., modification of the drug to introduce a reactive group and/or a degradable spacer prior to the polymeranalogous reaction [137], or the use of several polymeranalogous reactions, the first to convert the reactive group at the end of the side-chain (active ester to aliphatic amine) followed by attachment of drug via carboxylic groups [23].

3.1.1 Polymerizable Drug Derivatives

This method may be used when the structure of the drug is stable during the polymerization process. An advantage of this method is the avoidance of side-reactions which sometimes occur during polymeranalogous reactions. Examples of earlier studies may be found in reviews [20, 140], some recent examples follow.

Ulbrich et al. [141] synthesized a polymerizable derivative of daunomycin, N-methacryloylglycylphenylalanylleucylglycyldaunomycin (MA-Gly-Phe-Leu-Gly-DNM). In this monomer the polymerizable methacryloyl group is separated from the drug by a tetrapeptide spacer which is degradable by the lysosomal thiol proteinase, cathepsin B [26]. Copolymerization of MA-Gly-Phe-Leu-Gly-DNM with HPMA and MA-Gly-Gly-ONp yielded an HPMA copolymer with bound drug and attachment sites (active ester groups) for the binding of targeting moieties.

Polymerizable derivatives of an antidiabetic drug, N-(4-amino-benzenesulfonyl)-N'-urea, where the active substance is bound to a methacrylate moiety via spacers of various lengths containing amide or urethane links were synthesized and copolymerized with HPMA [142].

Ouchi et al. synthesized 1,2-mono-O-isopropylidene-3-[3-(5-fluorouracil-1-yl)propionyl]-6-O-acryloyl-α-D-glucofuranose and copolymerized it with acrylamide [143], Ozaki et al. synthesized 1-(meth)acryloyloxymethyl-5-fluorouracils and copolymerized with a number of comonomers, such as acrylic acid, methacrylic acid, methyl acrylate, and methyl methacrylate [144]. All the above mentioned polymer bound drugs possessed biological activity.

3.1.2 Conjugation of Drugs by Polymeranalogous Reactions

A prerequisite for such an attachment is the match of reactive groups on the polymeric carrier and the drug. Usually a polymer precursor is synthesized which contains reactive groups at side-chain termini, e.g., reactive esters. Groups which contain complementary groups, e.g., aliphatic amino groups may be easily bound by aminolysis forming an amide group [102]. Anticancer drugs bound to HPMA copolymers discussed in Sect. 5 are prepared by this route (see also Fig. 10).

Another example is a polymeric carrier containing side-chains terminated in COOH groups. An ester or amide bond between the drug may be formed when the drug contains an OH or NH_2 group. The carboxylic group on the polymeric carrier and the reactive group on the drug may be directly linked using carbodiimides as condensing agents. For example, drugs bearing NH_2 groups (isoniazid, procaine, histamine) were covalently coupled to poly(α, β-aspartic acid) using N-ethyl-N'-(3-dimethylaminopropyl)carbodiimide as the condensing agent [145].

However, in many cases the structure of the drug does not match the structure of the reactive groups on the polymeric carrier. In this case, the structure of at least one reactant has to be modified before conjugation. Derivatives of 5-fluorouracil, namely 1-(3-bromopropionyl)-5-fluorouracil, 1-chlorocarbonyl-5-fluorouracil, and 1-(4-bromobutyl)-5-fluorouracil were synthesized to permit the drug attachment to poly(L-cysteine) [146] (Fig. 6).

Reactive derivatives of anthracycline antibiotics were synthesized and attached to polymeric carriers. To bind adriamycin to poly(glutamic acid) and poly(aspartic acid) via ester bonds, Zunino et al. [147] used the reaction of alkylbromides (14-bromodaunorubicin) with the carboxylic group of the carrier.

A new derivative of adriamycin, namely (6-maleimidocaproyl)hydrazone of adriamycin was synthesized and conjugated to mAbs via a Michael addition

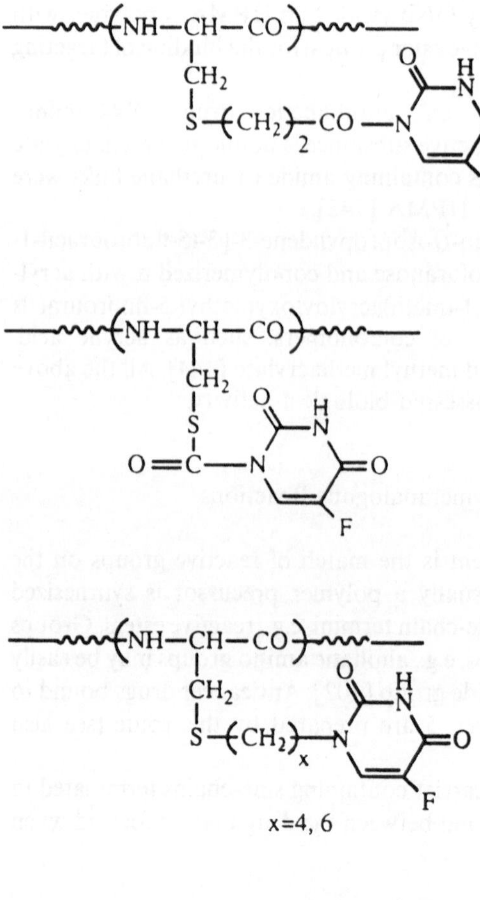

Fig. 6. Examples of attachment of 5-fluorouracil to cysteine residues in poly(amino acid) carriers (according to [146])

reaction to thiol-containing mAbs [28]. The conjugates released adriamycin under acidic conditions that mimic the lysosomal environment, while they were relatively stable at neutral pH.

A convenient synthetic method was developed for the site-selective conjugation of methotrexate (MTX) to peptide carriers (somatostatin and LHRH analogs) [148]. The selective protection of MTX with benzotriazol-1-yloxy-tris(dimethylamino)phosphonium hexafluorophosphate (BOP reagent) permits the selective coupling of either the α- or the γ-carboxyl group of the glutamic acid moiety in MTX to a free amino group in the peptide carrier (Fig. 7). Both of the above mentioned methods [28, 148] can be used for the conjugation to polymeric carriers.

On the other hand, polymeric carriers can also be modified to introduce reactive groups. Polysaccharides such as dextran and inulin may be activated [149] by periodate oxidation to create aldehyde groups, by succinic anhydride activation to create carboxylic groups, or by p-nitrophenyl chloroformate activation to create reactive ester groups.

Nukui et al. [133] synthesized poly(glutamic acid)-adriamycin conjugates. ADR was conjugated via amide bonds either directly or via an oligopeptide

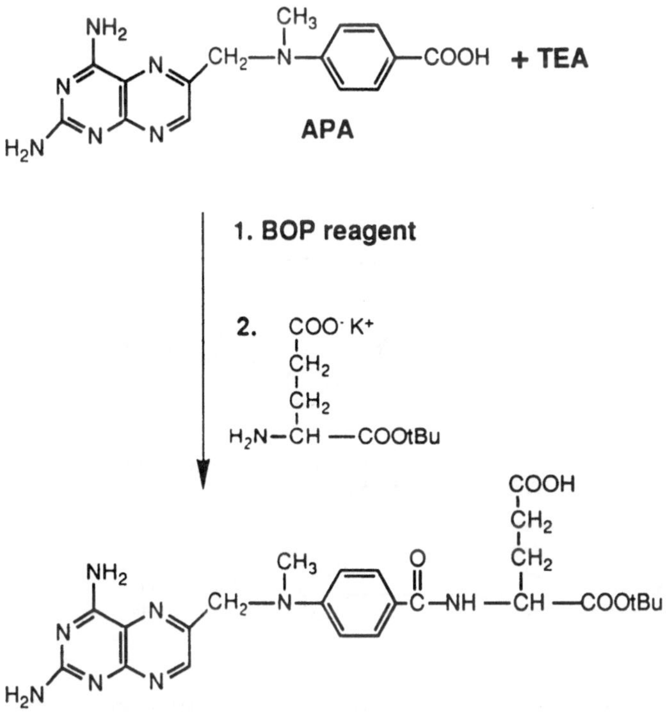

Fig. 7. Synthesis of selectively protected methotrexate (MTX) derivative. The synthesis of *tert*-butyl ester of the α-carboxyl group of the glutamic acid moiety in MTX from APA (4-amino-4-deoxy-N^{10}-methylpteroic acid), an MTX precursor (according to [148])

spacer. Direct conjugation was carried out using *N*-ethoxycarbonyl-2-ethoxy-1,2-dihydroquinoline (EEDQ) as the coupling agent. The spacer (Gly-Gly-Gly-Leu) was introduced after preactivating the carrier with saccharin and *N*,*N'*-carbonyldiimidazole. ADR was attached to the terminal amino acid residue in the presence of EEDQ. The noncovalently associated ADR was removed by ion-exchange chromatography.

One of the very important developments in the last decade or so was that a large group of heterobifunctional reagents became commercially available. Such reagents contain two different types of reactive functional moieties. Their advantage when compared to homobifunctional reagents (two identical functional groups) is in the fact that many reactions with macromolecules (polymeric carriers, antibodies) do not have to be performed with a large excess of the reagent to avoid intramolecular reactions. Examples of heterobifunctional reagents are shown in Fig. 8. Excellent reviews appeared recently dealing with this class of compounds [157–159]. However, the emphasis is on their use for the

NAME	STRUCTURE	REACTIVE TOWARD	REF.
1. N-Succinimidyl 3-(2-pyridyldithio propionate (SPDP)	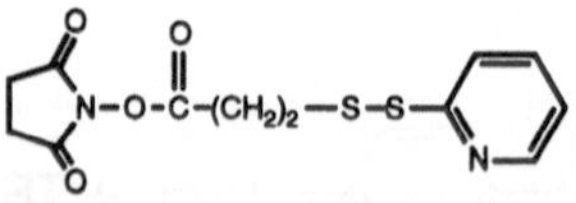	Amino, Sulfhydryl	150
2. N-Succinimidyl maleimido-acetate (AMAS)	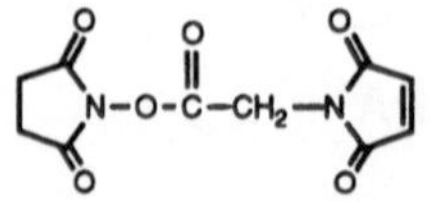	Amino, Sulfhydryl	151
3. N-Succinimidyl 4-(p-maleimidophenyl) butyrate (SMPB)	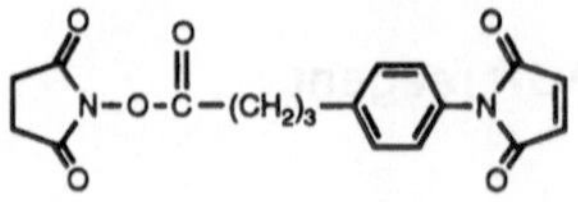	Amino, Sulfhydryl	152
4. Pyridyl-2,2'-dithiobenzy-diazoacetate (PDD)	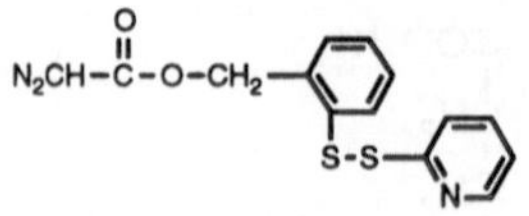	Carboxyl, Sulfhydryl or Amino	153
5. 1-(Aminooxy)-4-{(3-nitro-2-pyridyl)dithio}butane	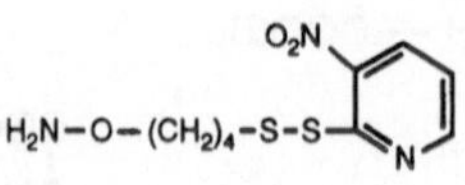	Carbonyl, Sulfhydryl	154
6. N-Succinimidyl-2-{4-azido-phenyl)dithio}acetate (NHS-APDA)	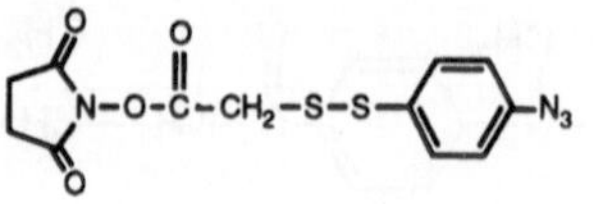	Amino anchor, photoactivatable	155
7. 4-Azidophenylmaleimide (APM)	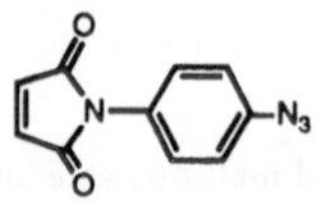	Sulfhydryl anchor, photoactivatable	156

Fig. 8. Typical heterobifunctional reagents

modification of biological macromolecules. Their use in polymeranalogous reactions on synthetic macromolecules will undoubtedly increase in the future. Some examples of the use of heterobifunctional reagents in the binding of biologically active molecules include the binding of Bowman-Birk protease inhibitor to poly(D-lysine) and poly(L-glutamate) with N-succinimidyl 3-(2-pyridyldithio)propionate (SPDP) [160], and the coupling of polylysine protein-binding polyhedral boron derivative conjugates to antibodies with N-maleimidobenzoyl sulfosuccinimide ester (sulfo-MBS) [161].

3.2 Attachment of Targeting Moieties

The binding methods used depend on the structure of the targeting moiety. Carbohydrates, hormones, and antibodies (or their fragments) are frequently used to direct polymer conjugates to specific cell subsets.

Some cell receptors recognize both carbohydrate and the N-acylated aminosugar. For example, the asialoglycoprotein receptor on hepatocytes recognizes both galactose and N-acetylgalactosamine. To incorporate galactose into HPMA copolymers, a monomer with protected OH groups, namely 1,2,3,4-di-O-isopropylidene-6-O-methacryloyl-α-D-galactopyranose was synthesized, copolymerized with HPMA and the protecting (isopropylidene) groups removed by formic acid [162]. To synthesize polymer conjugates containing N-acylated galactosamine is an easier task. Reactive HPMA copolymer precursors, containing side-chains terminated in p-nitrophenyl esters are aminolyzed with galactosamine, a reaction which can be performed in DMSO at room temperature [163].

Antibodies may be bound directly to anticancer drugs. However, extensive modification of mAbs with drugs may lead to loss of solubility and antibody activity [158]. Using a polymer intermediate increases the specific toxicity of the conjugates [164]. Abs contain a large number of lysine residues. Their ε-amino groups may be used for attachment to polymeric carriers [23, 24]. Oxidizing the carbohydrate moieties on the Fc fragment near the hinge region with sodium periodate produces aldehyde groups [165] suitable for binding to the carrier. For example, anti-Thy 1.2 antibodies were oxidized and bound to HPMA copolymers containing side-chains terminating in aliphatic amino groups. The azomethine bond was stabilized by reduction with NaCNBH$_3$ [23]. Conjugates containing antibodies bound via oxidized carbohydrate moieties are usually more active when compared to conjugates containing antibodies bound via lysine [23, 165, 344] or aspartic and glutamic acid side-chains [165].

When using amino groups on the polymeric carrier (or drug) for attachment to aldehyde groups in oxidized saccharide residues of antibodies, oligomers of the latter may be formed by the reaction of amino groups of lysine residues of one antibody molecule with the aldehyde groups on the other. To avoid this side-reaction, hydrazides may be used and the coupling reaction performed at a lower pH where the reactivity of amino groups is minimal [166].

An example of the synthesis of an adriamycin-dextran-antibody (anticarcino-embryonic mAb) is shown in Fig. 9 [54]. The synthetic design called for the modification of all three components, i.e., drug, polymeric carrier, and antibody before conjugation.

Recently, a modified method for the synthesis of drug-dextran-antibody conjugates was developed [167]. Aminopropyldextran was selectively mono-functionalized at its reducing terminus via reductive amination with 2-(4-nitrophenyl)ethylamine. The nitro group was eventually converted into a reactive isothiocyanato group used for the attachment of the polymer-drug

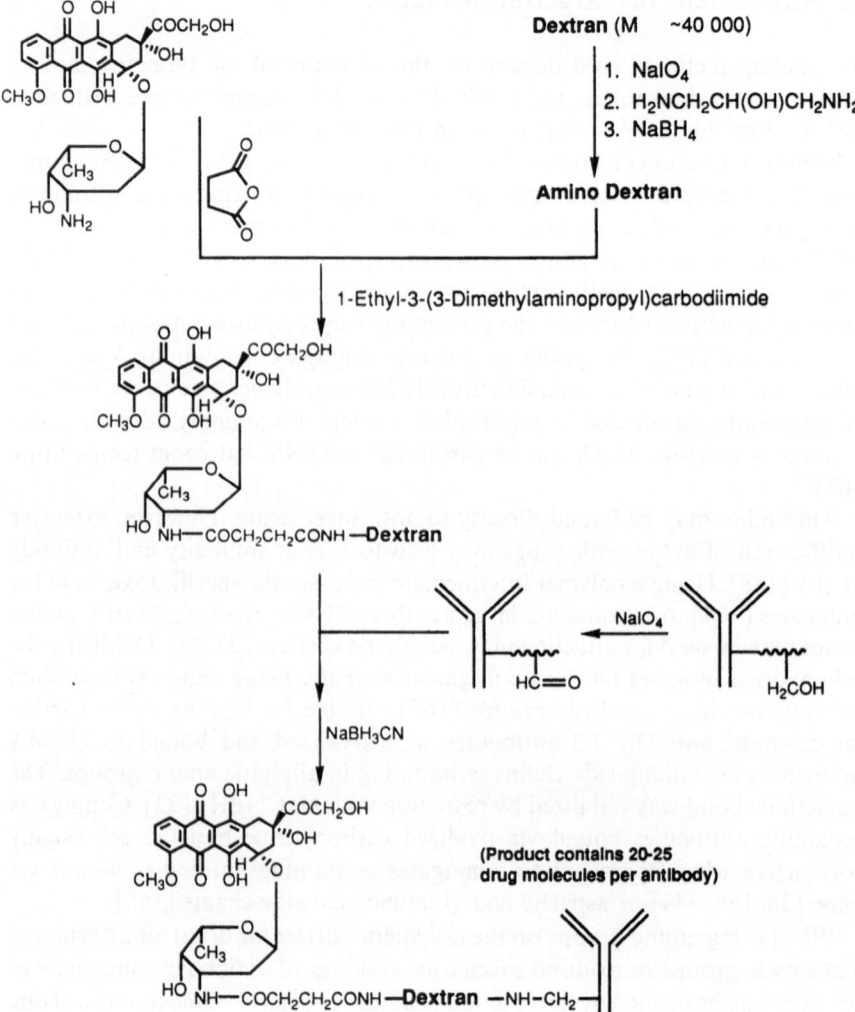

Fig. 9. Synthesis of a targetable dextran-adriamycin conjugate. Note that amino groups were introduced into the dextran structure, whereas the amino group of adriamycin was converted into a carboxylic group. According to [54]

conjugate to the targeting molecule [167]. A similar approach was used by Kato et al. [168] who synthesized daunomycin-poly(glutamic acid) conjugates with a single masked thiol group. The latter was used for attachment to an anti-α-fetoprotein antibody.

Synthesis of an adriamycin-HPMA copolymer-galactosamine conjugate is shown in Fig. 10. Attachment/release points at the side-chain termini are created by incorporation of tailor-made comonomers into the copolymer structure.

3.3 Choice of Synthetic Design Based on the Site of Action

Binding of drugs to polymeric carriers renders the conjugate lysosomotropic. However, the targeting moiety attached to the polymer-drug conjugate (or only to drug) may influence the site of action. It is well known that some antibodies do not internalize immediately, in fact a prolongation on the cell surface may occur. There are antibodies which do not internalize and stay for a long period at the cellular surface [170]. Systems for the delivery of anticancer drugs to the cell

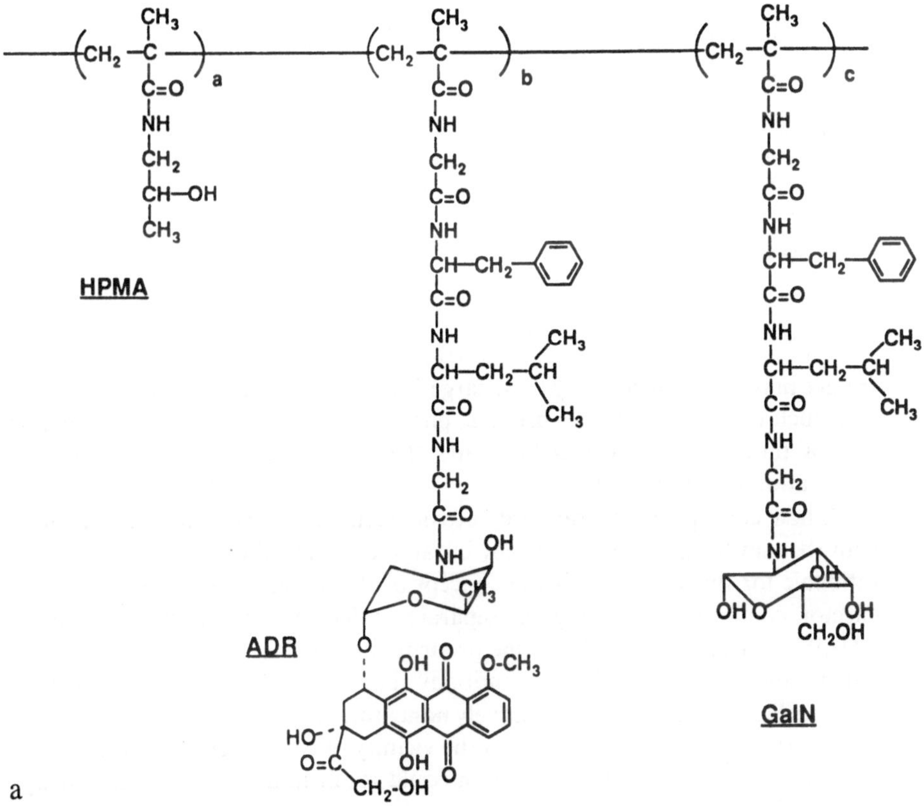

a

Fig. 10a. Structure of a targetable polymeric drug based on HPMA copolymer carrier.

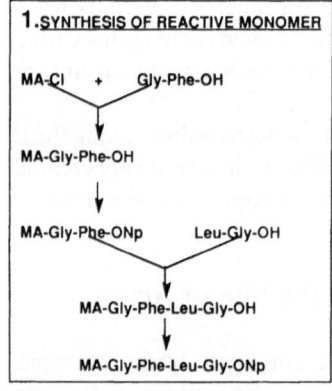

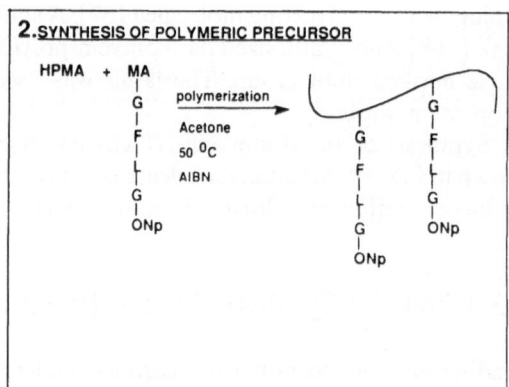

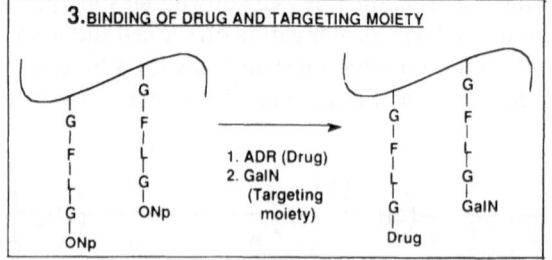

MA .methacryloyl
ONp p-nitrophenyloxy
ADR .adriamycin
GalN .galactosamine

Fig. 10b. A scheme of its synthesis. First, a reactive comonomer. *N*-methacryloylglycyl-phenylalanylleucylgycine *p*-nitrophenyl ester was synthesized. In the second step a polymeric precursor is prepared by copolymerization of the reactive comonomer with *N*-(2-hydroxypropyl)methacrylamide. After purification and characterization of the polymeric precursor, the anticancer drug (adriamycin) and targeting moiety (galactosamine) are attached by consecutive aminolysis. For details see, e.g., [169]

surface may be designed using these targeting moieties. Suitable drugs to be used with such types of antibodies are, e.g., photosensitizers, which can be activated after a time lag by external light and their site of action is at the site of localization (see Sect. 5.4).

A new concept of generating cytotoxic agents at cancer sites was designed with the aim to increase the efficacy of targeted chemotherapy [171, 172]. The rationale for this approach is the concept of the delivery components and the effector components residing on separate molecules. In this method, called ADEPT (antibody directed enzyme prodrug therapy), an enzyme which has no human analogue is linked to an antibody that binds to antigens expressed on tumor cells. After a time lag, a relatively noncytotoxic prodrug is administered. It is converted into a cytotoxic drug in the vicinity of cancer cells or within tumor masses (Fig. 11). Enzymatic activation could result in a large number of drugs generated per conjugate molecule, and the drug formed may be able to diffuse to tumor regions that are not accessible to the conjugate [172].

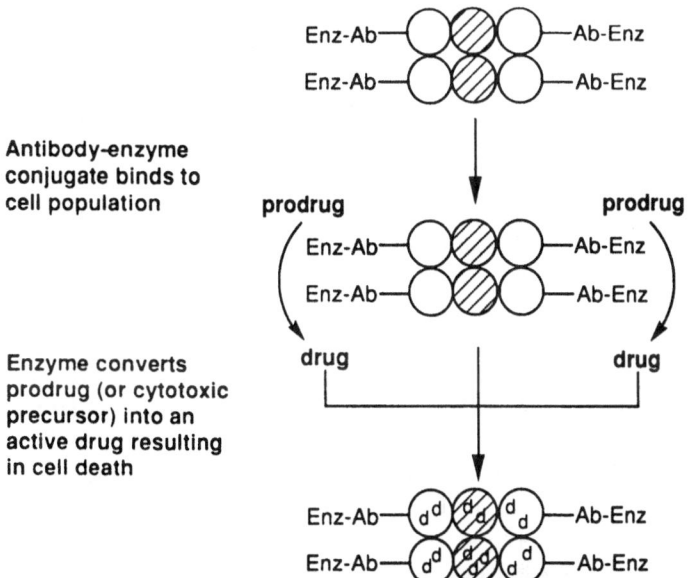

Antibody-enzyme conjugate binds to cell population

Enzyme converts prodrug (or cytotoxic precursor) into an active drug resulting in cell death

Fig. 11. Principle of antibody directed enzyme prodrug therapy. Hatched circles represent cells that have not bound antibody-enzyme conjugate. Internalized drug is abbreviated with the letter "d". (From [172] with permission)

This field was recently reviewed [172]. Enzymes that have been used for ADEPT are, e.g., alkaline phosphatase, carboxypeptidase A, cytosine deaminase, β-lactamase; prodrugs and corresponding drugs were, e.g., adriamycin phosphate and adriamycin, methotrexate-alanine and methotrexate, 5-fluorocytosine and 5-fluorouracil, and vinca-cephalosporin and vinca alkaloid [172 and references therein].

The main unsolved problem connected with this approach is the immunogenicity of the conjugates. Immune responses developed in patients treated with anti-hCG or anti-CEA monoclonal antibody-carboxypeptidase G2 conjugates followed by the administration of a modified benzoic acid mustard {p-N-bis-(2-chloroethyl)aminobenzoyl glutamic acid} [173].

4 Biological Properties of Conjugates

4.1 Methods of Evaluation

In this section some of the tests used to evaluate the biological activity of anticancer drugs are listed. The intent is to give the chemical audience a feeling of the methods suitable to evaluate their conjugates and the background to effectively discuss the testing protocols with biological coworkers. However, there are two important points to take into account.

First, the majority of published literature on testing protocols, and discussion of their suitability, focuses on screening protocols, i.e., the rapid screening of a number of new chemical entities to recognize those which are active in the particular test and warrant further studies. The situation with polymer bound drugs is in the majority of cases quite different. A drug with a known anticancer activity is chosen for the synthesis of conjugates with the aim to increase its therapeutic activity and/or to decrease the nonspecific side-effects. The testing protocols may be chosen based on the mechanism of action of the drug.

Second, extreme care has to be devoted to the purification of drug-polymer conjugates. Due to the different mechanism of cellular uptake of free and polymer bound drugs, the presence of free drug in the conjugates may render the evaluation of their biological activity in vitro meaningless. The results obtained from in vitro studies must be carefully assessed due to possible positive or negative aberrations which may arise. For example, doxorubicin is an amphipathic compound which not only self-associates in aqueous solution, but can also strongly interact with drug delivery systems. Consequently, free doxorubicin may exist in a delivery system in which the drug is thought to be only covalently bound. This free fraction of drug may alter the behavior of the delivery system, effectively producing false positive anticancer effects in vitro [36].

4.1.1 In Vitro Evaluation

The in vitro methods by which macromolecular anticancer drug conjugates are evaluated for efficacy are similar to those used to screen for the anticancer activity of low molecular weight compounds. A thorough review of in vitro models used for the evaluation of antitumor therapy was published several years ago [174].

Requirements of in vitro models are listed below [174].
1. Tumor sample must closely represent in vivo tumor.
2. Metabolism of tumor sample must approximate in vivo tumor.
3. Conjugate must be in active form or must be able to be activated by model.
4. Must be reproducible, inexpensive, fast and simple to perform.
5. Easily adapted to various histologic tumor types.
6. Drug concentrations in the system must be achievable in clinical setting.
7. In vitro drug scheduling must parallel in vivo kinetics and dosage schedules.
8. Must have a strong positive correlation between in vitro and clinical results.
9. In vitro model must be more sensitive than selective to avoid exclusion of a potentially active agent.

Numerous methodologies have been developed. Generally, in vitro methods used to show the anticancer effect of compounds are based upon the determination of the number of viable cells or cell colonies present in a treated population of cells relative to the number of cells or colonies present in an untreated population. Those methods used for polymeric conjugate efficacy studies shall be discussed in this section.

4.1.1.1 Radiolabeled Nucleic Acid Precursors

The use of radiolabeled nucleosides as markers for anticancer activity has become a popular method due to the commercial availability of such compounds. The technique is based upon the knowledge that cells rendered unable to replicate or killed by the anticancer agent are unable to effectively incorporate nucleic acid precursors into their DNA or RNA structure. Therefore, a decrease in cell viability correlates with a decrease in radioactivity relative to a control cell population. Although specific procedures differ, the basic technique involves the incubation of tumor cells in the presence of the radiolabeled compound with or without anticancer agent followed by scintillation counting to determine the radioactivity of the samples.

The use of radiolabeled nucleic acid precursors, primarily tritiated thymidine or (deoxy)uridine, however, is subject to numerous drawbacks. Nevertheless, the use of radiolabeled precursors have provided some promising clinical correlation data [174].

The ability of an antibody-dextran-adriamycin conjugate to inhibit the growth of tumor cells using [methyl-^{3}H]-thymidine was studied by Sheldon et al. [175]. Goff et al. [176] has shown that chlorin e_6-polyglutamic acid-OC125 (anti-ovarian carcinoma monoclonal antibody) exhibited high cytotoxycity against a number of human ovarian carcinoma cells ex vivo, whereas little to no cytotoxicity against nonovarian cancer cells was observed. Radiolabeled nucleic acid precursors were also used to determine the nonspecific toxicity toward hematopoietic precursors in bone marrow of free and HPMA copolymer bound adriamycin [177]. Other examples of evaluation of the activity of macromolecular drug conjugates are available [178–182].

The inhibition of RNA synthesis may be followed by the use of [^{3}H]-deoxyuridine or [^{3}H]-uridine in much the same manner as those used for the DNA assays above [183–186].

4.1.1.2 Colorimetric MTT Assay

Although the radiolabeled nucleic acid precursor assays described above produce accurate and reproducible quantification of the number of viable cells in a sample, Mosmann [187] sought to develop a more rapid assay capable of handling large numbers of samples. A colorimetric technique was developed based upon the tetrazolium salt, 3-(4,5-dimethylthiazol-2-yl)-2,5-diphenyl tetrazolium bromide, or MTT. Early studies of MTT by Slater et al. [188] determined that when the MTT tetrazolium salt interacts with the dehydrogenase enzymes of mitochondria, the salt is reduced forming a colored formazan type product (detectable spectrophotometrically).

The MTT assay was initially developed as a quantitative assay for cell survival and proliferation, not as an in vitro assay for chemosensitivity testing. Further study was required to ascertain if the method accurately predicted the in vivo antitumor activities of anticancer agents. Shimoyama et al. [189] studied the predictability of the MTT assay with respect to a clonogenic assay (Sect. 4.1.1.3.) and showed excellent reproducibility and a close correlation to the in vivo predictability rate of the clonogenic assay. Another study [190] also showed that the MTT assay closely approximated (90%) the clinical activity of anticancer agents. Many authors have since utilized the MTT assay to determine the efficacy of polymeric anticancer drug conjugates.

The activity of HPMA copolymer anticancer drug conjugates against Walker sarcoma [38] and a human hepatocarcinoma cell line [23, 43, 191], the in vitro efficacy of conjugates of mitomycin C with monoclonal antibody recognizing small cell lung carcinoma [192], and of genetically engineered (anti-transferrin receptor) antibody-tumor necrosis factor hybrid molecule [193] have been determined through the MTT assay.

4.1.1.3 Clonogenic Assays

It has been suggested that the ability of tumor cells to form colonies in vitro following exposure to antitumor agents may be the best method by which to evaluate an anticancer compounds's activity [194]. Prior to the 1970s, such assays were primarily limited to cultures of patient tumor tissue; however, culture techniques now allow tumor cell lines to be used for such studies.

Hamburger and Salmon [195] provided a simple clonogenic assay capable of evaluating many tumor types. This assay, and modifications thereof [196], convert solid tumor masses into single cell suspensions through mechanical and/or enzymatic means and grow them in the presence of an agar underlayer to prevent fibroblast attachment to the vessel surface [174]. It is this suspension which is utilized to determine the efficacy of anticancer conjugates.

Rakestraw et al. [197] recently demonstrated the efficacy of chlorin e_6 immunoconjugates through a clonogenic assay. Other examples of the use of clonogenic assays may be found [198, 199].

4.1.1.4 Direct Cell Suspension Count

Cultures consisting of cell suspensions may be used directly to determine the efficacy of macromolecular antitumor agents. For example, suspensions of L1210 leukemia cells were used to show the activity of HPMA copolymer anticancer drug conjugates [39]. The cells, while in exponential growth, were diluted in culture medium to produce a cell density of approximately

1×10^4 cells/ml and allowed to culture for 24 h after which drug conjugate was added. The cultures were then mixed thoroughly and incubated for an additional 72 h. A Coulter counter was then used to count the number of cells present in treated and nontreated cultures. The number of viable cells was assessed microscopically through the use of trypan blue (Sect. 4.1.1.5). Penetration of the dye was an indicator of cell death.

Hirano et al. [200] cultured mouse P338 leukemia cells with conjugates of adriamycin or daunomycin and divinyl ether and maleic anhydride (DIVEMA) copolymers. The cells were incubated with the conjugate for 48 h in growth medium followed by counting via a Coulter counter. The activity of the conjugate was ascertained through comparison with a control group. It seems from the literature that fluid tumors, such as leukemia, are best suited for this type of assay. This may be true since solid tumors may form microcolonies while in suspension resulting in inaccurate quantification of the number of cells present in a given culture.

4.1.1.5 Assays Based on Physical Characteristics of Cells

The loss of certain cell characteristics, such as membrane integrity, has been exploited as a means by which to determine the viability of cells. Damage to the cell membrane resulting from interference with normal cell function by an anticancer macromolecule renders the membrane permeable to marker dyes, such as trypan blue [174] or fluorescent dyes [201], and allows radiolabeled proteins within the cytoplasm to be released [194].

Trypan blue exclusion from a cell appears to have evolved as the standard dye method used to test cell membrane integrity. However, it has been suggested that the trypan blue exclusion test does not reliably predict the cytotoxicity of drug conjugates [194]. Recently, the trypan blue exclusion test was compared with a fluorometric assay using simultaneous fluorometric staining with fluorescein diacetate and propidium iodide [202]. During the early phase of rapid cell (murine hybridoma) growth, differences were not significant. However, as the culture aged, the trypan blue dye exclusion assay significantly overestimated cell viability, thereby underestimating nonviable cell density and yielding an erroneous estimation of the overall viability of the culture [202].

The trypan blue dye exclusion method has been used to assay the antitumor effects of methotrexate-antibody conjugates [203, 204], ribosome inactivating proteins [205], nucleotide analogue-antibody conjugates [178], HPMA copolymers-chlorin e_6-antibody conjugates [23, 43], daunomycin-poly-L-glutamic acid-antibody conjugates [168], and ricin-antibody conjugates [206].

4.1.1.6 Vital Enzyme Inhibition Assay

If the mechanism of action of an antitumor conjugate is known, such as the inhibition of a vital enzyme, then the efficacy of the conjugate may be determined in vitro by measuring the conjugate's ability to inhibit the enzyme. This type of assay is limited to the antimetabolite class of anticancer compounds, and is further limited to those antimetabolites for which the target vital enzyme is known. The inhibition of dihydrofolate reductase by methotrexate is a prime example of a vital enzyme inhibition assay. Endo et al. [203] utilized this type of assay to determine the efficacy of their methotrexate-antibody conjugate with a monoclonal antibody (aMM46) against mouse mammary tumor MM46 cells.

4.1.1.7 Protein Synthesis Inhibition Assay

The ability of a tumor cell to manufacture proteins is a result of intact DNA, RNA and biochemical intracellular mechanisms. Interference with any one of these structures or processes will result in the inability of the cell to produce required proteins. Hence, quantitation of tumor cell protein synthesis over a period of time may constitute a marker allowing determination of the efficacy of a macromolecular drug conjugate. The technique is based on the fact that decreased cell viability in the presence of radiolabeled amino acids correlates to a decrease in radioactivity relative to a control cell population. For example, ^{3}H-leucine [175, 208], a mixture of [^{14}C]-labeled amino acids [205], and ^{75}Se-lenomethionine [54, 209] have been used to evaluate the activity of conjugates.

4.1.2 In Vivo Evaluation

If the results following the in vitro evaluation of a macromolecular anticancer agent are promising, in vivo tests are then warranted. Animal tumor models are the status quo for the in vivo study of anticancer agents. An animal tumor model should be reproducible, have a low rate of spontaneous cures or remissions, and a low and nonvariable immunogenicity in the host. For tests of mechanisms that also occur in human tumors, the parameters involved in these mechanisms should be similar to those of human tumors [210].

The efficacy of a macromolecular anticancer conjugate against an in vivo cancer model will be influenced by the conjugate's ability to cross compartmental barriers within the organism as mentioned in Sect. 2.2. Therefore, initial in vivo testing of a polymeric anticancer conjugate will often employ the same compartment of the animal, e.g., peritoneum, for both the cancer cell and the conjugate. The data resulting from such an experimental design must be viewed as preliminary since the conjugate is essentially in direct contact with the cancer cells. Much more useful information will result from experiments that use different compartments, e.g., peritoneum and vascular, since the conjugate must then cross compartmental barriers in order to exert its anticancer effects.

Another important factor associated with an accurate in vivo tumor model is the number of days separating cancer cell inoculation and the initiation of the anticancer conjugate treatment. For example, if the inoculation of the cancer cells and the start of treatment are on the same day, it is likely that the tumor model will not be accurate since the cancer cells would not be established in the host, i.e., the cancer is easily eradicated. However, if treatment with the anticancer conjugate begins many days following inoculation of the cancer cells into the host, the cancer will be well established and the efficacy of the model greatly enhanced.

There exists no scientific basis for the use of animal studies to predict the antitumor effect of a compound; however, a fairly good correlation has been shown between the antitumor effects seen in animal studies and those observed in the clinical setting [210]. In order to produce an animal model which possesses the requirements listed above, the variable factors present in the experimental procedure, and the characteristics of the tumor and animals must be addressed.

The test procedure must insure that the number of animals used in the study is sufficient to eliminate animal variation and the tumor samples, methods of tumor introduction into the test animal, and treatment protocol must be standardized to eliminate irreproducibility of results. Tumors should be non-immunogenic in order to eliminate the possibility of host defenses eradicating the tumor. This requirement is difficult to meet. The National Cancer Institute has attempted to eliminate this problem through the use of mouse tumors which are derived from mice of inbred strains and used in the same mouse strain or the F1 hybrid [210]. The problem of immunogenicity may be overcome through the use of rapidly proliferating tumors such as Walker carcinosarcoma [211], sarcoma 180, and Ehrlich ascites tumor cells [210]. These tumors are strongly immuno-genic but grow at such a rate that the immune response is overcome resulting in the growth of primary tumors. Metastasis tend not to grow from these tumors since the immune system is capable of eliminating such a small sample of cells unless the immune system is hindered by the anticancer compound. The tumor model system must also be resistant to the influence of nutritive starvation which may be caused by the decrease in appetite of the animals exposed to toxic compounds.

Far too many animal tumor models are used in laboratories today to describe in detail. Therefore, the general principles associated with the in vivo testing of macromolecular antitumor conjugates will be discussed along with a selection of widely used models to illustrate the point.

4.1.2.1 Initial Screening In Vivo Models

Initial in vivo testing of a macromolecular anticancer conjugate is usually carried out by utilizing a tumor type which is highly sensitive to the effects of anticancer drugs. Such a primary screen will have the greatest chance of detecting the anticancer activity of a conjugate, minimizing the possibility of a false negative outcome. The model used should also be easily handled and convenient to use.

The L1210 leukemia has historically been the primary tumor cell line used in animal models for initial anticancer activity screening. A retrospective study by Goldin [212] led to the acceptance of L1210 leukemia as fulfilling the initial screening model criterion.

Duncan et al. [36] administered L1210 cells into mice intraperitoneally (i.p.) or subcutaneously (s.c.) to study the in vivo effects of HPMA copolymer-doxorubicin conjugates containing optionally targeting moieties. The mice were inoculated with the cells on day 0 of the study and subsequently treated i.p or intravenously (i.v.) with the conjugates on days 1,2 and 3. The conjugates expressed higher antitumor activity than the free drug.

Leukemia P388 has also been used as a primary screen to detect the antitumor activity of macromolecular conjugates. For example, Ohya et al. [213] used P388 to determine the antitumor activities of polygalactosamine immobilized 5-fluoruracil, and Zunino et al. [214] of a poly-L-aspartic acid-daunomycin conjugate.

4.1.2.2 Secondary Screening In Vivo Models

Conjugates which show promising activity in initial screening tests, such as L1210 and P388 leukemias, are then exposed to a secondary animal tumor model. The purpose of the secondary testing is to determine if the anticancer activity is indeed promising enough to warrant further study. Secondary models consist of cancer cells which are classified as difficult to kill, therefore eliminating conjugates with weak activities from further study.

B16 melanoma, Walker sarcoma, and M5076 forming liver metastasis have been used in the preclinical evaluation of HPMA copolymer-adriamycin conjugates [36]. Other tumors useful for secondary screening are MS-2 sarcoma, NMU-1 murine lung adenocarcinoma, and murine adenocarcinoma Colon 26. These have been used by Zunino et al. [147] to determine the activity of poly (carboxylic acid) immobilized anthracyclines. Mice inoculated intramuscularly with Lewis lung carcinoma have been used by Pratesi et al. [215] to assess the effect of a poly-L-aspartic acid/doxorubicin conjugate.

4.1.2.3 Xenografts

Mouse tumors by nature tend to grow very rapidly. This may be thought to be advantageous to the researcher since the time required to produce tumors is directly proportional to the financial burden of the animal model. However, the rapid growth of the tumors imposes a greater tumor dependence upon the normal function of cellular metabolic processes and, as such, the mouse tumors are more sensitive to chemical disruption of the normal function of the cell than are the slower growing human tumors. Mouse tumors which have undergone many transplant passages are also more homogeneous than human tumors. Due to these and other major differences between mouse and human tumors, xenograft animal tumor models have been developed to approximate more closely the clinical situation and thus to increase the correlation between animal testing models and the real clinical response of anticancer conjugates. Unfortunately, this type of model has yet to show a greater correlation to clinical response than the mouse tumor models described in Sects. 4.1.2.2 and 4.1.2.3. massive immune response by the host test animal owing to the tremendous species differences between mice and humans. However, the employment of athymic nude mice as human tumor hosts bypasses this potential barrier. Such mice do not possess the ability to reject the tumor since their immune system is impaired. The tumors used in xenograft studies may be introduced into the mice through s.c. injection or through the transplantation of tumor sections. The general procedure may be described through the following example.

The antitumor effect of an anthracycline/dextran-immunoconjugate was studied by using a GW-39 colonic carcinoma xenograft [54]. The immunoconjugate was efficient for targeting in vivo. The conjugate possessed a greater antitumor activity than either the free drug or an irrelevant antibody conjugate. In this connection it is important to note that the incorporation of proper controls, such as immunoconjugates with an irrelevant antibody, is necessary.

HPMA copolymer-adriamycin conjugates were studied using a human colorectal carcinoma LS174T xenograft [36]. Adriamycin-dextran immunoconjugate activities against KB tumor xenograft [180] or human cervical carcinoma xenografts [183] have also been studied.

4.2 Decreased Toxicity and Immunogenicity of Conjugates

As a rule of thumb it may be assumed that binding of anticancer drugs and targeting moieties such as antibodies to polymeric carriers improves their biocompatibility. Using a nontoxic and nonimmunogenic polymeric carrier, the decrease of nonspecific toxicity of an attached drug (when compared to an unbound drug) may be mainly attributed to the change in body distribution. For example, anthracycline antibiotics have a nonspecific cardiotoxicity and bone marrow toxicity, limiting the dose which can be administered. In a recent study on patients with non-Hodgkin's lymphoma, the escalation of adriamycin dose in a multidrug regimen increased the nonspecific toxicity but did not improve the rate of response or survival [216]. Binding of anthracycline to polymeric carriers decreases their rate of uptake in the heart tissue. Consequently, a smaller fraction of drug administered as a macromolecular conjugate will localize in the heart. On the other hand, a low molecular weight drug bound to a polymeric carrier may act as a hapten, facilitating an immune response. On the contrary, conjugation of high molecular weight compounds, such as (targeting) antibodies to nonimmunogenic polymeric carriers, usually decreases the anti-antibody immune response due to the steric hindrance and impaired biorecognition of the conjugate by the immune system.

The above mentioned effects were observed with different inert polymeric carriers. The results obtained with HPMA copolymer-drug conjugates are discussed in Sect. 5.3., the decrease of the immunogenicity of proteins after PEG modification in Sec. 2.1.1.1.

Covalent attachment of adriamycin to poly (L-aspartic acid) resulted in decreased toxicity (as compared to free drug) in mice and rats after single and multiple administrations [215]. Similar results were obtained when adriamycin was attached to oxidized dextran [217] or HPMA copolymers [177]. Also, attachment of targeting antibodies to polymeric carriers decreases their immunogenicity. Anti-Thy 1.2 antibody [41], human immunoglobin and human transferrin [218] attached to HPMA copolymers possessed considerably lower immunogenicity when compared to the free protein. Maeda et al. [219] have shown that binding of poly(styrene-co-maleic anhydride) to neocarzinostatin (antitumor protein, mol. wt. 12 000) decreases its immunogenicity approximately by one order of magnitude when compared to unbound neocarzinostatin.

When a polymeric carrier is used which possesses intrinsic biological activity the situation is more complex due to the possibility of synergistic effects and a pronounced molecular weight dependence of biological properties of the polymeric carrier. Meijer et al. [220, 221] studied sugar-derivatized albumins as carriers of antiviral nucleoside analogs like AZT. Their finding that negatively charged neoglycoalbumins inhibit HIV replication is being used in the design of polymer-drug conjugates with a potential synergistic effect of drug and carrier.

4.3 Anticancer Effects

Examples of the anticancer activity of polymer-anticancer drug conjugates are shown in Table 1 and in Sect. 5. To further improve the activity of conjugates, suitable targeting systems have to be designed. There are several problems associated with the biorecognition of antibodies by tumor cells [222]: the heterogeneity of antigens expressed on tumor cells, the emergence of idiotypic variants, the presence of circulating antigens that can compete with tumor-bound antigens for antibody binding, and downregulation of antigens at the cell surface. Moreover, the effect of the antibody affinity and antigen density on the uptake of conjugates is an important factor. Recently, Sung et al. [223] proposed a model describing the uptake of monoclonal antibodies in solid tumors which could be used in the design of targetable polymeric anticancer drugs.

Fritzberg et al. [224] attempted to increase tumor cell retention and internalization by modifying antibodies (or their fragments) with synthetic peptides designed to provide secondary sites of attachment. Binding analogs of the peptide "GALA", an amphipathic peptide interacting with uncharged lipid bilayers, to Fab fragments of antitumor antibodies NR-ML-05 increased the biorecognition by FEMX human melanoma cells. This approach, tested only in vitro, is very interesting with regard to polymeric carriers, since both antibody fragments and peptides could be attached to the carrier macromolecules in different ways to optimize the binding.

5 *N*-(2-Hydroxypropyl)Methacrylamide Copolymers as Drug Carriers

Many factors must be considered during the development of an anticancer polymer conjugate. This section will discuss the systematic research and rationale associated with the design of soluble synthetic polymeric carriers of drugs, most particularly anticancer drugs. HPMA copolymers shall be used as an example to demonstrate the logical development of soluble polymeric drug carriers from design to clinical trials which started in the spring of 1994.

The early studies described below were performed at the Institute of Macromolecular Chemistry, Czechoslovak Academy of Sciences (IMC) in Prague in one of the author's (J.K.) laboratory. In the late sixties and seventies, IMC was a suitable place to study hydrophilic biomedical polymers. Hydrogels were designed there by Wichterle and Lim [225] as well as soft contact lenses [226]. This was the driving force to start a detailed study of the relationship between the structure of crosslinked and soluble hydrophilic polymers and their biocompatibility [207, 227–231]. Based on these investigations, hydrogels have been used successfully in human medicine [232] and a new hydrophilic polymer, poly[*N*-(2-hydroxypropyl)methacrylamide] [233, 234] was chosen as a candidate for a soluble polymeric drug carrier. Oligopeptide side-chains were designed as drug attachment/release sites [235] and it was shown that oligopeptide sequences attached to HPMA copolymers were degradable in vivo and thus had potential as attachment/release sites [236]. The chemistry of polymeranalogous reactions at the end of oligopeptide side-chains terminated in *p*-nitrophenyl esters with compounds containing aliphatic amino groups was studied in detail [237–239]. Insulin [240] and ampicillin [241] were attached to HPMA copolymers by aminolysis of reactive polymeric precursors, whereas polymer conjugates containing *N*-(4-aminobenzenesulfonyl)-*N*′-butylurea were prepared by copolymerization of HPMA and polymerizable derivatives of the drug [142]. At this time it became clear that the structure of the HPMA copolymer carrier had to be optimized to permit the biorecognition of the polymer-drug conjugate both at the cellular level (cell surface antigens) and intracellularly on a molecular level (lysosomal enzymes). The prerequisite for such a detailed interdisciplinary study was to attract enthusiastic coworkers from the biological field. At the beginning of the eighties, we were lucky to initiate collaborations with J.B. Lloyd and R. Duncan from the University of Keele, and B. Říhová from the Institute of Microbiology in Prague.

5.1 Optimization of the Bond Between the Drug and the Carrier

It was clear from the fate of macromolecules in the living organism and the restricted permeability of the lysosomal membrane that the bond between the carrier and the drug has to be cleavable in the lysosomal compartment of the cell. There were two main options how to design the bond between the drug and the HPMA carrier: to take advantage of the change in pH in the prelysosomal and lysosomal compartments, or to match the specificity of one of the different groups of enzymes present in the lysosomal compartment. The pH option [29] was abandoned because there was no potential to vary widely the rate of drug release in the lysosomal compartment. Consequently, oligopeptide, oligoglyco-side, and oligonucleotide sequences were considered as potential spacers between the carrier macromolecule and the drug. It appeared that the biochemical community collected the most information on the proteolytic group of enzymes. Consequently, we have chosen to use oligopeptide side-chains as attachment/release points.

The active site of enzymes performs the two-fold function of binding a substrate and catalyzing a reaction. The efficiency of these actions determines the overall activity of the enzyme toward a particular substrate, i.e., the specificity of the enzyme. Since the active site of proteolytic enzymes is quite large and capable of combining with a number of amino acid residues, it is convenient to subdivide the binding site to subsites [242]. A subsite is defined as a region on the enzyme surface which interacts with one amino acid residue (P_j) of the substrate. The substrates are lined up on the enzyme in such a way that the CO–NH group to be hydrolyzed always occupies the same place (the catalytic site). The amino acid residues occupy the adjacent subsites, those towards the NH_2 end occupy subsites S_1, S_2 etc., while those towards the COOH end occupy subsites S_1', S_2', etc. The effects of structure on properties can be conveniently described as S_1-P_1; S_2-P_2 etc. interactions [242]. Usually four amino acid residues toward the NH_2 end (i.e., towards the polymeric backbone) and two amino acid residues toward the COOH end take part in the interactions, which determine the formation of the enzyme-substrate complex and ultimately the rate of release. A schematic of an interaction of a polymeric drug with an enzyme active site is shown in Fig. 12.

The relationship between the structure of the oligopeptide sequences and the rate of enzymatically catalyzed release of a drug or drug model was studied in detail. Over 50 different oligopeptide sequences were introduced into HPMA copolymers and their degradability by different enzymes studied. First, model enzymes, chymotrypsin [235, 238, 239, 243, 244], trypsin [245], and papain

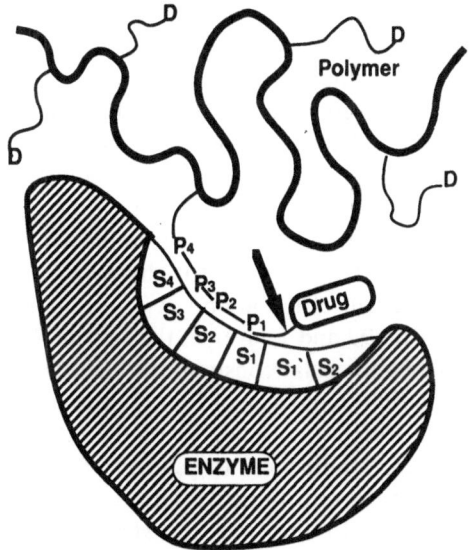

Fig. 12. Example of a proteolytic enzyme-polymeric substrate interaction. For explanation of symbols see text

[246] were used with the aim of answering two questions: a) what is the influence (steric hindrance) of the macromolecular backbone on the formation of the enzyme-substrate complex; b) is it possible to predict from the data on the crystal structure of inhibitor-enzyme complexes the structure of suitable polymeric substrates. After this basic information was obtained, we proceeded to the study of intracellular (lysosomal) enzymes. A mixture of intracellular enzymes isolated from rat liver [247, 248] was used to evaluate the degradability of the polymeric conjugates. It was shown that cysteine (thiol) proteinases are responsible for the cleavage of such polymer-drug conjugates [248]. Subsequently, polymers were designed to match the specificity of individual cysteine proteinases: cathepsin B [249], cathepsin L, cathepsin H, and artificial mixtures of lysosomal enzymes [250]. The stability of the oligopeptide side-chains in blood plasma and serum was determined [251]. Based on these results it was possible to control the degradability of HPMA copolymers by a particular enzyme as well as in the in vivo system [169, 252].

5.1.1 Main Factors Influencing the Degradability of Oligopeptide Side-Chains

These studies permitted the recognition of the following four main factors as influencing the rate of release of a drug bound to a polymeric carrier via oligopeptide bonds [253]: a) structure and length of the oligopeptide sequence; b) structure of the drug itself; c) structure of the main chain; d) drug loading. These factors are important in the design of polymer-drug conjugates and are explained using the following examples.

The influence of the length of the oligopeptide sequence is clear from Fig. 12. If the oligopeptide is too short, part of the carrier macromolecule has to be accommodated into the active site which is energetically unfavorable. The influence of the detailed structure of the oligopeptide sequence on the rate of release is shown in Fig. 13. Three polymeric substrates were compared with a general structure P–Gly–X–Phe–NAp, where P is the HPMA copolymer backbone, and NAp is p-nitroanilide. How to modify the structure to optimize the S_2–P_2 interactions in the chymotrypsin-substrate complex? From the X-ray diffraction studies at 2 Å resolution of crystals of inhibitor-chymotrypsin complexes [254] it is known that the side-chain of the amino acid residue in position P_2 of the substrate interacts with Ile-99 and Leu-97 in the subsite S_2. From the results of cleavage (k_{cat}/K_M is the bimolecular constant; the higher the value, the faster the cleavage) shown in the right hand side of Fig. 13, it is clear that the change from H (in Gly) to CH_3 (in Ala) has no influence on the rate of cleavage. The methyl group in alanine is hydrophobic, but too short to form a van der Waals contact with Ile-99. If an isobutyl group (in Leu) is introduced, the rate of cleavage increases three times [243]. In this way the structure of oligopeptide sequences can be optimized.

The prerequisite is the knowledge of the interactions taking place in the enzyme-substrate complex. This was known for the model enzymes. However, for cathepsin B, the most important enzyme in the lysosomal compartment from the drug release point of view, only limited information was available [26]. It was known that the amino acid residues in positions P_2 and P_3 should be

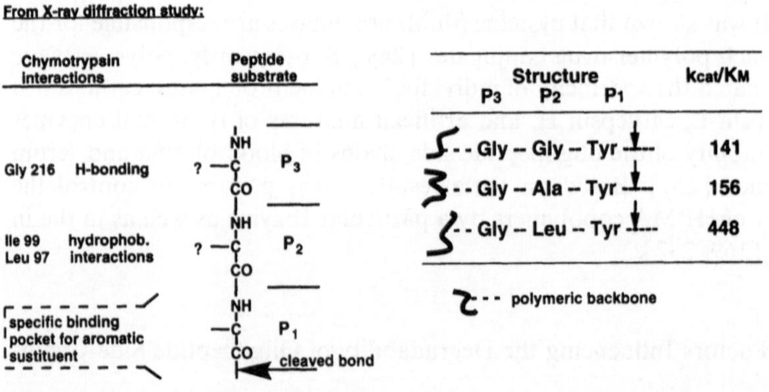

Fig. 13. Principles of design of oligopeptide side-chains in polymer-drug conjugates as demonstrated by the interaction of chymotrypsin with N-(2-hydroxypropyl)methacrylamide copolymers containing oligopeptide side-chains terminated in p-nitroaniline (leaving group serving as a drug model; not shown). The optimal amino acid residue in the P_2 position of the substrate was chosen based on the structure of the active site of the enzyme [254]. The bimolecular rate constant k_{cat}/K_M was used to characterize the rate of release. The higher the value, the faster the release of the leaving group. Data from [243]

hydrophobic. Additional information on the optimal structure of the oligopeptide sequence had to be obtained [249] from the cleavage study of numerous newly synthesized substrates (Fig. 14).

The active site of proteinases usually accommodates two amino acid residues of the substrate towards the COOH end. Unfortunately, in polymeric prodrugs the drug molecule occupies these positions (see Fig. 12) if the sequence is designed to be cleaved between the distal amino acid residue and the drug in order to release the unmodified drug. Consequently, different drugs bound to the very same polymeric carrier will have different rates of release [169] since their structure influences the interaction with the active site of the enzyme. It would be relatively easy to design such an oligopeptide sequence that would ensure a constant rate of release of a number of attached drugs. A hexapeptide is a possibility; however, a dipeptidyl derivative of the drug would be the first product of cleavage. By the consequent action of lysosomal aminopeptidases the parent drug would be regenerated. This approach is complicated from the regulatory point of view. It would require the determination of the biological properties of all three cleavage products, i.e., AA-AA-drug, AA-drug, and drug.

The influence of the structure of the main chain on the chymotrypsin catalyzed release of p-nitroaniline is shown in Fig. 15. Oligopeptide p-nitroanilides were attached to HPMA copolymers (as side-chains) and to polyethylene glycol (end-point attachment). From the values of k_{cat}/K_M it is evident that all oligopeptide p-nitroanilides attached to PEG were cleaved faster than those attached to HPMA copolymers. It appeared that the PEG substrates fit better into the active site of chymotrypsin due to the linearity and flexibility of the PEG molecule, and the type of spacer attachments [255].

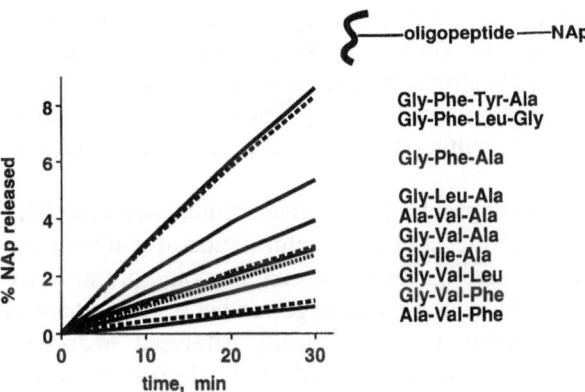

Fig. 14. Initial interval of cleavage of HPMA copolymer based polymeric substrates by lysosomal cysteine proteinase cathepsin B (isolated from bovine spleen). Only the cleavage of the bond between the distal amino acid residue and p-nitroaniline was monitored. Conditions of cleavage: [Cathepsin B] $= 1.9 \times 10^{-7}$ M; [NAp] $= 1.2 \times 10^{-3}$ M; [EDTA] $= 1 \times 10^{-3}$ M; [Cys] $= 2.5 \times 10^{-2}$ M; 0.1 M phosphate buffer; pH $= 6.0$; 40 °C. Data from [249]

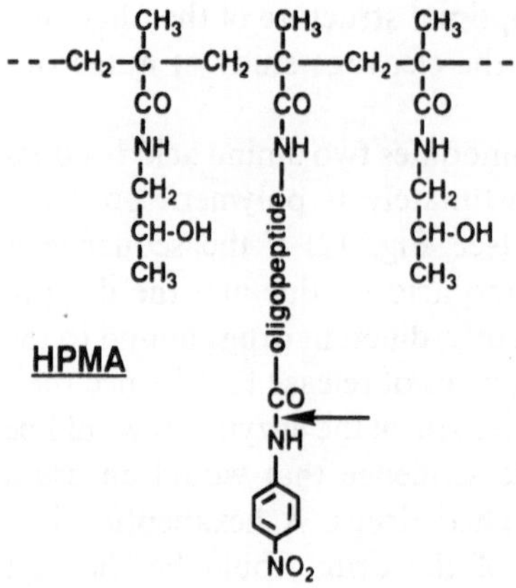

Oligopeptide sequence	k_{cat}/K_M HPMA	k_{cat}/K_M PEG
-Phe↓-NAp	~ 0	80
-Gly-Val-Phe↓-NAp	245	9550
-Ala-Gly-Val-Phe↓-NAp	14 200	123 000

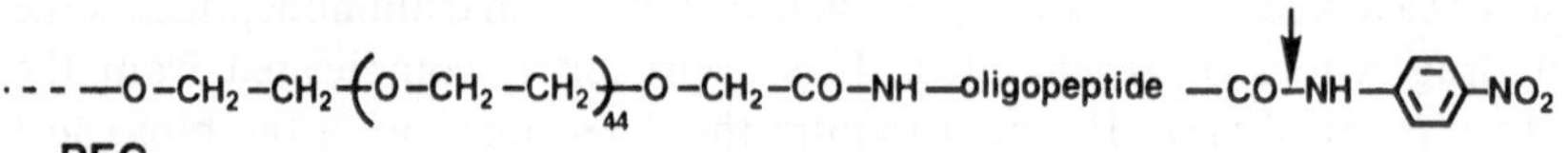

Fig. 15. The influence of the structure of the main chain on the chymotrypsin catalyzed release of *p*-nitroaniline. The cleaved bond is denoted by an arrow. Data from [255]

Solution properties of polymer-drug conjugates may influence their biological properties. HPMA copolymers have a hydrophobic backbone and hydrophilic side-chains. Oligopeptide side-chains terminated in drug are hydrophobic to match the specificity of lysosomal cysteine proteinases. As mentioned in Sect. 2.2.3, an increase in the amount of hydrophobic side-chains may lead to the formation of aggregates. The problem is schematically depicted in Fig. 16. The comparison of HPMA copolymers containing Gly-Leu-Phe-Nap side-chains showed that the higher the content of side-chains, the higher the association number. The aggregate structure renders the accessibility of oligopeptide side-chains by enzymes more difficult [132]. The diffusion of the latter into the micellar core is restricted resulting in the decreased rate of cleavage (Fig. 16).

The oligopeptide sequence was optimized for cysteine proteinases, especially cathepsin B [169, 249]. Due to a limited amount of information at that time, the main conclusions reached were based on the traditional mapping of the active site of cathepsin B [26, 249], i.e., the synthesis of a large number of substrates and determination of their susceptibility to enzymatically catalyzed hydrolysis (Fig. 14). The sequence Gly-Phe-Leu-Gly was used in the design of HPMA copolymer based anticancer drugs used in clinical trials [169]. It is interesting to note that just recently data on the structure of the active site of cathepsin B have been published [256, 257]. Data on the X-ray structure of cathepsin B may now be used in molecular dynamic simulations for the rational design of effective prodrugs.

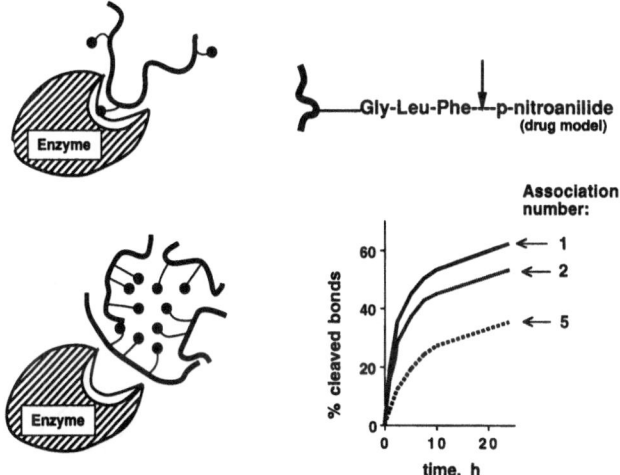

Fig. 16. The influence of solution properties of polymer-drug conjugates on their biorecognition. HPMA copolymers containing different amounts of side-chains terminated in *p*-nitroaniline were evaluated. The higher the number of side-chains per macromolecule, the higher the association number (average number of macromolecules in one aggregate). Aggregation renders the recognition of the macromolecule by chymotrypsin more difficult, decreasing the rate of drug model release. Based on data from [132]

5.2 Targetability of HPMA Copolymers

5.2.1 Principles of Design

The choice and design of a targeting system has to be based on a sound biological rationale. The design of the first targetable HPMA copolymer was based on the observation [258] that small changes in the structure of glycoproteins lead to dramatic changes in the fate of the modified glycoprotein in the organism (Fig. 17). When a glycoprotein (ceruloplasmin) is administered into rats, a long intravascular half-life is observed. However, when the terminal sialic acid is removed from ceruloplasmin, the asialoglycoprotein (asialoceruloplasmin) formed contains side-chains exposing the penultimate galactose units. The intravascular half-life of the latter is dramatically shortened due to the biorecognition of the molecule by the asialoglycoprotein receptor on the hepatocytes.

This receptor recognizes galactose and *N*-acetylgalactosamine moieties [259]. To determine if a synthetic macromolecule containing one of these units would be biorecognizable in vivo, we synthesized HPMA copolymers with *N*-methacryloylglycylglycine *p*-nitrophenyl ester and attached galactosamine by

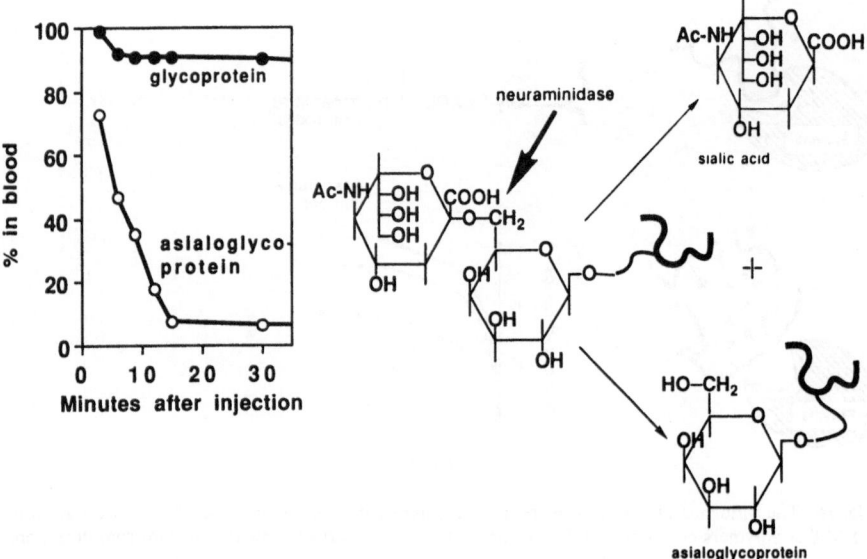

Fig. 17. Comparison of blood clearance of glycoproteins and asialoglycoproteins. Removal of terminal sialic acid moieties by enzyme (neuraminidase) exposes the penultimate galactose moieties. The modified macromolecules are removed from the bloodstream very fast due to their biorecognition by the asialoglycoprotein receptor on hepatocytes. Based on data from [250]

aminolysis [163]. The copolymers contained approx. 1 mol % of N-meth-acryloyltyrosinamide to permit radioiodination. These copolymers behaved similarly to the asialoglycoproteins and were biorecognizable in vivo (Fig. 18). Their clearance from the bloodstream was related to the N-acylated galactosamine content (1-11 mol %) of the HPMA copolymer [260, 261]. Separation of the rat liver into hepatocytes and non-parenchymal cells indicated that the polymer is largely associated with hepatocytes, and density-gradient subcellular fractionation of the liver at various times after administration confirmed that the HPMA copolymers were internalized by liver cells and transported, with time, into the secondary lysosomes [261, 262]. The kinetics of binding, the internalization and subcellular trafficking of galactosamine containing HPMA copolymers by a human hepatoma cell line HepG$_2$ was recently studied by confocal fluorescence microscopy [263]. It was very important to find that HPMA copolymers containing side-chains terminated in galactosamine and adriamycin (for structure, see Fig. 10) also preferentially accumulated in the liver (Fig. 19), i.e., it appeared that nonspecific hydrophobic interactions with cell membranes did not interfere with the biorecognition by hepatocytes [264]. However, it has to be recognized that macromolecules captured by receptor-mediated pinocytosis are internalized effectively at a low dose. Receptor saturation occurs at high i.v. doses decreasing the specificity of targeting [265].

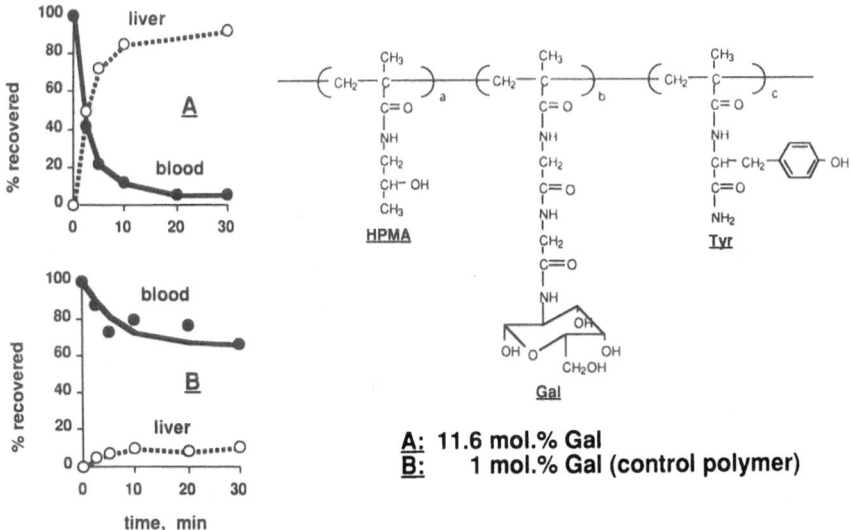

Fig. 18. Structure and blood clearance of synthetic HPMA copolymers mimicking the structure of glycoproteins and asialoglycoproteins. Based on data from [261]

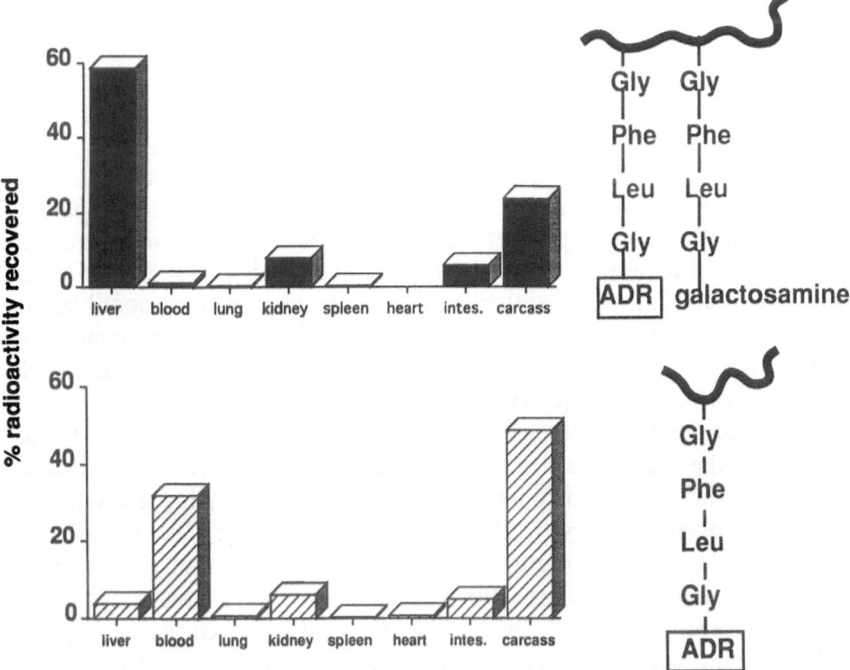

Fig. 19. Body distribution of adriamycin–HPMA copolymer conjugates 1 h after intravenous administration to rats. The distribution of a targetable conjugate containing the galactosamine moiety is compared with a conjugate without targeting moiety. Based on data from [264]

5.2.2 Structural Factors Influencing the Biodistribution of HPMA Copolymers

As mentioned in Sect. 2.2.3, the biodistribution of HPMA copolymers depends on many factors. Molecular weight influences the uptake in the isolated tissue of yolk sac [266] as well as the elimination in vivo [124, 125, 267, 268]. Nonspecific increase in the rate of polymer uptake can be achieved by incorporation of positively charged or hydrophobic comonomers into the HPMA copolymer structure, such as methacryloyloxyethyltrimethylammonium chloride [22], N-methacryloyltyrosinamide [21], or N-[2-(4-hydroxyphenyl)ethyl]acrylamide [267]. The incorporation of hydrophobic moieties may influence the solution properties of the HPMA copolymer conjugates [132, 134, 269]. The interaction with the cellular surface may depend on the association number and the stability of the micelles.

However, the greatest changes in the biodistribution of HPMA copolymers were achieved, as expected, by the attachment of targeting moieties. A number of targeting moieties were used: (N-acylated) galactosamine [23, 39, 40, 43, 113, 191, 260–265, 270–272], (6-O-bound) galactose [162, 272], (N-acylated) fucosylamine [39, 40, 264], anti-Thy 1.2 antibodies and anti Ia^k antibodies [23, 24, 41, 273–275], B 72.3 antibodies [276], transferrin and anti-transferrin receptor (B3/25) antibodies [277], anti-CD3 antibodies [344] melanocyte stimulating hormone [37], and secretin [278, 279].

5.3 Biocompatibility of HPMA Copolymer Conjugates

A prerequisite for developing a biocompatible drug carrier is to start with a biocompatible polymer. This does not imply that the conjugates will also be biocompatible, but it will considerably improve the chances. The biocompatibility of poly[N-(2-hydroxypropyl)methacrylamide] was evaluated using the evaluation procedures developed for blood plasma expanders [124, 280–285]. The results of these studies indicated that the homopolymer was biocompatible. The next step was to evaluate the biocompatibility of HPMA copolymers containing oligopeptide side-chains [286] and attached haptens {arsanilic acid (ARS), fluorescein isothiocyanate (FITC)} as drug models [287]. It was shown [286–288] that the homopolymer was not recognized as a foreign macromolecule in any of the five inbred strains of mice used, and no detectable antibodies were found against it. Attachment of oligopeptide side-chains gives rise to compounds possessing very weak immunogenic activity. The intensity of response depends on the structure of the side-chain, dose, and genetic background of the mice. Neither the homopolymer nor the HPMA copolymers possessed mitogenic activity. Similar results were obtained when haptens, namely arsanilic acid (ARS) and fluorescein isothioisocyanate (FITC) were attached to HPMA copolymers [287–289]. The intensity of response increased with increasing molecular weight of the conjugate. While the ARS-HPMA copolymer conjugate behaved as a high dose tolerogen, tolerogenicity was not observed with the FITC conjugate. In summary, HPMA copolymers behaved as thymus independent antigens with low immunogenicity [24]. The immune response towards HPMA copolymers was about 4 orders of magnitude lower when compared with human gamma globulin [26]. However, the results obtained strongly indicated that the biocompatibility of each drug-HPMA copolymer conjugate would have to be evaluated independently.

HPMA homopolymer and HPMA copolymer conjugates did not have a prominent effect on the complement system in vitro [290]. An important observation was the fact that the attachment of antibodies of HPMA copolymers decreases the former's immunogenicity [41, 218]. Also, attachment of anticancer drugs to HPMA copolymer carriers decreases the drug's toxicity. The effects of free daunomycin (DNM) and DNM bound to HPMA copolymer-anti-Thy 1.2 antibody conjugates on the depletion of hematopoietic stem cells from bone marrow were compared [24]. Conjugation of DNM to the targetable carrier substantially decreased its toxicity. The histopathological evaluation has shown that the free drug was irritable to Kupffer cells while HPMA copolymer conjugates had no such effect [24].

Finally, the biocompatibility of two HPMA copolymer conjugates designated for Phase I/Phase II clinical trials was evaluated [177]. HPMA based polymeric prodrugs containing adriamycin bound to the polymeric backbone via glycylphenylalanylleucylglycine side-chains, and optionally galactosamine bound via the same sequence, were evaluated for immunogenicity after i.v., s.c., and oral administration in two inbred strains of mice. It was found that antibodies were produced in very small amounts (Table 4). Attachment of adriamycin to HPMA copolymers considerably decreased its toxicity against hematopoietic precursors in bone marrow as determined by the in vivo colony-forming unit-spleen assay (Table 5) and its ability to inhibit [^{3}H]thymidine incorporation by mouse splenocytes and human peripheral blood lymphocytes in vitro [177]. These results were corroborated during the preclinical evaluation of the HPMA copolymer conjugates (Table 6) [36]. The decreased cardiotoxicity of HPMA copolymer-adriamycin conjugates was recently evaluated in rats using a dose of 4 mg/kg [291]. Whereas those animals to which free drug was administered exhibited a time dependent decrease in cardiac output and these animals subsequently died, animals receiving the polymer conjugate showed no change in cardiac output and no overt signs of toxicity [291].

5.4 Examples of HPMA Copolymer-Anticancer Drug Conjugates

The overview of HPMA copolymer-anticancer drug conjugates and their antitumor activities can be found in Table 1. As with other carrier systems, attachment of an anticancer drug to the HPMA copolymer carrier changes the drug's pharmacokinetics. For example, ADR attached to HPMA copolymers via a Gly-Phe-Leu-Gly sequence showed a 15-times longer intravascular half-life and a 100-fold lower initial peak level in the heart tissue when compared to free

Table 4. Immunogenicity of HPMA copolymer-ADR conjugates (from [177] with permission)

Sample	Dose Copolymer (μg)	Dose ADR	Route of Application	Antibody titre A/J mice 3rd day	A/J mice 6th day	C57BL/10ScSn mice 3rd day	C57BL/10ScSn mice 6th day
P-Gly-Phe-Leu-Gly-ADR	300	25.5	i.v.	1/64	1/128	1/32	1/128
			s.c.	1/64	1/64	1/128	1/128
			orally	1/64	1/128	1/64	1/128
	100	8.5	i.v.	1/16	1/16	1/64	1/256
			s.c.	1/128	1/128	1/128	1/256
			orally	1/64	1/128	1/32	1/64
P〈 Gly-Phe-Leu-Gly-ADR / Gly-Phe-Leu-Gly-gal	300	21.9	i.v.	1/256	1/256	1/128	1/128
			s.c.	1/256	1/256	1/64	1/128
			orally	1/128	1/256	1/64	1/128
	100	7.3	i.v.	1/64	1/128	1/32	1/69
			s.c.	1/64	1/64	1/32	1/69
			orally	1/32	1/32	1/64	1/69
None (control)	–	–	–	1/32	1/32	1/16	1/16

Ten mice per group were immunized every 3rd day (five times). On the 3rd and 6th days after the last treatment the mice were exsanguinated and the sera stored at −70 °C. Numbers represent an average of ten individually tested sera. P = HPMA copolymer backbone; ADR = adriamycin (doxorubicin); gal = galactosamine bound via the NH₂ group

Table 5. Spleen colony-forming units (CFUs) detected in irradiated recipient mice after injection of bone marrow harvested from mice injected with free ADR or HPMA copolymer ADR conjugates (from [177] with permission)

Sample	Immunization protocol	Route of Application	Number of CFUs ± SE per spleen detected on day	
			3rd	6th
*P-Gly-Phe-Leu-Gly-ADR		i.v.	–	26 ± 4
	A	s.c.	–	29 ± 3
		orally	–	30 ± 4
		i.v.	26 ± 5	27 ± 4
	B	s.c.	25 ± 3	28 ± 4
		orally	27 ± 2	27 ± 5
*P⟨ Gly-Phe-Leu-Gly-ADR / Gly-Phe-Leu-Gly-gal		i.v.	–	30 ± 3
	A	s.c.	–	28 ± 5
		orally	–	30 ± 5
		i.v.	23 ± 3	29 ± 5
	B	s.c.	26 ± 1	27 ± 3
		orally	27 ± 4	27 ± 3
ADR		i.v.	–	14 ± 2
	A	s.c.	–	18 ± 3
		orally	–	28 ± 5
		i.v.	5 ± 1	13 ± 3
	B	s.c.	11 ± 3	19 ± 3
		orally	25 ± 5	27 ± 5
None (control)	–	–	–	25 ± 4

A = Immunization daily five times with 300 μg of polymer (22–25 μg of ADR per immunization). On the 6th day after the last treatment, cells from bone marrow were isolated
B = Immunization every 3rd day, five times with 300 μg of polymer (22–25 μg of ADR per immunization). On the 3rd and 6th days after the last treatment, cells from bone marrow were isolated
Data are expressed as the average of the triplicate ± SE
* P = HPMA copolymer backbone; ADR = adriamycin (doxorubicin); gal = galactosamine bound via the NH_2 group

ADR [292]. Delivery to the cell interior, however, is not sufficient for anticancer activity. Daunomycin, or adriamycin are active only when attached to the carrier via an enzymatically degradable spacer [24]. The release from the carrier in the lysosomal compartment clears their diffusive pathway to the cell nucleus where they intercalate into the DNA. On the contrary, photosensitizers, such as chlorin e_6 [23, 42, 43, 191], or phthalocyanines [293] which act as energy transfer agents from external light to oxygen, are also active when covalently bound to the HPMA copolymer carrier. Because the lifetime of singlet oxygen is short, the site of action of photosensitizers is largely determined by its location [294]. Most probably polymer bound photosensitizers contribute to the damage of the lysosomal membrane, resulting in the release of lysosomal enzymes into the cytoplasm, and ultimately cell death.

Table 6. Treatment of mice bearing L1210 leukemia (intraperitoneally) with HPMA copolymer-ADR conjugates (from [36] with permission)

Treatment	Dose[a] (mg/kg)	T/C[b] (%)	Toxic Deaths	Long-term survivors[c]
Intraperitoneal administration (3 doses days 1, 2, 3)				
ADR	2.5	168	0/30	1/30
	5.0	214	3/40	0/40
	10.0	80	40/40	0/40
[d]P-Gly-Phe-Leu-Gly-ADR	2.5	113	0/10	0/10
	5.0	138	0/10	0/10
	10.0	170	0/30	1/30
	20.0	231	0/10	0/10
	30.0	> 430	2/30	8/30
	50.0	110	20/20	0/20
P Gly-Phe-Leu-Gly-ADR / Gly-Phe-Leu-Gly-gal	2.5	125	0/10	0/10
	5.0	138	0/10	0/10
	10.0	203	0/20	2/20
	20.0	256	0/10	2/10
	30.0	> 762	0/20	17/20
	50.0	113	10/10	0/10
Intravenous administration (single dose)				
ADR	13	167	0/10	0/10
	16.9	167	0/10	0/10
	22	183	0/10	0/10
[d]P-Gly-Phe-Leu-Gly-ADR	25.0	125	0/10	0/10
	50	150	0/10	0/10
	75	183	0/10	0/10
	100	233	5/10	0/10

[a] Represents the equivalent ADR dose
[b] Ratio of median survival of the test group (T) to that of untreated control (C) expressed as a percentage
[c] Animals surviving until 60 days
[d] P = HPMA copolymer backbone; ADR = adriamycin (doxorubicin); gal = galactosamine bound via the NH_2 group

HPMA copolymer-anticancer drug conjugates have been shown to be active against a number of tumor models (Table 1). An HPMA copolymer-daunomycin conjugate selectively accumulated in solid tumor. In rats bearing subcutaneous Walker sarcoma, a four-fold increase in the area under the curve for the polymer conjugate was found [295]. A similar accumulation of HPMA copolymer-chlorin e_6 conjugates was found in mice bearing Neuro 2A neuroblastoma [42].

Recently, a preclinical study of two adriamycin-HPMA copolymer conjugates designed for Phase I/II clinical trial was published [36]. The conclusions reached in previous studies were corroborated with large groups of animals (Table 6). HPMA copolymer-ADR conjugates were active against a number of solid tumor models: M5076, P388, B16 melanoma, Walker sarcoma, and the xenograft LS174T.

Two further points are worth discussion: tumor accumulation and tumor targeting. As mentioned above, the threshold for glomerular filtration of HPMA copolymers is about 50 kD. Since the whole molecular weight distribution has to be under this threshold, HPMA copolymers with M_W 20kD are being used. However, the accumulation of macromolecules in solid tumors increases with molecular weight [16]. This discrepancy could be solved if HPMA carriers were used which contain synthetic macromolecules (M_W below the glomerular threshold) joined into higher molecular weight carriers by oligopeptide sequences. Such HPMA copolymers can be easily prepared by crosslinking of polymeric precursors with diamines below the gel point (i.e., the resulting polymer is water soluble) [128, 238, 239]. In fact, we proposed and studied this concept a long time ago [102, 128], but from the regulatory point of view decided to pursue simpler systems in the first phase.

Targeting is an excellent concept, however, its efficiency is not always satisfactory. Other approaches are being used such as the use of photosensitizers in polymer-anticancer drug conjugates which permits double targeting and combination therapy.

The combination of light and certain absorbing molecules, called photosensitizers, in the presence of oxygen, can lead to rapid cell destruction [294, 296, 297]. It is widely believed that the generation of singlet oxygen is ultimately responsible for the majority of such phototoxic effects, although other reactions, e.g., formation of radicals, do indeed occur [298]. A reactive excited state of molecular oxygen, singlet oxygen, lies only 90 kJ mol^{-1} above the triplet [294]. Because the lifetime of singlet oxygen is short ($\sim 10^{-6}$ s in aqueous environment) its site of action is largely determine by its location. Polymer bound photosensitizers will end up in the lysosomal compartment of target cells where they will be inactive without light. A double targeting effect can be achieved by biorecognition (incorporation of targeting moieties) followed by the subsequent localized application of light [293]. A number of polymer bound photosensitizers were synthesized and their biological activity evaluated [23, 43, 176, 191, 197, 198, 299–302].

A novel concept of combination chemotherapy and photodynamic therapy (PDT) using HPMA copolymer bound drugs [279] was used in the treatment of experimental Neuro 2A neuroblastoma (solid tumor). Mice were administered (i.v.) a mixture of HPMA copolymer bound adriamycin (Gly-Phe-Leu-Gly spacer was used) and HPMA copolymer bound meso-chlorin e$_6$ (Mce$_6$; Gly-Phe-Leu-Gly or Gly spacers were used). A time lag was allowed for optimal uptake in the tumor and for adriamycin to begin to take effect, after which light (650 nm) was applied to activate the photosensitizer. Tumor cures were obtained (Table 7) with the combination therapy that could not be achieved by either chemotherapy or PDT alone [42]. It is interesting to note that results obtained when using two different spacers to bind meso-chlorin e$_6$ were in good agreement with the time dependence of drug localization inside the tumor [42]. Similar experiments are underway to treat ovarian carcinoma [303].

Table 7. Combined chemotherapy and photodynamic therapy of A/J mice bearing Neuro 2A neuroblastoma tumors using HPMA copolymer conjugates (from [42])

Samples	Time Lag[a] (h)	Irradiation (650 nm)	Long-term survivors
[b]P-Gly-Phe-Leu-Gly-ADR (8.2 mg/kg) + P-Gly-Mce$_6$ (4 mg/kg)	48	500 mW/cm^2 5 min	5/15
[b]P-Gly-Phe-Leu-Gly-ADR (8.2 mg/kg) + P-Gly-Phe-Leu-Gly-Mce$_6$ (1.5 mg/kg)	48	500 mW/cm^2 10 min	0/5
[b]P-Gly-Phe-Leu-Gly-ADR (8.2 mg/kg) + P-Gly-Phe-Leu-Gly-Mce$_6$ (1.5 mg/kg)	24	500 mW/cm^2 10 min	4/5

[a] Time lag between the administration of conjugates and irradiation of mice
[b] P = HPMA copolymer backbone; ADR = adriamycin (doxorubicin); Mce$_6$ = meso chlorin e$_6$

6 Alternative Delivery Systems: Liposomes, Nanoparticles, Microspheres, Immunotoxins

Many other delivery systems have been employed in attempts to improve the delivery of anticancer drugs. The major types include liposomes, microspheres, nanoparticles and immunotoxins. This review concentrates on soluble polymeric carriers. However, basic characteristics of alternate delivery systems have been included to permit the reader the comparison of different delivery systems.

6.1 Liposomes

The scientific community was attracted to the study of liposomes due to the relatively simple procedure of their preparation. Moreover, if prepared from natural phospholipids, they are biocompatible, and possess low cytotoxicity, low immunogenicity, and biodegradability [304]. Liposomes, however, have two main disadvantages: the structural instability both in vitro and in vivo, and low cell specificity [304]. To increase the stability, the structure of the phospholipid layer has been modified to include artificial lipids and/or cholesterol. Polymerizable vesicles have also been prepared [305]. It is obvious that the biocompatibility of these modified systems has to be addressed.

It is well established that size, charge, and chemical composition of liposomes affect their fate in vivo [306]. To manipulate their biodistribution and/or drug release, liposomes of different structures have been prepared, including those sensitive to changes in pH [307, 308] or temperature [309, 310]. Compared to soluble polymers discussed in previous chapters, liposomes, when applied i.v., are captured by a substantially greater extent by the specialized cells (macrophages) of the reticuloendothelial system (RES) [311]. The removal of liposomes from the bloodstream takes place by nonspecific endocytosis (phagocytosis) [312].

The delivery of drugs to the RES (also referred to as the mononuclear phagocyte system) is an area where liposomes have a potential for success. They were successfully used in the delivery of antimonial drugs to treat experimental leishmaniasis infection [313]. Very good results were obtained in the treatment of systemic fungal infections in cancer patients using liposomal amphotericin B [314, 315].

Specific targeting moieties have been attached to the liposome surface to promote biorecognition. For example, Sunamoto and Iwamoto [316] modified polysaccharides with palmitoyl chains and incorporated the conjugates into the liposomal structure to modify biorecognition. Numerous immunoliposomes have been prepared containing covalently attached antibodies or antibody fragments. However, nonspecific recognition by the RES interferes with the recognition by target cells [317]. Consequently, sterically stabilized liposomes [314] were designed as better candidates for targeting. The long circulating or Stealth® liposomes with a PEG modified surface [318] have long circulation half-lives, dose independent pharmacokinetics, and the ability to move through the lymph after subcutaneous injection [319]. Antibodies or their fragments may be attached to Stealth® liposomes [320]. However, more data have to be collected before a final judgement can be made on their clinical potential.

Liposomes may also be used as a controlled depot release of an encapsulated drug in the blood compartment, thereby reducing toxic side-effects of the drug by avoiding toxic peak concentrations of free drug in the bloodstream [304], and decreasing the accumulation of the toxic drug in such areas as heart, kidney, and gastrointestinal tract [321]. Phase I clinical trials of adriamycin (doxorubicin) encapsulated liposomes [321] have shown that the delivery system was well tolerated and produced less side-effects than the free drug. In an exciting recent trial, Nabel et al. [322] injected liposome-encapsulated DNA (HLA-B7 gene) into tumors of five patients with advanced skin cancer. It was encouraging to find that the genes were able to be expressed and boost the immune system of recipients.

6.2 Nanoparticles

Nanoparticles were developed as an alternative to liposomes. Nanoparticles are spherical particles with a diameter typically between 20 and 400 nm (this value varies depending on the author). Contrary to liposomes, they are solid porous spheres with a dense polymeric structure [8]. Their use as a delivery system for anti-cancer agents has been extensively reviewed [8, 9, 323, 324]. Early studies of nanoparticles were performed on non-biodegradable polyacrylamide [325] or poly(methyl methacrylate) [326] nanoparticles. The risk of lysosomal overloading [8] initiated the development of biodegradable nanoparticles based on poly (alkyl α-cyanoacrylates) [324]. Drugs can be combined with nanoparticles during or after polymerization. Following administration into the bloodstream, nanoparticles are rapidly cleared by the RES, typically by macrophages in the liver and spleen. Poly(alkyl α-cyanoacrylate) nanoparticles are then degraded, the rate of polymer degradation depending on the structure of the alkyl. The rate of drug release can be correlated with the polymer's bioerosion [327]. As in other polymeric delivery systems, the attachment of drugs to nanoparticles changes their body distribution and toxicity. Poly(isohexyl α-cyanoacrylate) nanoparticles loaded with [^{14}C]-adriamycin were administered i.v. to mice and body distribution evaluated [328]. Adriamycin blood clearance was decreased during the initial interval and the concentrations in heart and kidney were substantially lower when compared to the administration of free drug.

In summary, nanoparticles may be used in the delivery of drugs to the reticuloendothelial system, but it is a challenge to modify their biorecognition by surface modification with either antibodies or PEG chains. Poly(hexyl α-cyanoacrylate) nanoparticles with a surface coated by monoclonal antibodies to target to mice bearing human xenografts were accumulated mainly in the liver and spleen due to either displacement or incorrect orientation of the antibody [329]. The adsorption of PEG containing block copolymers was also not successful since there was no significant difference in the liver and spleen uptake between the coated and uncoated nanoparticles [330]. These observations may be due to the loss of the coating layer either by exchange of the coating macromolecules with plasma proteins [330] or due to the degradation of the matrix [324]. Recently, it was shown that covalent attachment of PEG chains to crosslinked albumin nanospheres (220 nm) decreased their biorecognition by the RES by a statistically significant amount [331].

6.3 Microspheres

Microspheres are particles ranging between 1 and 100 μm. They are typically formed from degradable polymeric materials such as albumin, polysaccharides, or poly(α-hydroxy acids) by precipitation or phase-separation emulsion techniques [6, 332]. The relatively large diameters of microspheres make their extravasation into the tumor mass difficult and the uptake of microspheres by the RES is very rapid.

In the circulation, microspheres loaded with drug act as a depot injection which produces a slow, continuous delivery of drug into the bloodstream. Although this is not a targeting mechanism of drug delivery, the large size of microspheres does allow a certain amount of regional targeting through physical entrapment. Most organ systems possess small capillary beds which will entrap particles such as microspheres. The infusion of microspheres directly into arteries or veins which preferentially run into a desired organ system will result in the entrapment of the microspheres which will release drug in that organ.

Chemoembolization with microencapsulated drugs has been in clinical use since 1978 [6]. Using biodegradable starch microspheres containing anticancer drugs which occlude selected arteries, anticancer drugs can be locally released upon the degradation of starch by serum amylases [333, 334].

Several attempts were made to influence the biodistribution of microspheres. The site of embolization depends on the site of administration and on the diameter of the microspheres [335]. Block copolymers have been used to coat the microspheres with the aim to prevent the recognition by the RES [9, 336–338] or to target them to the bone marrow [339]. Interesting results in changing the body distribution of coated microspheres have been obtained. However, it appears that covalent attachment of hydrophilic polymer chains, e.g., PEG chains, is the better way [340].

Poly(lactide-co-glycolide) microspheres developed by Takeda Chemical Industries containing luteinizing hormone releasing hormone (LHRH) analog are used in human medicine for prostate cancer treatment. These microspheres release the drug for 28 days after subcutaneous administration [11, 341–343]. It is interesting to note that for biocompatibility reasons, the Japanese team synthesized the copolymer by a step-growth process, avoiding the metal containing catalysts necessary for the ring-opening polymerization [345].

6.4 Immunotoxins

Immunotoxins are toxins with altered cell surface receptor specificity and consequently altered cell type specificity. These compounds consist of toxin domains and binding domains derived from growth factors, cytokines or monoclonal antibodies [346]. Several protein toxins have been used to construct immunoconjugates: ricin and diphtheria toxin inactivate protein synthesis enzymatically [347], cholera toxin provides alterations in cAMP, and tetanus and botulinum toxins inhibit neurotransmitter release [346]. The most frequently studied were the conjugates of ricin A-chain subunit with a number of monoclonal antibodies [348]. However, during in vivo studies antibodies against the immunoconjugates were detected [349–352]. Recently, these conjugates were produced as recombinant fusion proteins [10, 353, 354]. Preclinical testing has shown that several recombinant immunotoxins exhibit curative antitumor activities in nude mouse models of human cancer [355].

The inherent immunogenicity of immunotoxins [349–352] and the low cytotoxicity of conjugates of mAbs with the majority of commonly used anticancer drugs (without a polymer intermediate) [356] raised the interest in compounds possessing two or three orders of magnitude higher cytotoxicity than common anticancer drugs. For example, Chari et al. [357] bound maytansinoid to several monoclonal antibodies via disulfide-containing linkers. The conjugates showed a high antigen-specific cytotoxicity in vitro (50% inhibiting concentration, 10-40 pM), low systemic toxicity in mice, and good pharmacokinetic behavior. Hinman et al. [356] evaluated calicheamicins, antitumor antibiotics capable of producing double-stranded DNA breaks at subpicomolar concentrations. This group of compounds contains a methyl trisulfide group, which upon reduction causes a molecular rearrangement of the enediyne bicyclo moiety generating a diradical which attacks the DNA [356, 358]. A hydrazide of calicheamicin was prepared and conjugated to an oxidized (internalizing) anti-polyepithelial mucin antibody and the potential for the treatment of solid tumors was demonstrated.

7 Conclusions and Future Prospects

The principles for the synthesis and design of polymer conjugates demonstrated in this review have shown the great diversity and potential of polymers and polymer conjugates for the treatment of cancer. Although many polymer conjugates have been synthesized and studied, to date, few anticancer polymer conjugates have become approved for clinical use. The problem of effectively targeting conjugates to a specific subset of cells, e.g., cancer cells, must be solved prior to a major breakthrough in cancer therapy. Polymer conjugates shall then become prime candidates for anticancer drug delivery owing to the extensive study of their chemistry and their favorable biocompatibility within the body.

Challenging opportunities exist for the use of polymers as agents in the future treatment and prevention of cancer. Vaccines have been under investigation for the prevention of cancer, and biodegradable polymers could play a major role in the delivery of these vaccines into the body. Gene therapy of cancer [359] is another area in which polymer-conjugates may be utilized to overcome the problems of metabolic stability of oligonucleotides in the biological environment, their transmembrane passage, and their distribution within various cell compartments [360, 361]. For example, poly (L-lysine)-asialoorosomucoid conjugate has been studied as a carrier [362] and shown to deliver antisense oligonucleotides efficiently to NIH 3T3 cells. The synthetic part of the conjugate serves as the antisense oligonucleotide binding site, whereas the asialoorosomucoid part is biorecognizable by the asialoglycoprotein receptor expressed on the cell surface. Another important area of polymer conjugates will be in the area of basic studies of the function of organism's genes. By suppressing a selected gene with a (polymer delivered) antisense oligonucleotide, important information on its function may be obtained.

In summary, the results obtained so far and reviewed here bode well for the future of polymers in the delivery of anticancer agents. A further systematic study of their advantages and limitations is necessary to evaluate fully their therapeutic potential. The final judgement, however, will come from the results of clinical trials.

Acknowledgements. We thank Dr. P. Kopečková for valuable discussions and for the design of the majority of figures. The research in the authors' laboratory was supported in part by NIH grant CA 51578 and by NIH Biotechnology Training Grant GM 08393.

8 References

1. Ehrlich P (1906) Studies in immunity. Plenum Press, New York
2. De Duve C, DeVarsy Th, Poole B, Trouet A, Tulkens P, Van Hoof F (1974) Biochem Pharmacol 23: 2495
3. Jatzkewitz H (1954) Hoppe-Seyler's Z Physiolog Chem 297: 149
4. Morell AG, Irvine RA, Sternlieb I, Scheinberg IM, Ashwell G (1968) J Biol Chem 243: 155
5. Ringsdorf H (1975) J Polym Sci Polym Symp 51: 135
6. Zee-Cheng RKY, Cheng CC (1989) Meth and Find Exp Clin Pharmacol 11: 439
7. Papahadjopoulos D, Allen T, Gabizon A, Mayhew E, Matthay K, Huang SK, Lee KD, Woodle MC, Lasic DD, Redemann C, Martin FJ (1991) Proc Natl Acad Sci USA 88: 11460
8. Couvreur P, Roblot-Treupel L, Poupon MF, Brasseur F, Puisieux F (1990) Adv Drug Deliv Rev 5: 209
9. Wright JJ Illum L (1992) in: Donbrow M (ed) Microcapsules and nanoparticles in medicine and pharmacy, CRC Press, Boca Raton, p 281
10. Pastan I, Chaudhary VK, FitzGerald DJ (1992) Ann Rev Biochem 61: 331
11. Debruyne FM, Denis L, Lunglmeyer G, Mahler C, Newling DW, Richards B (1988) J Urol 140: 775
12. Krinick NL, Kopeček J (1991) in: Juliano RL (ed) Targeted drug delivery. Springer, Berlin, p 105 (Handbook of experimental pharmacology, Vol 100)
13. Duncan R (1992) Anti-Cancer Drugs 3: 175
14. Seymour LW (1992) Crit Rev Ther Drug Carrier Syst 9: 135
15. Maeda H, Seymour LW, Miyamoto Y (1992) Bioconjugate Chem 3: 351
16. Sezaki H, Takakura Y. Hashida M (1989) Adv Drug Delivery Rev 3: 247
17. Hoes CJT, Feijen J (1989) in: Roerdink FDH, Kroon AM (eds) Drug carrier systems, John Wiley, New York, p 57
18. Kopeček J Duncan R (1987) J Controlled Rel 6: 315
19. Lloyd JB (1987) in: Johnson P, Lloyd-Jones JG (eds) Drug delivery systems fundamentals, Ellis Horwood, p 95
20. Duncan R, Kopeček J (1984) Adv Polym Sci 57: 51
21. Duncan R, Cable HC, Rejmanová P, Kopeček J, Lloyd JB (1984) Biochim Biophys Acta 799: 1
22. McCormick LA, Seymour LCW, Duncan R, Kopeček J (1986) J Bioact Comp Polym 1: 4
23. Krinick NL, Říhová B, Ulbrich K, Strohalm J, Kopeček J (1990) Makromol Chem 191: 839
24. Říhová B Kopečková P, Strohalm J, Rossmann P, Větvička V, Kopeček J (1988) Clin Immunol Immunopathol 46: 100
25. McGraw TE, Maxfield FR (1991) in: Juliano RL (ed) Targeted drug delivery. Springer, Berlin, p 11 (Handbook of experimental pharmacology, Vol 100)
26. Kopeček J (1984) Biomaterials 5: 1984
27. Kopeček J (1990) J Controlled Rel 11: 279
28. Willner D, Trial PA, Hofstead SJ, King HD, Lasch SJ, Braslawsky GR, Greenfield RS, Kaneko T, Firestone RA (1993) Bioconjugate Chem 4: 521
29. Shen WC, Ryser HJP (1981) Biochem Biophys Res Commun 102: 1048
30. Shen WC, Ryser HJP, LaManna L (1985) J Biol Chem 260: 10905
31. Feener EP, Shen WC, Ryser HJP (1990) J Biol Chem 265: 18780
32. Thorpe PE, Wallace PM, Knowless PP, Relf MG, Brown ANF, Watson GJ, Knyba RE, Wawrzynczak EJ, Blakey DC (1987) Cancer Res 47: 5924
33. Goddard P, Petrak K (1989) J Bioact Compat Polym 4: 372
34. Chiu HC, Kopečková P, Deshmane S, Kopeček J (1994) J Controlled Rel., to be submitted
35. Chiu HC, Koňák Č, Kopečková P, Kopeček J (1994) J Bioact Compat Polymers, in press
36. Duncan R, Seymour LW, O'Hare KB, Flanagan PA, Wedge S, Hume IC, Ulbrich K, Strohalm J, Subr V, Spreafico F, Grandi M, Ripamonti M, Farao M, Surato A (1992) J Controlled Rel 19: 331
37. O'Hare KB, Duncan R, Strohalm J, Ulbrich K, Kopečková P (1993) J Drug Targeting 1: 217
38. Duncan R, Hume IC, Yardley HJ, Flanagan PA, Ulbrich K, Šubr V, Strohalm J (1991) J Controlled Rel 16: 121
39. Duncan R, Kopečková-Rejmanová P, Strohalm J, Hume I, Cable HC, Pohl J, Lloyd JB, Kopeček J (1987) Br J Cancer 55: 165

40. Duncan R, Kopečková P, Strohalm J, Hume IC, Lloyd JB, Kopeček J (1988) Br J Cancer 57: 147
41. Říhová B, Kopeček J (1985) J Controlled Rel 2: 289
42. Krinick NL, Sun Y, Joyner D, Spikes JD, Straight RC, Kopeček J (1994) J Biomat Sci Polym Ed., 5: 303
43. Říhová B, Krinick NL, Kopeček J (1993) J Controlled Rel 25: 71
44. Hoes CJT, Potman W, van Heeswijk WAR, Mud J, de Grooth BG, Greve J, Feijen J, (1985) J Controlled Rel 2: 205
45. Hoes CJT, Grootoonk J, Duncan R, Hume IC, Bhakoo M, Bouma JMW, Feijen J, (1993) J Controlled Rel 23: 37
46. Hurwitz E, Stancovaki I, Wilchek M, Shouval D, Takahashi H, Wands JR, Sela M, (1990) Bioconjugate Chem 1: 285
47. Abraham A, Nair MG, Kisliuk RL, Gaumont Y, Galivan J (1990) J Med Chem 33: 711
48. Yokoyama M (1989) PhD Thesis, University of Tokyo p 172
49. Yokoyama M, Okano T, Sakurai Y, Ekimoto H, Shibazaki C, Kataoka K (1991) Cancer Res 51: 3229
50. Tanaka T, Kaneo Y, Iguchi S (1991) Bioconjugate Chem 2: 261
51. Ohkawa K, Hatano T, Tsukada Y, Matsuda M (1993) Br J Cancer 67: 274
52. Song Y, Onishi H, Nagai T (1993) Biol Pharm Bull 16: 48
53. Noguchi A, Takahashi T, Yamaguchi T, Kitamura K, Takakura Y, Hashida M, Sezaki H (1992) Bioconjugate Chem 3: 132
54. Shih LB, Goldenberg DM, Xuan H, Lu H, Sharkey RM, Hall TC (1991) Cancer Res 51: 4192
55. Trouet A, Jolles G (1984) Seminars in Oncol. 11 (Suppl 3): 64
56. Oda T, Sato F, Yamamoto H, Akagi M, Maeda H (1989) Anticancer Res 9: 261
57. Konno T, Maeda H (1987) in: Okada K, Ishak KG (eds) Neoplasma of the Liver, Springer, New York, p 343
58. Hirano T, Ringsdorf H, Zaharko DS (1980) Cancer Res 40: 2263
59. Seymour LW (1991) J Bioact Compat Polym 6: 178
60. Ambrose EJ, Easty DM, Jones PCT (1958) Br J Cancer 12: 439
61. McGuire JJ, Russell CA (1990) Leukemia 4: 48
62. Larsen B, Thorling EB (1969) Acta Path Microbiol Scandinav 75: 229
63. Saiki I, Murata K, Matsuno K, Ogawa R, Nishi N, Tokura S, Azuma I (1990) Jpn J Cancer Res 81: 660
64. Saiki I, Murata J, Kakajima M, Tokura S, Azuma I (1990) Cancer Res 50: 3631
65. Zaharko DS, Corey JM (1984) Cancer Treat Rep 68: 1255
66. Ardalan B, Paget GE (1986) Cancer Res 46: 5473
67. Rosenblum MG. Hortobagyi GN (1986) Cancer Chemother Pharmacol 18: 247
68. Bocci V (1987) Pharmac Ther 34: 1
69. Thomas H (1992) Drugs of today 28: 311
70. Wada H, Imamura I, Sako M, Katagiri S, Tarui S, Nishimura H, Inada Y (1990) Ann N Y Acad Sci, 613: 95
71. Fuertges F, Abuchowski A (1990) J Controlled Rel 11: 139
72. Delgado C, Francis GE, Fisher D (1992) Crit Rev Ther Drug Carrier Syst 9: 249
73. Katre NV, Knauf MJ, Laird WJ (1987) Proc Natl Acad Sci USA 84: 1487
74. Chiu HC, Zalipsky S, Kopečková P, Kopeček J (1993) Bioconjugate Chem 4: 290
75. Wang QC, Pai LH, Debinski W, FitzGerald DJ, Pastan I (1993) Cancer Res 53: 4588
76. Kyriadikis DA, Tsirka SAE, Tsavdaridis IK, Iliadis SN, Kortsaris AH (1990) Mol Cell Biochem 96: 137
77. Ashihara Y, Kono T, Yamazaki S, Inada Y (1978) Biochem Biophys Res Comm 83: 385
78. Hershfield MS, Buckley RH, Greenberg ML, Melton AL, Shiff R, Hatem C, Kurtzberg J, Markert ML, Kobayashi RH, Hobayashi AL, Abuchowski A (1987) N Engl J Med 316: 589
79. Davis S, Abuchowski A, Park YK, Davis FF (1981) Clin Exp Immunol 46: 649
80. Beauchamp CO, Gonias SL, Menapace DP, Pizzo SV (1983) Analytical Biochem 131: 25
81. Pyatak PS, Abuchowski A, Davis FF (1980) Res Comm Chem Pathol Pharmacol 29: 113
82. Veronese FM, Largajolli R, Boccu E, Banassi CA, Shiavon O (1985) Appl Biochem Biotechnol 11: 141
83. Nishimura H, Ashihara Y, Matsushima A, Inada Y (1979) Enzyme 24: 261
84. Katre NV, Knauf MJ, Laird WJ (1987) Proc Natl Acad Sci USA 84: 1487
85. Rajagopalan S, Gonias SL, Pizzi SV (1985) J Clin Invest 75: 413

86. Iwashita Y, Yabuki A, Yamaji KY, Iwasaki K, Okami T, Hirata C, Kosaka K (1988) Biomat Art Cells Art Org 16: 271
87. Tanaka H, Sateke-Ishikawa R, Ishikawa M, Matsuki S, Asano K (1991) Cancer Res 51: 3710
88. Zalipsky S, Lee C (1992) in: Harris JM (ed) Poly (ethylene glycol) chemistry: Biotechnical and biomedical applications, Plenum, New York, p 347
89. Ingham KC (1984) Methods Enzymol. 104: 351
90. Topchieva IN (1980) Russ Chem Rev 49: 260
91. Andrade JD, Nagaoka S, Cooper S, Okano T, Kim SW (1987) ASAIO J 10: 75
92. Lee JH, Kopeček J, Andrade JD (1989) J Biomed Mater Res 23: 351
93. Lee JH, Kopečková P, Kopeček J, Andrade JD (1990) Biomaterials 11: 455
94. Poznansky MJ, Shandling M, Salkie MA, Elliott, J, Lau E (1982) Cancer Res 42: 1020
95. Lääne A, Aaviksaar A, Haga M, Chytrý V, Kopeček J (1985) Makromol Chem Suppl 9: 35
96. Kamei S, Kopeček J (1995) Pharmaceutical Res, in press
97. Takei YG, Aoki T, Sanui K, Ogata N, Okano T, Sakurai Y (1993) Bioconjugate Chem 4: 42
98. Takei YG, Aoki T, Sanui K, Ogata N, Okano T, Sakurai Y (1993) Bioconjugate Chem 4: 341
99. Ito Y, Kotoura M, Chung DJ, Imanishi Y (1993) Bioconjugate Chem 4: 358
100. Drobník J, Rypáček F (1984) Adv Polym Sci 57: 1
101. Drobník, (1989) Adv Drug Deliv Rev 3: 229
102. Kopeček J (1982) in: Benoit H, Rempp P (eds) IUPAC macromolecules, Pergamon, Oxford. p 305
103. Kopeček J, Duncan R (1987) in: Illum L, Davis SS (eds) Polymers in controlled drug delivery, John Wright, Bristol, p 152
104. Joyner WL, Kern DF (1990) Adv Drug Deliv Rev 4: 319
105. Pitha J (1980) in: Donaruma IG, Ottenbrite RM, Vogl D (eds) Anionic polymeric drugs. Wiley, New York, p 277
106. Petrak K, Goddard P (1989) Adv Drug Deliv Rev 3: 191
107. Maeda H, Matsumura Y (1989) Crit Rev Ther Drug Carrier Syst 6: 193
108. Senger DR, Galli SJ, Dvorak AM, Perruzzi CA, Harvey VS, Dvorak HF (1983) Science 219: 983
109. Dvorak HF, Nagy JA, Dvorak JT, Dvorak AM (1988) Am J Pathol 133: 95
110. Nugent LJ, Jain RK (1984) Am J Physiol 246: H129
111. Matsumura Y, Maeda H (1986) Cancer Res 6: 6387
112. Magerstadt M, Antibody conjugates and malignant disease, CRC Press, Boca Raton, FL, 1991
113. Seymour LW, Duncan R, Kopečková P, Kopeček J (1987) J Bioact Compat Polym 2: 97
114. Bos E, Boon P, Kaspersen F, McCabe R (1991) J Controlled Rel 16: 101
115. Flavell DJ, Cooper S, Morland B, French R. Flavell SU (1992) Br J Cancer 65: 545
116. Fanger MW, Morganelli PM, Guyre PM (1992) Crit Rev Immunol 12: 101
117. Baxter LT, Jain RK (1991) Microvascular Res 41: 5
118. Dilman RO (1990) Antibody, Immunoconjug Radiopharm 3: 1
119. Yokota T, Milenic DE, Whitlow M, Schlom J (1992) Cancer Res 52: 3402
120. Wright A, Shin SU. Morrison SL (1992) Crit Rev Immunol 12: 125
121. LoBuglio AF, Wheeler RH, Trang J, Haynes A, Rogers K, Harvey EB, Sun L, Ghrayeb J, Khazaeli MB (1989) Proc Natl Acad Sci USA 86: 4220
122. Hale G, Clark MR, Marcus R, Winter G, Dyer MJS, Philips JM, Reichmann L, Waldmann H (1988) Lancet p 1394
123. Trail PA, Willner D, Lasch SJ, Henderson AJ, Hofstead S, Casazza AM, Firestone RA, Hellstrom I, Hellstrom KE (1993) Science 261: 212
124. Šprincl L, Exner J, Štěrba O, Kopeček J (1976) J Biomed Mater Res 10: 953
125. Seymour LW, Duncan R, Strohalm J, Kopeček J (1987) J Biomed Mater Res 21: 1341
126. Arturson G. Granath K, Thoren L, Wallenius G (1964) Acta Chir Scand 127: 543
127. Yamaoka T, Tabata Y, Ikada Y (1993) Drug Delivery 1: 75
128. Kopeček J (1981) in: Williams DF (ed) Systemic aspects of biocompatibility, Vol II, CRC Press, Boca Raton, p 159
129. Hespe W, Meier AM, Blakwater YJ (1977) Drug Res 27: 1158
130. Takakura Y, Fujita T, Hashida M, Sezaki H (1990) Pharmaceutical Res 7: 339
131. Duncan R, Pratten MK, Cable HC, Ringsdorf H, Lloyd JB (1981) Biochem J 196: 49
132. Ulbrich K, Koňák Č, Tuzar Z, Kopeček J (1987) Makromol Chem 188: 1261
133. Nukui M, Hoes K, van den Berg H, Feijen J (1991) Makromol Chem 192: 2925
134. Koňák Č, Kopečková P, Kopeček J (1992) Macromolecules 25: 5451
135. Yokoyama M, Inoue S, Kataoka K, Yui N, Okano T, Sakurai Y, (1989) Makromol Chem 190: 2041

136. Yokoyama M, Kwon GS, Okano T, Sakurai Y, Ekimoto, Okamoto K, Mashiba H, Seto T, Kataoka K (1993) Drug Delivery 1: 11
137. Trouet A, Masquelier M, Baurain R, Deprez-de Campeneere D (1982) Proc Natl Acad Sci USA 79: 626
138. Schechter B, Neumann A, Wilchek M, Arnon R (1989) J Controlled Rel 10: 75
139. Howe-Grant ME, Lippard SJ (1980) in: Sigel H (ed) Metal ions in biological systems, Vol 11, Marcel Dekker, New York, p 63
140. Kopeček J (1977) Polymers in Medicine (Wroclaw) 7: 191
141. Říhová B, Strohalm J, Plocová D, Ulbrich K (1990) J Bioact Compat Polym 5: 249
142. Obereigner B, Burešová M, Vrána A, Kopeček J (1979) J Polym Sci Polym Symp. 66: 41
143. Ouchi T, Fujie H, Jokei S, Sakamoto Y. Chikashita H, Inoi T, Vogl O (1986) J Polym Sci Part A: Polym Chem 23: 2059
144. Ozaki S, Ohnishi J, Watanabe Y, Nohda T, Nagase T, Akiyama T, Uehara N, Hoshi A (1989) Polymer J 21: 955
145. Giammona G, Giannola LI, Carlisi B, Bajardi ML (1989) Chem, Pharm Bull 37: 2245
146. Yang F, Zhuo R (1990) Polymer J 22: 572
147. Zunino F, Pratesi G, Micheloni A (1989) J Controlled Rel 10: 65
148. Nagy A, Szoke B, Schally AV (1993) Proc Natl Acad Sci USA 90: 6373
149. Schacht E (1987) in: Illum L, Davis SS (eds) Polymers in drug delivery. Wright, Bristol, p 131
150. Carlsson J, Drevin H, Axen R (1978) Biochem J 173: 723
151. Kitagawa T, Shimozono T, Aikawa T, Yoshida T, Nishimura H (1981) Chem Pharm Bull 29: 1130
152. Kitagawa T, Aikawa T (1976) J Biochem 79: 233
153. Henkin J (1977) J Biol Chem 252: 4293
154. Webb RR II, Kancko E (1990) Bioconjugate Chem 1: 96
155. Vanin EF, Ji TH (1981) Biochemistry 20: 6754
156. Trommer WE, Kolkenbrock H, Pfleiderer G (1975) Hoppe-Seyler's Z Physiol Chem 356: 1455
157. Wong SS (1993) Chemistry of protein conjugation and crosslinking, CRC Press, Boca Raton, Florida
158. Pietersz GA (1990) Bioconjugate Chem 1: 89
159. Brinckley M (1992) Bioconjugate Chem 3: 2
160. Persiani S, Yeung A, Shen W-C, Kennedy A (1991) Carcinogenesis 12: 1149
161. Alam F, Soloway A, Barth RF, Mafune N, Adams DM, Knoth WH (1989) J Med Chem 32: 2326
162. Chytrý V, Kopeček J, Leibnitz E, O'Hare K, Scarlett L, Duncan R (1987) New Polym Mater 1: 21
163. Duncan R, Kopeček J, Rejmanová P, Lloyd JB (1983) Biochim Biophys Acta 755: 518
164. Pietersz GA, Cunningham Z, McKenzie FC (1988) Immunol Cell Biol 66: 43
165. Rodwell JD, Alvarez VL, Lee C, Lopes AD, Goers JWF, King HD, Powsner HJ, McKearn TJ (1986) Proc Natl Acad Sci USA 83: 2632
166. O'Shannesy DJ (1990) J Chromatography 510: 13
167. Mann JS, Huang JC, Keana JFW (1992) Bioconjugate Chem 3: 154
168. Kato Y, Umemoto N, Kayama Y, Fukushima H, Takeda Y, Hara T, Tsukada Y (1984) J Med Chem 27: 1602
169. Kopeček J, Rejmanová P, Strohalm J, Ulbrich K, Říhová B, Chytrý V, Lloyd JB, Duncan R (1991) US Patent 5,037,883
170. Matzku S, Moldenhauer G, Kalthoff H, Canevari S, Colnaghi M, Schuhmacher J, Bihl H (1990) Br J Cancer 62 Suppl X: 1
171. Bagshawe KD (1989) Br J Cancer 60: 275
172. Senter PD, Wallace PM, Svensson HP, Vrudhula VM, Kerr DE, Hellstrom I, Hellstrom KE (1993) Bioconjugate Chem 4: 3
173. Bagshawe KD, Sharma SK, Springer CJ, Antoniw P, Rogers GT, Burke PJ, Melton R, Sherwood R (1991) Antibody Imm Radiopharm 4: 915
174. Cowan JD, Von Hoff DD (1987), in: Muggia FM, Rozencweig M (eds) Clinical evaluation of antitumor therapy, Nijhoff, Boston, p 33
175. Sheldon K, Marks A, Baumal R (1989) Anticancer Res 9: 637
176. Goff BA, Bamberg M, Hasan T (1991) Cancer Res 51: 4762
177. Říhová B, Bilej M, Větvička V, Ulbrich K, Strohalm J, Kopeček J, Duncan R (1989) Biomaterials 10: 335
178. Hurwitz E, Kashi R. Arnon R, Wilchek M, Sela M (1985) J Med Chem 28: 137

179. Smyth MJ, Pietersz GA, McKenzie IFC (1987) Cancer Res 47: 62
180. Aboud-Pirak E, Hurwitz E, Bellot F, Schlessinger J, Sela M (1989) Proc Natl Acad Sci USA 86: 3778
181. Dillman RO, Shawler DL, Johnson DE, Meyer DL, Koziol JA, Frincke JM (1986) Cancer Res 46: 4886
182. Pietersz GA, Smyth MJ, McKenzie IFC (1988) Cancer Res 48: 926
183. Yeh MY. Roffler SR, Yu MH (1992) Int J Cancer 51: 274
184. Kanellos J, Pietersz GA, Cunningham Z, McKenzie IFC (1987) Immunol. Cell Biol 65: 483
185. Kanellos J, Pietersz GA, McKenzie IFC (1985) J Natl Cancer Inst 75: 319
186. Hurwitz E, Levy R, Maron R, Wilchek M, Arnon R, Sela M (1975) Cancer Res 35: 1175
187. Mosmann T (1983) J Immunol Methods 65: 55
188. Slater TF, Sawyer B, Strauli U (1963) Biochim Biophys Acta 77: 383
189. Shimoyama Y, Kubota T, Watanabe M, Ishibiki K, Abe O (1989) J Surg Oncol 41: 12
190. Suto A, Kubota T, Shimoyama Y, Ishibiki K, Abe O (1989) J Surg Oncol 42: 28
191. Krinick NL, Říhová B, Ulbrich K, Andrade JD, Kopeček J (1988) Proc SPIE 997: 70
192. Kubota T, Yamamoto T, Takahara T, Furukawa T, Ishibiki K, Kitajima M, Shida Y, Nakatsubo H (1992) J Surg Oncol 51: 75
193. Hoogenboom HR, Raus JCM, Volckaert G (1991) Biochim. Biophys Acta 1096: 345
194. Roper PR, Drewinko B (1976) Cancer Res 36: 2182
195. Hamburger AW, Salmon SE (1977) Science 197: 461
196. Scott Jr. CF, Goldmacher VS, Lambert JM, Jackson JV, Malntyre GD (1987) J Natl Cancer Inst 79: 1163
197. Rakestraw SL, Ford WE, Tompkins RG, Rodgers MAJ, Thorpe WP, Yarmush ML (1992) Biotechnol Prog 8: 30
198. Oseroff AR, Ohuoha D, Hasan T, Bommer JC, Yarmush ML (1986) Proc Natl Acad Sci USA 83: 8744
199. Triton TR, Yee G (1982) Science 217: 248
200. Hirano T, Ohashi S, Morimoto S, Tsuda K, Kobayashi T, Tsukagoshi S (1986) Makromol Chem 187: 2815
201. Jones KH, Senft JA (1985) J Histochem Cytochem 33: 77
202. Altman SA, Randers L. Rao G (1993) Biotechnol Prog 9: 671
203. Endo N, Takeda Y, Umemoto N, Kishida K, Watanabe K, Saito M, Kato Y, Hara T (1988) Cancer Res 48: 3330
204. Umemoto N, Kato Y, Endo N, Takeda Y, Hara T (1989) Int J Cancer 43: 677
205. Ovadia M, Hager CC, Oeltmann TN (1990) Anticancer Res 10: 671
206. Youle RJ, Neville Jr, DM (1980) Proc Natl Acad Sci USA 77: 5483
207. Kopeček J, Šprincl L (1974) Polymers in Medicine (Wroclaw) 4: 109
208. Teece R, Fraioli R, De Fabritis P, Sandrelli A, Savarese A, Santoro L, Cuomo M, Natali PG, (1991) Int J Cancer 49: 310
209. Brooks CG, Rees RC, Robins RA (1978) J Immunol Meth 21: 111
210. Van Putten LM (1987), in: Muggia FM, Rozencweig M (eds) Clinical evaluation of antitumor therapy, Nijhoff, Boston p 17
211. Ueda Y, Munechika K, Kikukawa A, Kanóh Y, Yamanouchi K, Yokoyama K (1989) Chem Pharm Bull 37: 1639
212. Goldin A (1966) Cancer Chemotherapy Reports 50: 173
213. Ohya Y, Huang TZ, Ouchi T, Hasegawa K, Tamura J, Kadowaki K, Matsumoto T, Suzuki S, Suzuki M (1991) J Controlled Rel 17: 259
214. Zunino F, Giuliani F, Saui G, Dasdia T, Gambetta R (1982) Int J Cancer 30: 465
215. Pratesi G, Savi G, Pezzoni G, Bellini O, Penco S, Tinelli S, Zunino F (1985) Br J Cancer 52: 841
216. Meyer RM, Quirt IC, Skillings JR, Cripps MC, Bramwell VHC, Weinerman BH, Gospodarowitz MK, Burns BF, Sargeant AM Shepherd LE, Zee, B. Hryniuk WM (1993) N Engl J Med 329: 1770
217. Bernstein A, Hurwitz E, Maron R, Arnon R, Sela M, Wilchek M (1978) J Natl Cancer Inst 60: 379
218. Flanagan PA, Duncan R, Říhová B, Šubr V, Kopeček J (1990) J Bioact Compat Polym 5: 151
219. Maeda H, Matsumoto T, Konno T, Iwai K, Ueda M (1984) J Protein Chem 3: 181
220. Meijer DKF, Jansen RW, Pauwels R, Schols D, de Clercq E (1992) 2nd Eur Symp on Controlled Drug Delivery, Noordwijk aan Zee, The Netherlands, April 1-3, Abstracts p 167
221. Meijer DKF, Jansen RW, Molema G (1992) Antiviral Res 18: 215
222. Myers KJ, Ron Y (1992) Pharmaceutical Technology p 26 (January)

223. Sung C, Shockley TR, Morrison PF, Dvorak HF, Yarmush ML, Dedrick RL (1992) Cancer Res 52: 377
224. Anderson DC, Manger R, Schroeder J, Woodle D, Barry M, Morgan AC, Fritzberg AR (1993) Bioconjugate Chem 4: 10
225. Wichterle O, Lím D (1960) Nature 185: 117
226. Wichterle O (1968) US Patent 3,361, 858
227. Šprincl L, Vacík J, Kopeček J, Lím D (1971) J Biomed Mater Res 5: 197
228. Šprincl L, Vacík J, Kopeček J (1973) J Biomed Mater Res 7: 123
229. Kopeček J, Šprincl L, Bažilová H, Vacík J (1973) J Biomed Mater Res 7: 111
230. Šprincl L, Kopeček J, Lím D (1973) Calc Tissue Res 13: 63
231. Kopeček J, Šprincl L, Lím D (1973) J Biomed Mater Res 7: 179
232. Voldřich Z, Tománek Z, Vacík, J Kopeček J (1975) J Biomed Mater Res 9: 675
233. Kopeček J, Bažilová H (1973) Europ. Polym J 9: 7
234. Bohdanecký M, Bažilová H, Kopeček J (1974) Europ Polym J 10: 405
235. Drobník J, Kopeček J, Labský J, Rejmanová P, Exner J, Saudek, V, Kálal J (1976) Makromol Chem 177: 2833
236. Kopeček J, Cífková I, Rejmanová P, Strohalm J, Obereigner B, Ulbrich K (1981) Makromol Chem 182: 2941
237. Rejmanová P, Labský J, Kopeček J (1977) Makromol Chem 178: 2159
238. Kopeček J (1977) Makromol Chem 178: 2169
239. Kopeček J, Rejmanová P (1979) J Polym Sci Poly Symp 66: 15
240. Chytrý V, Vrána A, Kopeček J (1978) Makromol Chem 179: 329
241. Solovskij MV, Ulbrich K, Kopeček J (1983) Biomaterials 4: 44
242. Schechter I, Berger A (1967) Bichem Biophys Res Commun 27: 157
243. Kopeček J, Rejmanová P, Chytrý V (1981) Makromol Chem 182: 799
244. Rejmanová P, Obereigner B, Kopeček J (1981) Makromol Chem 182: 1899
245. Ulbrich K, Strohalm J, Kopeček J (1981) Makromol Chem 182: 1917
246. Ulbrich K, Zacharieva EI, Obereigner B, Kopeček J (1980) Biomaterials 1: 199
247. Duncan R, Cable HC, Lloyd JB, Rejmanová P, Kopeček J (1982) Biosci Rep 2: 1041
248. Duncan R, Cable HC, Lloyd JB, Rejmanová P, Kopeček J (1983) Makromol Chem 184: 1997
249. Rejmanová P, Pohl J, Baudyš M, Kostka V, Kopeček J (1983) Makromol Chem 184: 2009
250. Šubr V, Kopeček J, Pohl J, Baudyš M, Kostka V (1988) J Controlled Rel 9: 133
251. Rejmanová P, Kopeček J, Duncan R, Lloyd JB (1985) Biomaterials 6: 45
252. Kopeček J, Rejmanová P, Duncan R, Llyod JB (1985) Ann N Y Acad Sci 446: 93
253. Kopeček J (1987) Proc Intern Symp Control Rel Bioactive Mater 14: 125
254. Blow DM (1974) Bayer Symp 5: 473
255. Ulbrich K, Strohalm J, Kopeček J (1986) Makromol Chem 187: 1131
256. Musil D, Zucic D, Turk D, Engh RA, Mayr I, Huber R, Popovic T, Turk V, Towatari T, Katunuma N, Bode W (1991) EMBO J 10: 2321
257. Yamamoto A, Kaji T, Tomoo K, Ishida T, Inoue M, Murata M, Kitamura K (1992) J Mol Biol 227: 942
258. Van den Hamer CJA, Morell AG, Scheiberg IH, Hickman J, Ashwell G (1970) J Biol Chem 245: 4397
259. Ashwell G, Harford J (1982) Annu Rev Biochem 51: 531
260. Duncan R, Lloyd JB, Rejmanová P, Kopeček J (1985) Makromol Chem Suppl 9: 3
261. Duncan R, Seymour LCW, Scarlett L, Lloyd JB, Rejmanová P, Kopeček J (1986) Biochim Biophys Acta 880: 62
262. Wedge SR, Duncan R, Kopečková P (1991) Br J Cancer 63: 546
263. Omelyanenko V, Kopečková P, Rathi RC, Kopeček J (1994) Biochim Biophys Acta to be submitted
264. Duncan R, Hume IC, Kopečková P, Ulbrich K, Strohalm J, Kopeček J (1989) J Controlled Rel 10: 51
265. Seymour L, Ulbrich K, Wedge SR, Hume IC, Strohalm J, Duncan R (1991) Br J Cancer 63: 859
266. Cartlidge SA, Duncan R, Llyod JB, Rejmanová P, Kopeček J (1986) J Controlled Rel 3: 55
267. Goddard P, Williamson I, Brown J, Hutchinson LE, Nicholls J, Petrak K (1991) J Bioact Compat Polym 6: 4
268. Cartlidge SA, Duncan R, Lloyd JB, Kopečková-Rejmanová P, Kopeček J (1987) J Controlled Rel 4: 253
269. Ambler LE, Brookman L, Brown J, Goddard P, Petrak K (1992) J Bioact Compat Polym 7: 233

270. Cartlidge SA, Duncan R, Lloyd JB, Kopečková-Rejmanová P, Kopeček J (1987) J Controlled Rel 4: 265
271. Ulbrich K, Zacharieva EI, Kopeček, J Hume IC, Duncan R (1987) Makromol Chem 188: 2497
272. O'Hare KB, Hume IC, Scarlett L, Chytrý V, Kopečková P, Kopeček J, Duncan R (1989) Hepatology 10: 207
273. Říhová B, Kopeček J, Kopečková-Rejmanoý P, Strohalm J, Plocová D, Semorádová H (1986) J Chromatogr Biomed Appl 376: 221
274. Říhová B, Větvička V, Strohalm J, Ulbrich K, Kopeček J (1989) J Controlled Rel 9: 21
275. Říhová B, Vereš V, Fornüsek L, Ulbrich K, Strohalm J, Větvička V, Bilej M, Kopeček J (1989) J Controlled Rel 10: 37
276. Seymour LW, Flanagan PA, Al-Shamkani A, Subr V, Ulbrich K, Cassidy J, Duncan R (1991) Select. Cancer Therapeut 7: 59
277. Flanagan PA, Kopečková P, Kopeček J, Duncan R (1989) Biochim Biophys Acta 993: 83
278. Krinick NL, Combination polymeric drugs as anticancer agents, Ph D Thesis, University of Utah, 1992
279. Kopeček J, Krinick NL (1993) US Patent 5,258,453
280. Štěrba O. Paluska E, Jozová O, Špunda J, Nezvalová J, Šprincl L, Kopeček J, Činátl J (1975) Čas Lék Českých 114: 1268
281. Korčáková L, Paluska E, Hašková V, Kopeček J (1976) Z. Immun Forsch 151: 219
282. Paluska E, Činátl J, Korčáková L, Štěrba O, Kopeček J Hrubá A, Nezvalová j, Staněk R (1980) Folia biologica (Prague) 26: 304
283. Štěeba O, Uhlířová Z, Petz R, Viktora L, Jirásek A, Kopeček J (1980) Čas Lék. Českých 119: 994
284. Činátl J, Štěrba O, Paluska E, Polednová V, Kopeček J (1980) Čs Farmacie 29: 134
285. Paluska E, Hrubá A, Štěrba O, Kopeček J (1986) Folia Biologica (Prague) 32: 91
286. Říhová B, Kopeček J, Ulbrich K, Pospíšil M, Mančal P (1984) Biomaterials 5: 143
287. Říhová B, Ulbrich K, Kopeček J, Mančal P (1983) Foila Microbiologica (Prague) 28: 217
288. Říhová B, Kopeček J, Ulbrich K, Chytrý V (1985) Makromol Chem Suppl 9: 13
289. Tlaskalová-Hogenová H, Kopeček J, Ulbrich K, Rýpáček F, Pospíšil M (1985) Makromol Chem Suppl 9: 137
290. Šimečková J, Říhová B, Plocová D, Kopeček J (1986) J Bioact Compat Polym 1: 20
291. Yeung TK, Hopewell JW, Simmonds RH, Seymour LW, Duncan R, Bellini O, Grandi M, Spreafico F, Strohalm J, Ulbrich K (1991) Cancer Chemother Pharamacol 29: 105
292. Seymour LW, Ulbrich K, Strohalm J, Kopeček J, Duncan R (1990) Biochem Pharmacol. 39: 1125
293. Gu ZW, Spikes JD, Kopečková P, Kopeček J (1993) Collect Czech Chem Commun 58: 2321
294. Bayley H, Gasparro F, Edelson R (1987) Trends Pharmacol Sci 8: 138
295. Cassidy J, Duncan R, Morrison GJ, Strohalm J, Plocová D, Kopeček J, Kaye SB (1989) Biochem Pharmacol 38: 875
296. Gomer CJ, Rucker N, Ferratio A, Wong S (1989) Radiat Res 120: 1
297. Henderson BW, Dougherty TJ (1992) Photochem Photobiol 55: 145
298. Spikes JD (1989) in: Smith KC (ed) The science of photobiology, 2nd ed., Plenum New York, p. 79
299. Kopeček J, Říhová B, Krinick NL (1991) J Controlled Rel 16: 137
300. Kopeček J, (1991) Ann N Y Acad Sci 618: 335
301. Jiang FN, Allison B, Liu D, Levy JG (1992) J Controlled Rel 19: 41
302. Wöhrle D, Krawczyk G, Paliuras M (1988) Makromol Chem 189: 1013
303. Peterson MC et al. (1994) in preparation
304. Sata T (1990) Polysaccharide coated liposomes, Ph D Thesis, University of Nagasaki
305. Gregoriadis G (1993) Liposome technology, Vol I-III, 2nd ed., CRC Press, Boca Raton, FL
306. Nässander UK (1991) Liposomes, immunoliposomes, and ovarian carcinoma, PhD Thesis, Unviersity of Utrecht
307. Connor J Huang L (1986) Cancer Res 46: 3431
308. Maeda M, Kumano A, Tirrell DA (1991) Ann New York Acad Sci 618: 362
309. Weinstein JN, Magin RL, Cysyk RL, Zaharko DS (1980) Cancer Res 40: 1388
310. Zou Y, Yamagishi M, Horikoshi I, Ueno M, Gu X, Perez-Soler R (1993) Cancer Res 53: 3046
311. Senior J (1987) Crit Rev Ther, Drug Carrier Syst 3: 123
312. Becker S (1988) Adv Drug Deliv Rev 3: 1
313. Alving CR (1988) Adv Drug Deliv Rev 2: 107

314. Lopez-Berenstein G, Fainstein V, Hopfer R, Mehta K, Sullivan MP, Keating M, Rosenblum MG, Mehta R, Luna M, Hersh EM, Reuben J, Juliano RL, Bodey GP (1985) J Infect Dis 150: 278
315. Sculier JP, Coune A, Meunier F, Brassine C, Laduron C, Hollaert C, Collete N, Heymans C, Klastersky J (1988) Eur J Cancer Clin Oncol 24: 527
316. Sunamoto J, Iwamoto K (1986) Crit Rev Ther Drug Carrier Syst 2: 117
317. Peeters PAM, Storm G, Crommelin DJA (1987) Adv Drug Deliv Rev 1: 249
318. Allen TM, Mehra T, Hansen C, Chin YC (1992) Cancer Res 52: 2431
319. Allen TM (1993) NATO ASI "Targeting of Drugs: Advances in Systems Constructs", Cape Sounion Beach, Greece, June 24 – July 5, 1993, Abstracts
320. Torchilin VP, Trubetskoy VS, Papisov MI, Bogdanov AA, Omelyanenko VG, Narula J, Khaw BA (1993) Proc Intern Symp Control Rel Bioact Mater 20: 194
321. Cowens JW, Creaven PJ, Greco WR, Brenner DE, Tung Y, Ostro M, Pilkiewicz F. Ginsberg R, Petrelli N (1993) Cancer Res 53: 2796
322. Nabel GJ, Nabel EG, Yang ZY, Fox BA, Plautz GE, Gao X, Huang L, Shu S, Gordon D. Chang AE (1993) Proc Natl Acad Sci USA 90: 11307
323. Douglas SJ, Davis SS, Illum L (1987) Crit Rev Ther Drug Carrier Syst 3: 233
324. Couvreur P, Vauthier C (1991) J Controlled Rel 17: 187
325. Birrenbach G, Speiser P (1976) J Pharm Sci 65: 1763
326. Kreuter J, Speiser P (1976) J Pharm Sci 65: 1424
327. Lenaerts V, Couvreur P, Christiaens-Leyh D, Joiris E, Roland M, Rollman B, Speiser P (1984) Biomaterials 5: 65
328. Verdun C, Brasseur F, Vranckx H, Couvreur P, Roland M (1990) Cancer Chemother Pharmacol 26: 13
329. Illum L, Jones PDE, Baldwin RW, Davis SS (1984) J Pharmacol Exp Ther 230: 733
330. Douglas SJ, Davis SS, Illum L (1986) Int J Pharm 34: 145
331. Müller BG, Kissel T (1993) Pharm Pharmacol Lett 3: 67
332. Donbrow M (1992) Microcapsules and nanoparticles in medicine and pharmacy, CRC Press, Boca Raton, FL
333. Pfeifle CE, Howell SB, Ashburn WL, Barone RM, Bookstein JJ (1986) Cancer Drug Deliv 3: 1
334. Edman P, Sjöholm I (1992) in: Donbrow M (ed) Microcapsules and nanoparticles in medicine and pharmacy, CRC Press, Boca Raton, FL, p 265
335. Juliano RL (1988) Adv Drug Del Rev 2: 31
336. Illum L, Davis SS (1984) FEBS Lett 167: 79
337. Illum L, Davis SS, Muller RH, Mak E, West P (1987) Life Sci 40: 367
338. Illum L, Davis SS (1987) Life Sci 40: 1553
339. Moghimi SM, Illum L, Davis SS (1990) Crit Rev Ther Drug Carrier Syst 7: 187
340. Harper GR, Davies MC, Davis SS, Tadros ThF, Taylor DC, Irving MP, Waters JA (1991) Biomaterials 12: 695
341. Okada H, Heya T, Ogawa Y, Shimamoto T (1988) J Pharmacol Exp Therap 244: 744
342. Ogawa Y, Okada H, Heya T, Shimamoto T (1989) J Pharm Pharmacol 41: 439
343. Okada H, Inoue Y, Heya T, Ueno H, Ogawa Y, Toguchi H (1991) Pharmaceutical Res 8: 787
344. Říhová B, Jegorov A, Strohalm J, Matha V, Rossmann, P, Fornůsek L, Ulbrich K (1992) J Controlled Rel 19: 25
345. Ogawa Y, Yamamoto M, Takada S, Okada H, Shimamoto T (1988) Chem Pharm Bull 36: 1502
346. Neville Jr DM, Scharff J, Srinivasachar K (1993) J Controlled Rel 24: 133
347. Sung C, Wilson D, Youle RJ (1991) J Biol Chem 266: 14159
348. FitzGerald D, Pastan I (1989) J Natl Cancer Inst 81: 1455
349. Wawrzynczak EJ (1991) Br J Cancer 64: 624
350. Byers VS, Rodvien R, Grant K, Durrant LG, Hudson KH, Baldwin RW, Scannon PJ (1989) Cancer Res 49: 6153
351. Vitetta ES, Stone M, Amlot P, Fay J, May R, Till M, Newman J, Clark P, Cunningham D, Ghetie V, Uhr JW, Thorpe PE (1991) Cancer Res 51: 4052
352. Oratz R, Speyer JL, Wernz JC, Hochster H, Meyers M, Mischak R, Spitler LE (1990) J Biol Respons Mod 9: 345
353. Kreitman RJ, FitzGerald D, Pastan I (1992) Int J Immunopharmac 14: 465
354. Pastan I, FitzGerald DJ (1991) Science 254: 1173
355. FitzGerald D, Pastan I (1993) NATO ASI "Targeting of Drugs: Advances in Systems Constructs", Cape Sounion Beach, Greece June 24 - July 5, 1993, Abstracts

356. Hinman LM, Hamann PR, Wallace R, Menendez AT, Durr FE, Upeslacis J (1993) Cancer Res 53: 3336
357. Chari RVJ, Martell BA, Gross JL, Cook SB, Shah SA, Blättler WA, McKenzie SJ, Goldmacher VS (1992) Cancer Res 52: 127
358. Lee MD, Dunne TS, Chang CC, Siegal MM, Morton GO, Ellestad GA, McGahren WJ, Borders DB (1992) J Am Chem Soc 114: 985
359. Anderson WF (1994) Human Gene Therapy 5: 1
360. Degols G, Machy P, Leonetti JP, Leserman L, Lebleu B (1992) in: Murray JAH (ed) Antisense RNA and DNA, Wiley-Liss, New York, p 255
361. Stein CA, Cheng YC (1993) Science 261: 1004
362. Bunnell BA, Askari FK, Wilson JM (1992) Somat Cell Mol Genet 18: 559

Biomedical Membranes from Hydrogels and Interpolymer Complexes

C. L. Bell and N. A. Peppas*
School of Chemical Engineering, Purdue University, West Lafayette,
IN 47907-1283, USA

Biomedical applications of hydrogel membranes require understanding of their structural characteristics and diffusive behavior. Thus, the subject of this review is the analysis of the response of such
biomembranes to their surrounding environment. This responsive behavior may be due to the
presence of certain functional groups along the polymer chains or specific interactions between
polymer chains (complexation). This behavior is particularly important in the use of these physiologically responsive materials in membrane applications. We begin with an introduction to the
structural characteristics and behavior of hydrogel membranes followed by a discussion of the types
of environmentally responsive behavior seen with hydrogels. The subject of interpolymer complexation is then treated with emphasis on complexation due to hydrogen bonding and how this type of
behavior may be used to produce responsive membranes. Finally, the theories of transport in
membranes are reviewed.

* Correspondence

List of Symbols and Abbreviations

Roman Symbols

a_1	activity of swelling agent
c	concentration
D	diffusivity
D_a	diffusivity through amorphous polymer portion of semicrystalline polymer
D_c	diffusivity through crystalline polymer portion of semicrystalline polymer
G	shear modulus
D_{eff}	effective diffusivity of solute in porous membrane
D_{im}	diffusivity of solute i in polymer membrane
D_{iw}	diffusivity of solute i in pure water
$D_{2,1}$	diffusivity of solute in pure water
$D_{2,13}$	diffusivity of solute in water swollen polymer
D_∞	diffusivity in bulk solution
E	elastic modulus
E_d	activation energy for diffusion
F_s	drag force between solute and solvent
f_∞	friction factor
G	shear modulus
g	lag coefficient
H	membrane hydration
I	ionic strength
i	ionization
J	solute flux
K	ratio of pore to bulk friction coefficients
K_p	partition coefficient in pores
k	Boltzmann constant
K'	partition coefficient
L	mechanochemical compliance
$\bar{M}_c$	number average molecular weight between crosslinks
$\bar{M}_n$	number average molecular weight of uncrosslinked polymer
M_t	mass released at time t
M_∞	mass released as t approaches ∞
N	Avogadro's number
$\bar{N}_s$	average solute flux
n	diffusional exponent
n_1	moles of swelling agent
P	permeability coefficient
$P_{2,13}$	permeability coefficient of solute in water swollen polymer
Q	equilibrium volume swelling ratio

q	equilibrium weight swelling ratio
q_s	cross-sectional area of solute
R	gas constant
r_p	pore radius
r_s	solute radius
T	absolute temperature
t	time
U_s	net solute velocity
V	volume
$V_{f,1}$	free volume of water in swollen membrane
V_0	molecular volume of unswollen network
$\bar{V}_1$	molar volume of swelling agent
V'_1	free volume of water
V_1	donor cell volume
V_{13}	free volume in swollen membrane
x_1	mole fraction of swelling agent

Greek Symbols

α_s	linear deformation factor
γ_1	activity coefficient of swelling agent
δ	membrane thickness
ΔG	total free energy change
ΔG_{el}	elastic free energy
ΔG_{ion}	ionic free energy
ΔG_{mix}	free energy of mixing
ΔS_{el}	entropy change due to deformation
ε	porosity
ζ	ratio of pore to bulk diffusivity
λ	ratio of r_s to r_p
$\lambda_{2,1}$	solute difusional jump length in water
$\lambda_{2,13}$	solute diffusional jump length in swollen polymer
μ	chemical potential
v	Poisson's ratio
v_e	effective number of crosslinks per unit chain
ρ	density
ρ_x	crosslinking density
τ	tortuosity
υ_a	amorphous polymer volume fraction in semicrystalline polymer
υ_c	crystalline polymer volume fraction in semicrystalline polymer
υ_1	volume fraction of swelling agent
υ_2	volume fraction of polymer
υ_s	size of solute
$\upsilon_{2,s}$	equilibrium polymer volume fraction
$\upsilon_{2,r}$	polymer volume fraction after crosslinking but before swelling

$\bar{\upsilon}_{2,s}$	specific volume of polymer
$\Phi(v)$	free volume contributions
$\phi(q_s)$	sieving mechanism parameter
χ	sieving coefficient
χ_1	Flory polymer-solvent interaction parameter

1 Structural Characteristics and Behavior of Hydrogels

Hydrogels are hydrophilic polymer networks capable of imbibing large amounts of water yet insoluble because of the presence of crosslinks, crystalline regions, or entanglements. The hydrophilicity of these materials is due to the presence of hydrophilic functional groups such as -OH, -COOH, -CONH$_2$, -CONH, and -SO$_3$H along the polymer chains [1]. Hydrogels are used in a variety of applications especially in the biomedical field. Because of the high water content of these materials, hydrogels are very similar to natural tissue and often exhibit good biocompatibility. These characteristics have allowed hydrogels to be used in such biomedical applications as drug delivery systems, biosensors, contact lenses, catheters, and wound dressings. Also, because of the water content of hydrogels, these materials are good candidates for use as membranes. The water in the gels allows for solute diffusion through the gels while the crosslinks, which define the specific mesh size of the network, determine whether a solute of a certain size is able to move through the space between the polymer chains [2].

1.1 Neutral Hydrogels

In order to describe the hydrophilicity of a particular hydrogel, the extent to which the material swells in a particular solvent is an important parameter. This parameter may be a function of such factors as the degree of crosslinking in the polymer network, the ionic nature of the polymer network, the nature of the swelling agent, or even the temperature of the swelling agent. The degree of swelling of a hydrogel is usually described as the equilibrium volume swelling ratio, Q, which is the ratio of the volume of the swollen gel to the volume of the dry polymer or as the equilibrium weight swelling ratio, q, which is the ratio of the weight of the swollen gel to the weight of the dry polymer [3]. Knowledge of these general swelling parameters of a particular hydrogel is important in evaluating the material for use in a desired application.

The basic parameter that describes the structure of a hydrogel network is the molecular weight between crosslinks, $\bar{M}_c$. This describes the average molecular weight of polymer chains between two consecutive junctions. These junctions may be chemical crosslinks, physical entanglements, crystalline regions, or even polymer complexes [2]. Additional parameters derived from the molecular weight between crosslinks include the crosslinking density, ρ_x, and the effective number of crosslinks per original chain, ν_e. These are defined in Eqs. (1) and (2).

$$\rho_x = \frac{1}{\bar{\upsilon}\,\bar{M}_c} \tag{1}$$

$$\nu_e = \frac{\bar{M}_n}{\bar{M}_c} - 1 \tag{2}$$

In these equations, $\bar{\upsilon}$ is the specific volume of the polymer and $\bar{M}_n$ is the initial molecular weight of the uncrosslinked polymer [4].

Several theories have been proposed to calculate the molecular weight between crosslinks in a hydrogel membrane. Probably the most widely used of these theories is that of Flory and Rehner [5]. This theory deals with neutral polymer networks and assumes a Gaussian distribution of polymer chains and tetrafunctional crosslinking within the polymer network.

If a polymer network is placed in contact with a solvent (see Fig. 1a), a thermodynamic swelling force is exerted as the macromolecular chains interact with the solvent and assume an elongated or solvated state. This swelling force is then counteracted by an elastic, retractive force. A state of equilibrium swelling is reached when these two forces are equal. The change in free energy in this process may be described as a combination of the free energy of mixing and the elastic free energy,

$$\Delta G = \Delta G_{mix} + \Delta G_{el}. \tag{3}$$

The free energy of mixing is as follows, keeping in mind that there are no individual polymer molecules in the crosslinked polymer system,

$$\Delta G_{mix} = kT(n_1 \ln \upsilon_1 + \chi_1 n_1 \upsilon_2). \tag{4}$$

Here, n_1 is the moles of the swelling agent, υ_1 and υ_2 are the corresponding volume fractions of swelling agent and polymer, χ_1 is the Flory interaction parameter, and k and T are the Boltzmann constant and absolute temperature, respectively.

The elastic free energy, assuming no appreciable change in internal energy in the network, may be described as

$$\Delta G_{el} = - T \Delta S_{el} \tag{5}$$

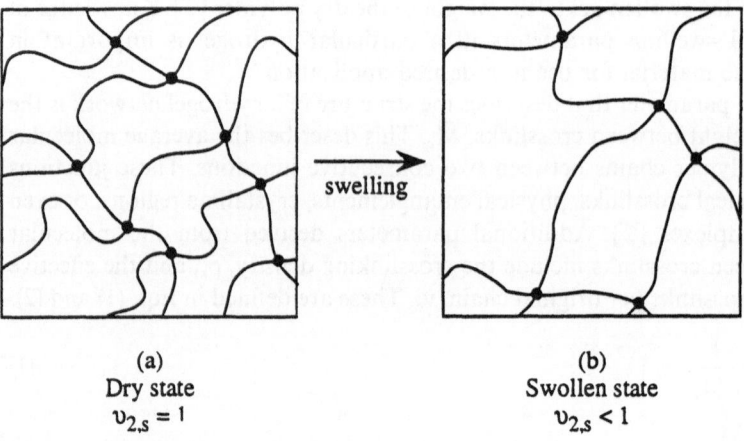

(a)
Dry state
$\upsilon_{2,s} = 1$

(b)
Swollen state
$\upsilon_{2,s} < 1$

Fig. 1. Swelling expansion of a gel membrane

where ΔS_{el} is the change in entropy due to the deformation. If α_s is the linear deformation factor and isotropic swelling is assumed ($\alpha_s = \alpha_x = \alpha_y = \alpha_z$) the elastic free energy is given by

$$\Delta G_{el} = (kTv_e/2)(3\alpha_s^2 - 3 - \ln\alpha_s^3) \tag{6}$$

with v_e being the effective number of chains in the network.

The chemical potential of the solvent in the swollen polymer is

$$\mu_1 - \mu_1^0 = N\left(\frac{\partial \Delta G_{mix}}{\partial n_1}\right)_{T,P} + \left(\frac{\partial \Delta G_{el}}{\partial \alpha_s}\right)_{T,P}\left(\frac{\partial \alpha_s}{\partial n_1}\right)_{T,P} \tag{7}$$

where N equals Avogadro's number. This theory assumes the crosslinks have been introduced in the unswollen polymer so the relationship of α_s in terms of n_1 is

$$\alpha_s^3 = \frac{1}{v_2} = \left(V_0 + \frac{n_1 V_1}{N}\right)\bigg/V_0. \tag{8}$$

Here, V_0 and $\bar{V}_1$ are the molecular volume of the unswollen network and the molar volume of the swelling agent. Evaluating the required derivatives and substituting into Eq. (7) yields

$$\mu_1 - \mu_1^0 = RT[\ln(1 - v_2) + v_2 + \chi_1 v_2^2 + V_1(v_e/V_0)(v_2^{1/3} - v_2/2)]. \tag{9}$$

Swelling equilibrium is defined as the point at which the activity of the solvent is unity, or $\mu_1 = \mu_1^0$, and the polymer volume fraction at this point is defined as $v_{2,s}$ (see Fig. 1b). At equilibrium, Eq. (9) becomes,

$$-[\ln(1 - v_{2,s}) + v_{2,s} + \chi_1 v_{2,s}^2] = V_1(v_e/V_0)(v_{2,s}^{1/3} - v_{2,s}/2). \tag{10}$$

The term v_e may be replaced by using the following relations:

$$v_e = v\left(1 - \frac{2\bar{M}_c}{\bar{M}_n}\right) \tag{11}$$

$$v = V/\bar{v}\bar{M}_c. \tag{12}$$

Here V is the total volume and $\bar{v}$ is the specific volume of the polymer. Substitution and rearrangement yields the final form of the Flory-Rehner model [6],

$$\frac{1}{\bar{M}_c} = \frac{2}{\bar{M}_n} - \frac{(\bar{v}/V_1)[\ln(1 - v_{2,s}) + v_{2,s} + \chi_1 v_{2,s}^2]}{[v_{2,s}^{1/3} - v_{2,s}/2]}. \tag{13}$$

As mentioned, the Flory-Rehner model describes the situation in which crosslinks are introduced in the dry state. Peppas and Merrill derived a model reported by Peppas and Barn-Howell [3] which accounts for the introduction of crosslinks in the swollen state as in the case of solution polymerization. The final form of the Peppas-Merrill model for determining the molecular weight between crosslinks is:

$$\frac{1}{\overline{M}_c} = \frac{2}{\overline{M}_n} - \frac{(\bar{v}/V_1)\left[\ln(1 - v_{2,s}) + v_{2,s} + \chi_1 v_{2,s}^2\right]}{v_{2,r}\left[\left(\frac{v_{2,s}}{v_{2,r}}\right)^{1/3} - \frac{1}{2}\left(\frac{v_{2,s}}{v_{2,r}}\right)\right]}. \tag{14}$$

Here, $v_{2,r}$ is defined as the polymer volume fraction after crosslinking but before swelling (the relaxed polymer volume fraction) and $v_{2,s}$ is the polymer volume fraction after equilibrium swelling (swollen polymer volume fraction).

Models which also describe the molecular weight between crosslinks for neutral polymer networks but use a non-Gaussian chain distribution have also been derived. These models would be useful in cases of highly crosslinked polymer networks. Examples of these types of models include those of Peppas and Lucht [7], Kovac [8], and Galli and Brummage [9].

1.2 Ionic Hydrogels

Hydrogels may be neutral or ionic in nature. If the polymer chains making up the network contain ionizable groups, the forces influencing swelling may be greatly increased due to localization of charges within the hydrogel [6]. Ionic polymer networks in aqueous salt solutions yield a far more complicated situation than that of neutral polymers. The equilibrium swelling ratios attained are often an order of magnitude larger than those of neutral networks, as intermolecular interactions such as coulombic, hydrogen-bonding, and polar forces are present [10]. The swelling of ionic hydrogels is governed by the two previously mentioned thermodynamic mixing and elastic-retractive forces as well as ionic interactions between charged polymer and free ions. The charged groups on the polymer have to be neutralized by mobile counterions which cause an osmotic force to develop and influence the swelling behavior. Also, electrostatic repulsion between fixed charges and between fixed charges and mobile ions inside the gel affect the swelling of ionic networks [11].

The swelling behavior of ionic hydrogels is unique because of the ionization of the pendant functional groups along the polymer chains. As a result of these ionizable groups, the swelling of ionic hydrogels may be sensitive to factors in the surrounding environment such as pH and ionic strength. For example, a hydrogel containing anionic groups may not be highly swollen at a low pH when the pendant groups are not ionized. Then, as pH is increased above the pK_a of the gel, the network swells to a high degree because of electrostatic repulsion produced by ionizing pendant groups. As a result, ionic hydrogels have received much attention as environmentally sensitive materials as will be discussed in Sect. 2.

Several factors affect the swelling of ionic hydrogels. Peppas and Khare [12] indicated some of the parameters that affect this swelling behavior.

1. Degree of ionization: an increase in the ionic content of a hydrogel increases the hydrophilicity and results in a higher degree of swelling.

2. Ionization equilibrium: fixed charges in the polymer network cause the formation of an electric double layer of fixed charges and counterions in the gel. Donnan equilibrium dictates that, at equilibrium, the chemical potential of the ions inside the gel is equal to that of the ions outside the gel (in the swelling agent). Also, the electroneutrality condition applies inside the gel and in the swelling medium. Donnan exclusion does not allow sorption of co-ions into the gel due to electroneutrality. Thus, the concentration of counterions is always higher in the gel than in the external solution. As a result, an electrostatic potential difference (Donnan potential) forms between the two phases. The efficiency of this co-ion exclusion, or an increase in the Donnan potential, increases with decreasing solution concentration and with increasing ionic content in the gel.

3. Nature of counterions: counterions of higher valence are more strongly attracted to the gel phase in ionic hydrogels. These ions are preferred by the gel because the concentration of counterions needed inside the gel decreases.

As mentioned above, the equilibrium between swollen gel and swelling medium resembles that of Donnan membrane equilibrium. The swelling force from the fixed charges in the gel may be identified as a swelling or osmotic pressure [6]. This affects the determination of structural characteristics such as the molecular weight between crosslinks of ionic networks. As a result, there are now three contributions to the free energy of the swollen gel as opposed to the two contributions discussed in Sect. 1.1. These three contributions are then that of the mixing, elastic-retractive, and ionic free energies as shown below,

$$\Delta G = \Delta G_{mix} + \Delta G_{el} + \Delta G_{ion}. \tag{15}$$

Differentiating Eq. (15) with respect to the number of molecules of swelling agent yields an expression for the total chemical potential of the system,

$$\mu_1 - \mu_1^0 = (\Delta\mu_1)_{mix} + (\Delta\mu_1)_{el} + (\Delta\mu_1)_{ion}. \tag{16}$$

Here, μ_1 is the chemical potential of the swelling agent in the gel and μ_1^0 is the chemical potential of pure swelling agent. At equilibrium swelling, the chemical potential of the swelling agent in the gel (μ_1) is equal to that of the swelling agent outside of the gel, μ_1^*. Thus,

$$(\Delta\mu_1^*)_{ion} - (\Delta\mu_1)_{ion} = (\Delta\mu_1)_{mix} + (\Delta\mu_1)_{el}. \tag{17}$$

This equates the difference between the ionic contributions from the chemical potentials outside (*) and inside the gel to the contributions of the two forces used to describe neutral networks. The contributions from the mixing and elastic portions of Eq. (17) may be described as discussed in Sect. 1.1 [13].

Brannon-Peppas and Peppas [13, 14] have derived models to describe this ionic contribution term for both anionic and cationic hydrogels. Here the chemical potential is described as

$$(\Delta\mu_1^*)_{ion} = \mu_1^* - \mu_1^\circ = RT \ln a_1^* = RT \ln \gamma_1 x_1^*. \tag{18}$$

Here, a_1^*, γ_1, and x_1^* are the activity, activity coefficient, and mole fraction of the swelling agent. Making the assumption that a highly swollen network may be approximated by a dilute polymer solution,

$$(\Delta\mu_1^*)_{ion} = -V_1 RT \sum_j c_j^* \tag{19}$$

$$(\Delta\mu_1)_{ion} = -V_1 RT \sum_j c_j \tag{20}$$

with the summations incorporating the mobile solute species. Thus,

$$(\Delta\mu_1^*)_{ion} - (\Delta\mu_1)_{ion} = V_1 RT \sum_j (c_j - c_j^*) \tag{21}$$

or

$$(\Delta\mu_1^*)_{ion} - (\Delta\mu_1)_{ion} = V_1 RT(c_+ + c_- - c_+^* - c_-^*). \tag{22}$$

Assuming Donnan equilibrium with an external solution containing c_s^* concentration of an electrolyte $M_{v+}A_{v-}$, one may write

$$a_+^{v+} a_-^{v-} = a_+^{*v+} a_-^{*v-} \tag{23}$$

$$\left(\frac{a_+}{a_+^*}\right)^{v+} = \left(\frac{a_-^*}{a_-}\right)^{v-}. \tag{24}$$

Assuming that the activities of the ionic species may be approximated by the their concentrations, the concentrations of ions are given by

$$c_+ = v_+ c_s \tag{25}$$

$$c_- = v_- c_s + \frac{ic_2}{z_-} \tag{26}$$

$$c_+^* = v_+ c_s^* \tag{27}$$

$$c_-^* = v_- c_s^*. \tag{28}$$

These expressions are for an anionic hydrogel and could easily be derived for a cationic gel. Here the term ic_2/z_- represents the concentration of anions along the polymer chains with i being the ionization. Substitution into Eq. (24) yields

$$\left(\frac{c_s}{c_s^*}\right)^{v+} = \left(\frac{c_s^*}{c_s + \frac{ic_2}{v_- z_-}}\right)^{v-}. \tag{29}$$

Using the concentrations above, a charge balance ($v = v_+ + v_-$), and the fact that

$$c_+^* + c_-^* = v c_s^* \tag{30}$$

the ionic contribution to the chemical potential becomes

$$(\Delta\mu_1^*)_{ion} - (\Delta\mu_1)_{ion} = V_1 RT\left[\frac{ic_2}{z_-} - v(c_s^* - c_s)\right]. \tag{31}$$

In their derivation, Brannon-Peppas and Peppas consider two cases. The first describes a case where the concentration of a 1:1 electrolyte in the external solution is small compared to the concentration of ions in the polymer. The result of this case is an ionic contribution given by

$$(\Delta\mu_1^*)_{ion} - (\Delta\mu_1)_{ion} = V_1 RT\left[\frac{ic_2}{z_-}\right]. \tag{32}$$

The following substitutions may be made to relate this contribution to more useful or measurable parameters,

$$c_2 = \frac{\upsilon_{2,s}}{\bar{\upsilon}} \tag{33}$$

$$i = \frac{K_a}{10^{-pH} + K_a}. \tag{34}$$

This yields

$$(\Delta\mu_1^*)_{ion} - (\Delta\mu_1)_{ion} = V_1 RT\left[\frac{K_a}{10^{-pH} + K_a}\right]\left[\frac{\upsilon_{2,s}}{z_-\bar{\upsilon}}\right]. \tag{35}$$

The second case described by these investigators was that in which the concentration of external electrolytes is greater than ic_2 and that $c_s^*-c_s$ is of the same magnitude as the number of ions along the polymer chain. The final result for this case yields the following expression:

$$(\Delta\mu_1^*)_{ion} - (\Delta\mu_1)_{ion} = \frac{V_1 RT}{4I}\left[\frac{K_a}{10^{-pH} + K_a}\right]^2\left[\frac{\upsilon_{2,s}}{\bar{\upsilon}}\right]^2 \tag{36}$$

where I is the ionic strength of the external solution.

It must be noted that Eqs. (35) and (36) are for the case in which the crosslinks in the polymer network were introduced in solution as with the Peppas-Merrill equation for neutral hydrogels and also that a Gaussian chain distribution is assumed. The complete equilibrium expressions accounting for the mixing, elastic-retractive, and ionic contributions to the chemical potential for anionic networks in the two cases described above are then

$$V_1\left[\frac{K_a}{10^{-pH} + K_a}\right]\left[\frac{\upsilon_{2,s}}{z_-\bar{\upsilon}}\right] = \left[\ln(1 - \upsilon_{2,s}) + \upsilon_{2,s} + \chi_1\upsilon_{2,s}^2\right]$$

$$+ \left(\frac{V_1}{\bar{\upsilon}\bar{M}_c}\right)\left(1 - \frac{2\bar{M}_c}{\bar{M}_n}\right)\upsilon_{2,r}\left[\left(\frac{\upsilon_{2,s}}{\upsilon_{2,r}}\right)^{1/3} - \frac{1}{2}\left(\frac{\upsilon_{2,s}}{\upsilon_{2,r}}\right)\right] \tag{37}$$

$$\frac{V_1}{4I}\left[\frac{K_a}{10^{-pH}+K_a}\right]^2\left[\frac{\upsilon_{2,s}}{\bar{\upsilon}}\right]^2 = \left[\ln(1-\upsilon_{2,s})+\upsilon_{2,s}+\chi_1\upsilon_{2,s}^2\right]$$
$$+\left(\frac{V_1}{\bar{\upsilon}\bar{M}_c}\right)\left(1-\frac{2\bar{M}_c}{\bar{M}_n}\right)\upsilon_{2,r}\left[\left(\frac{\upsilon_{2,s}}{\upsilon_{2,r}}\right)^{1/3}-\frac{1}{2}\left(\frac{\upsilon_{2,s}}{\upsilon_{2,r}}\right)\right]. \tag{38}$$

The expressions for cationic networks are derived similarly and are shown here [15].

$$V_1\left[\frac{K_b}{10^{pH-14}+K_b}\right]\left[\frac{\upsilon_{2,s}}{z-\bar{\upsilon}}\right] = \left[\ln(1-\upsilon_{2,s})+\upsilon_{2,s}+\chi_1\upsilon_{2,s}^2\right]$$
$$+\left(\frac{V_1}{\bar{\upsilon}\bar{M}_c}\right)\left(1-\frac{2\bar{M}_c}{\bar{M}_n}\right)\upsilon_{2,r}\left[\left(\frac{\upsilon_{2,s}}{\upsilon_{2,r}}\right)^{1/3}-\frac{1}{2}\left(\frac{\upsilon_{2,s}}{\upsilon_{2,r}}\right)\right]. \tag{39}$$

$$\frac{V_1}{4I}\left[\frac{K_b}{10^{pH-14}+K_b}\right]^2\left[\frac{\upsilon_{2,s}}{\bar{\upsilon}}\right]^2 = \left[\ln(1-\upsilon_{2,s})+\upsilon_{2,s}+\chi_1\upsilon_{2,s}^2\right]$$
$$+\left(\frac{V_1}{\bar{\upsilon}\bar{M}_c}\right)\left(1-\frac{2\bar{M}_c}{\bar{M}_n}\right)\upsilon_{2,r}\left[\left(\frac{\upsilon_{2,s}}{\upsilon_{2,r}}\right)^{1/3}-\frac{1}{2}\left(\frac{\upsilon_{2,s}}{\upsilon_{2,r}}\right)\right]. \tag{40}$$

Other models that have been proposed to describe the swelling behavior of ionic hydrogels include that of Prausnitz and collaborators [16] and Konak and Bansil [17].

Figure 2 shows the influence of pK_a of the anionic gel and pH of the surrounding solution on the equilibrium degree of swelling, i.e., the reciprocal of the equilibrium volume fraction, $\upsilon_{2,s}$, as predicted by Eq. (38). Clearly the nature of the surrounding fluid affects the swelling/deswelling behavior of an ionic gel. Similarly Fig. 3 shows the influence of the ionic strength of the surrounding solution, I, on the equilibrium degree of swelling for various gels.

1.3 Mechanical Behavior of Hydrogels

The stress-strain behavior of a hydrogel membrane can be described by the rubber elasticity theory [3]. In the unstrained state, the polymer chains are in randomly coiled configurations but are able to rearrange to accomodate deformation. Thus, when subjected to an external stress, the deformation may be accomodated through rearrangement and uncoiling of the polymer chains. Usually the chains become aligned parallel to the axis of the elongation. During this process a restoring force develops within the polymer because of the tendency of the network chains to return to their original conformations. Another characteristic essential to the elastic behavior of polymer networks is the presence of sufficient internal mobility. This is needed to allow the polymer chains to rearrange during deformation and also during recovery. Thus, this elastic behavior is not observed in glassy or crystalline polymers.

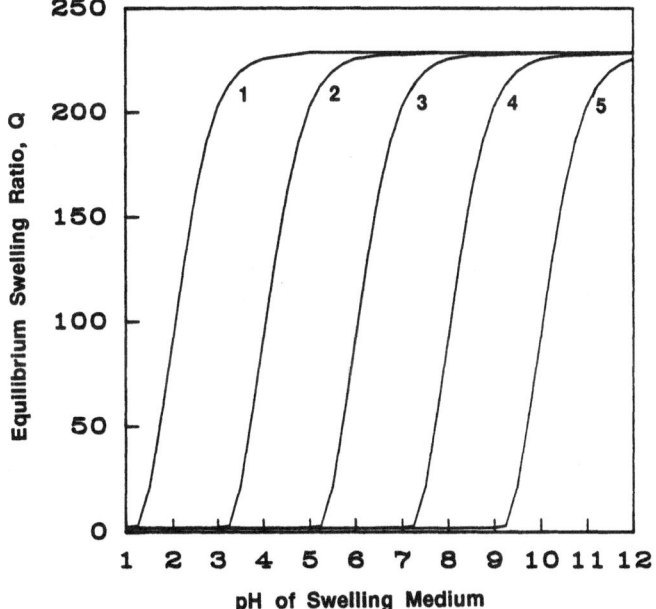

Fig. 2. Equilibrium swelling ratio of anionic polymer membranes with (1) $pK_a = 2.0$, (2) $pK_a = 4.0$, (3) $pK_a = 6.0$, (4) $pK_a = 8.0$, and (5) $pK_a = 10.0$, at comparable ionic strength, as a function of pH

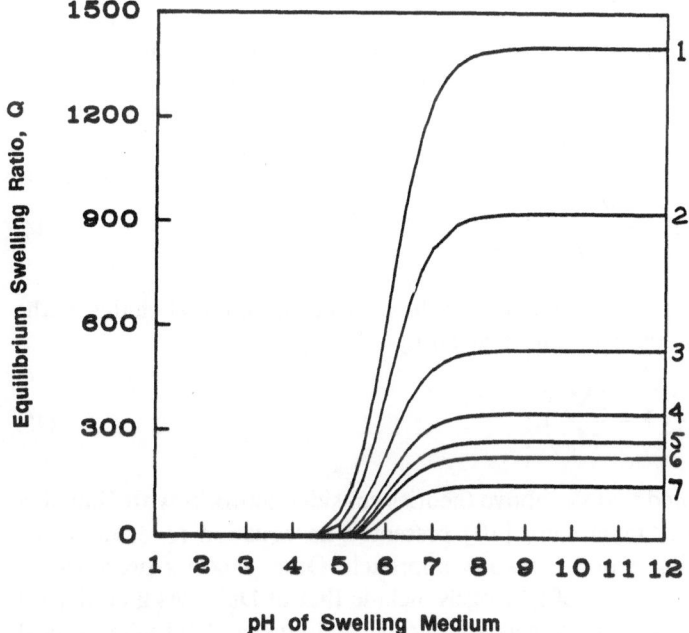

Fig. 3. Equilibrium swelling ratio of anionic polymer membranes, at comparable pK_a, as a function of pH with (1) $I = 0.05$, (2) $I = 0.1$, (3) $I = 0.25$, (4) $I = 0.5$, (5) $I = 0.75$, (6) $I = 1.0$, (7) $I = 2.0$

The shear modulus, G, describes the elastic behavior of hydrogels and is related [18] to the elastic modulus, E, by

$$E = 2G(1 + v). \tag{41}$$

Here, v is Poisson's ratio which is equal to 0.5 for elastic materials such as hydrogels. Rubber elasticity theory describes the shear modulus in terms of structural parameters such as the molecular weight between crosslinks. In the rubber elasticity theory, the crosslink junctions are considered fixed in space [19]. Also, the network is considered ideal in that it contained no structural defects. Known as the affine network theory, it describes the shear modulus as

$$G = \frac{\rho RT}{\bar{M}_c} \tag{42}$$

where ρ is the density of the polymer network. This theory was extended to describe swollen networks by including a term for the volume fraction of polymer in the swollen network, v_2,

$$G = \frac{\rho RT}{\bar{M}_c} v_2^{1/3}. \tag{43}$$

Thus the presence of a swelling agent is seen to reduce the modulus.

A subsequent theory [6] allowed for movement of the crosslink junctions through rearrangement of the chains and also accounted for the presence of terminal chains in the network structure. Terminal chains are those that are bound at one end by a crosslink but the other end is free. These terminal chains will not contribute to the elastic recovery of the network. This phantom network theory describes the shear modulus as

$$G = \frac{\rho RT}{\bar{M}_c} \left(1 - \frac{2\bar{M}_c}{M_n}\right). \tag{44}$$

Here, the term in brackets accounts for the presence of terminal chains. If the network is swollen, the modulus is given by

$$G = \frac{\rho RT}{\bar{M}_c v^{1/3}} \left(1 - \frac{2\bar{M}_c}{M_n}\right). \tag{45}$$

It should be noted that the above theories consider networks with Gaussian chain distributions. Oppermann [10] presented an equation to describe the shear modulus using a non-Gaussian approach. Other theories proposed to describe the shear modulus of hydrogels include that of Dubrovskii et al. [20] who investigated weakly ionic gels and Zrinyi and Horkay [21] who derived a phenomenological equation to describe the poly(vinyl acetate) networks investigated.

2 Environmentally Responsive Behavior

In recent years, there has been an explosion in the area of materials that respond in some physical manner to changes in their surrounding environment. Examples of this behavior include change in swelling ratio of a pH sensitive polymeric hydrogel due to a change in the pH of the swelling agent, or change in permeability of a thermosensitive membrane due to a change in temperature. These changes may be used in a variety of applications in such areas as separations, biosensors, superabsorbent materials, and controlled drug release [2]. Research on potential applications of these materials has become especially important in the areas of controlled drug delivery and membrane separations. Obviously, membranes that could separate solutes of different sizes under different conditions or drug delivery systems that could respond to changes in their environment, delivering drugs at varying rates as necessary, would be invaluable.

2.1 Materials Sensitive to the Nature of the Surrounding Solution

As early as 1950, researchers investigated the development and use of environmentally sensitive materials. Some of the earliest studies involved materials that responded to the nature of the surrounding solution. Factors influencing this type of behavior include the pH, ionic strength, and chemical composition of the surrounding solution. Kuhn et al. [22] noticed that the shape of ionizable polymeric molecules, specifically polyacids and polybases, depended on the degree of ionization of the molecular chain. They observed that by ionizing the carboxyl groups of a polyacid they could cause the originally coiled molecule to expand because of the electrostatic repulsion produced along the main chain. They studied this behavior using aqueous solutions of poly(methacrylic acid) (PMAA). About 50% ionization was sufficient to cause full length molecular extension, while this uncoiling caused a substantial increase in specific viscosity. This behavior was reversible, that is, when acids or neutral salts were added the extended molecules contracted and specific viscosity decreased. These results have recently been further confirmed by Liu et al. [23] who verified these conformational changes as a function of backbone ionization using a spectroscopic ruler technique. Another effect has been observed for PMAA in methanolic solutions on titration with lithium methoxide. Ionized molecules collapsed upon addition of the salt because the salt was more soluble in PMAA than in the methanol solvent [24].

Kuhn et al. [22] extended their studies to crosslinked networks of poly(acrylic acid) (PAA) and found that these gels swelled to a high degree in alkaline environment and contracted rapidly when acid was added to the surrounding solution. Dilations and contractions of the order of 300% were observed. They were reversible and could be repeated at will. A contracting and expanding gel was able to lift and lower a load, thus illustrating the concept of converting chemical energy to mechanical work.

During this time, Katchalsky [25] also worked with PMAA crosslinked with a low percentage of divinyl benzene. This polymer network also swelled to a high degree in basic solution and contracted or deswelled upon addition of acidic solution. This behavior was explained in a manner similar to that of Kuhn et al. [22]. When the polymeric acid chains were ionized by the presence of the basic solution they extended because of electrostatic repulsion of the carboxylate ions, and the gel could swell to a high degree. When an acidic solution was added, the carboxylate ions became neutralized and began to contract thus pushing out the imbibed water. Katchalsky et al. [26–28] presented an extensive theoretical treatment of the behavior observed in these solutions and gels. Thus, with these early investigations, the first synthetic examples of mechanochemical systems were produced, and the idea of synthetic materials responding to their surrounding environment was born.

Katchalsky and collaborators [29] later worked on the idea of using environmentally sensitive behavior to turn chemical energy into mechanical work. Upon finding that partially crosslinked collagen fibers could be used to lift and lower a weight several thousand times the weight of the fibers alone by treatment with lithium bromide, potassium thiocyanate or urea, they produced an "engine" which functioned on the basis of the resulting difference in chemical potential and directly produced mechanical work.

In more recent years, significant research has been done using materials that are sensitive to the chemical nature of their surrounding solution. Ohmine and Tanaka [30] observed that a discrete volume change occurred in ionized polyacrylamide gels upon varying the salt concentration in acetone-water solutions. The critical salt concentration depended on the valence of the positive ions in the salt. Without salt, the gels swelled in solutions of less than 65% acetone by volume and shrank in solutions above 65% acetone. At 65% acetone, the volume of the gel changed discontinuously by 350%. When NaCl was added to the solutions, this volume transition was observed at concentrations of acetone below 65%. With added salt, the gels showed volume transitions in solutions from 0–50% acetone, and as the percent of acetone in the solution approached 65% the amount of NaCl needed to cause the volume change dropped. At 50% acetone, 0.04 mol/l of NaCl was needed to observe the transition while at 60% acetone concentration, only 0.001 mol/l of NaCl was required. To investigate the effect of the valence of the salt ions, $MgCl_2$ was also added to the gel solutions. The gel collapsed at much lower concentrations of $MgCl_2$ than that of NaCl. This phenomenon was explained by using the Flory-Huggins theory for osmotic pressure of gel networks, and relating this behavior to the contraction and relaxation of muscles. Divalent Ca^{+2} ions play a role in muscle contraction and relaxation. Thus, the difference in the osmotic pressure of divalent and monovalent ions in the muscle may be important in muscle contraction.

Tanaka and collaborators [31] further investigated the polyacrylamide behavior. Based on the Flory-Huggins equation for osmotic pressure, they showed that one of the crucial parameters that caused the discontinuous collapse of the gels was the degree of ionization. In order to investigate this theory gels were selectively ionized by substituting molecules of an ester for acrylamide molecules. These gels were incubated at pH 9 for a short time. This time period was sufficient to hydrolyze virtually all of the ester molecules with no effect on the acrylamide groups leading to ionized acrylic acid. Thus the final gel was identical to a pure polyacrylamide gel that had been partially hydrolyzed and which had an ionization fraction which ideally equaled the mole fraction of the incorporated ester. As the molar fraction of incorporated ester, i.e. the degree of ionization, increased the volume collapse of the network became more discrete.

More recently, Kokufuta and Tanaka [32] have used this idea of environmental sensitivity to produce gels that undergo a discontinuous volume change due to the concentration of the products of an enzymatic reaction. The enzyme liver esterase, which hydrolyzes ethyl butyrate into ethanol and butyric acid, was immobilized in a poly(N-isopropyl acrylamide) P(NIPA) gel. As substrate molecules entered the gel, the enzyme-catalyzed reaction began to take place within the gel. The change in solvent composition within the gel, as substrate was converted into products, changed the phase equilibrium. At a certain concentration, there was a collapse of the gel and thus a termination of the reaction. Other aspects of this work included the study of the visible patterns which develop in gels as they swell or shrink [33] and development of gels which show a volume transition in organic solutions [34].

We have worked in the area of applications for pH sensitive materials, particularly in the area of the controlled release of drugs. Our group's work focuses on physiologically responsive hydrogels, and reviews of this topic have been presented [2, 35].

Brannon-Peppas and Peppas [36] have done work with pH sensitive hydrogels for use in controlled drug delivery. They have investigated the swelling behavior of poly(2-hydroxyethyl methacrylate) (PHEMA) and copolymers of HEMA with methacrylic acid (MAA), maleic anhydride (MAH), and isopropyl acrylamide (IPAc). Under cyclic pH conditions, that is when the gels were initially placed in a pH of 10 and then placed in a pH of 2 after 335 minutes and then back in a pH of 10 at 440 minutes, all of the gels showed increased swelling in the basic solution followed by a drop in swelling when placed in acidic solution and then an increase in swelling when returned to a basic medium (see also Fig. 4). The pure HEMA gels showed this dependence to a much lower degree than the copolymers of HEMA and IPAc and copolymers of HEMA and MAH. Copolymers of HEMA and MAA responded with the greatest volume changes of any of the copolymers studied.

This time-dependent response to pH of the swelling medium was modelled with a Boltzmann superposition type model shown in Eq. (46):

$$Q(t) = \left[1 + \int_0^t L(t - \tau) \frac{\partial I(\tau)}{\partial \tau} \, d\tau \right]^3.   \qquad (46)$$

This model describes the swelling ratio ($Q(t)$) in terms of a step function

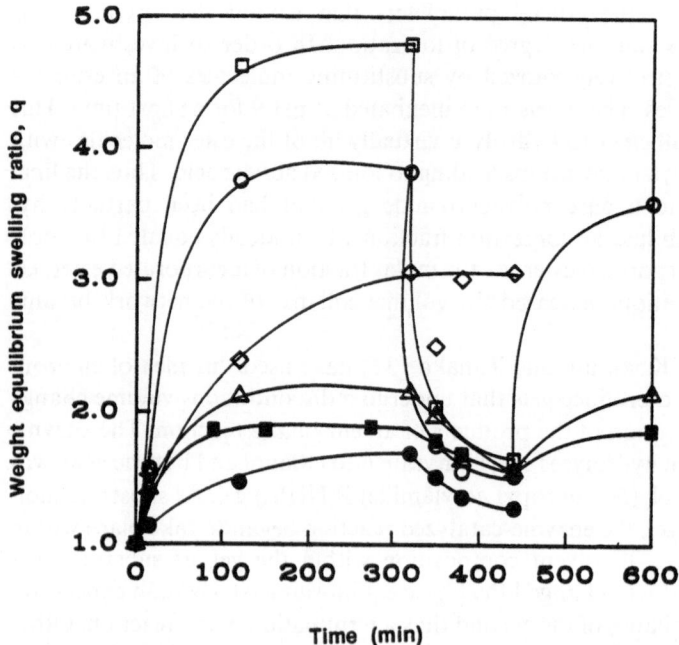

Fig. 4. Dynamic swelling of HEMA-containing polymers with 10 mol% MAA (○), 20 mol% MAA (□), 70 mol% NVP (◇), 40 mol% IPAc (△), 20 mol% MAH (●) and 100 mol% HEMA (■) at 25 °C. All samples were placed in pH 10.0 buffer at t = 0 min, pH 2.0 at t = 335 min and pH 10.0 buffer at t = 440 min

$(\partial I(\tau)/\partial\tau)$ which could be a change in ionic strength or H^+ concentration of the medium and a characteristic parameter of the polymer/swelling agent system termed the mechanochemical compliance because it describes how the system converts a chemical input (change in ionic strength, etc.) to a mechanical output (swelling or deswelling).

The equilibrium swelling behavior of these materials was also studied [37] and a model was derived which describes the swelling behavior of ionic networks. This model was based on the realization that when an ionic polymer is placed in a swelling agent, three contributions to the free energy govern the swelling behavior. These terms include the mixing, elastic-retractive, and ionic contributions. They may be described using thermodynamics and rubber elasticity theory. This model predicted the pH dependent swelling behavior seen in experiments fairly accurately with the predicted swelling being slightly higher than that seen in most experiments. Using this model, it was concluded that the swelling of these gels was due to a balance between the pH of the swelling medium, the pK_a of the network, and a number of structural parameters of the polymer gel. A full treatment of the equations was developed [38] that may be used to analyze the swelling of ionic gels. Equations were presented for use with anionic gels such that the ion concentration outside of the gel is much lower than that of the inside, for gels with comparable ionic concentrations inside and outside of the gel, for various distributions of chain lengths, and for different methods of crosslinking. Experimental results of the swelling of PHEMA gels, P(HEMA-*co*-MAA) gels were analyzed with these equations (see Fig. 5).

In addition, Brannon-Peppas and Peppas [39] studied the release of bioactive agents such as theophylline and proxyphylline from P(HEMA-*co*-MAH) and P(HEMA-*co*-MAA) gels. The drug was loaded into the polymer by allowing the polymer to swell in concentrated buffer solutions of the drugs. Because of increased swelling at higher pH values, they found that more drug was incorporated when loading was done at higher pH. Drug release from these gels was found to be constant over 2 h with release being faster from the P(HEMA-*co*-MAA) gels. In alkaline solutions, where the gel swelled to a higher degree, the drug release was thought to be an equal function of drug diffusion and macromolecular chain relaxation while release in acidic environment was a strong function of drug diffusion.

Khare and Peppas [40] used differential scanning calorimetry to investigate the structure of water in pH sensitive hydrogels of P(HEMA-*co*-MAA) and

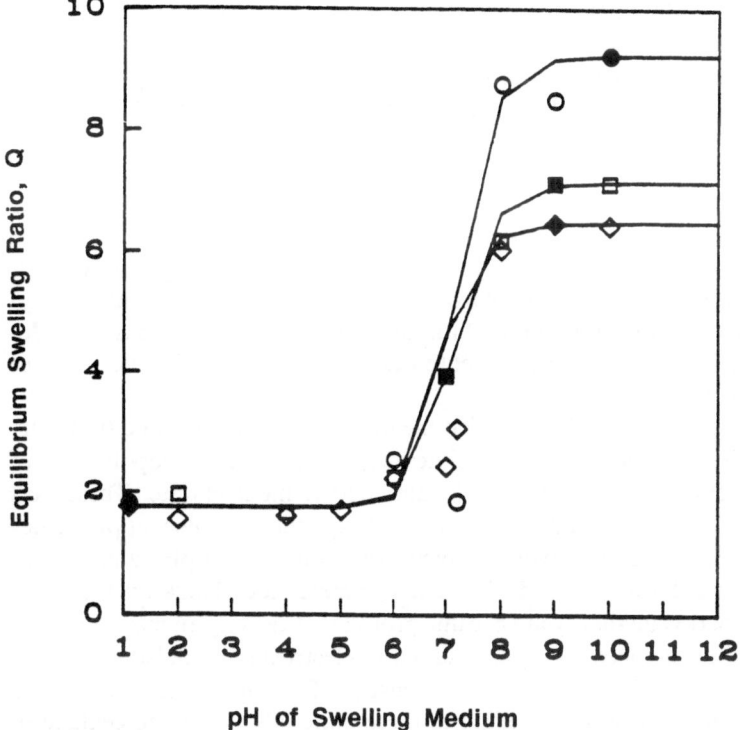

Fig. 5. Equilibrium swelling data for copolymers of P(HEMA-*co*-MAA) containing 10 mol% HEMA (○), 60 mol% HEMA (□), and 80 mol% HEMA (◇) as a function of the pH of the swelling medium along with predictions of that swelling behavior using Eq. (38)

P(HEMA-*co*-AA) as a function of solution pH and ionic strength. In a hydrogel, the water is essentially in three forms. Bound water is strongly associated with the polymer chains through hydrogen bonding or polar interactions. Free, or bulk, water has the same physical properties as normal water and is not attached to the macromolecular chains. Interfacial water has characteristics in between those of bound and free water and is characterized by ice-like configurations around the hydrophobic groups on the polymer chains. As mentioned, these anionic gels have been seen to swell to a much higher degree in basic solutions as opposed to acidic solutions. It was found [40] that the amount of bound water in the gels did not change as a function of pH but rather the increased swelling ratios at higher pHs were due to an increase in the amount of free water in the gels. Free water and total water contents were found to be highest in NaCl solutions as opposed to $CaCl_2$ solutions. This was attributed to the higher counterion valence of the $CaCl_2$. Every calcium ion was able to bind to two polymeric carboxylate ions and thus reduce the electrostatic repulsion within the gel more effectively than the sodium ions.

The pH sensitivity of cationic gels has also been investigated. Hariharan [41] synthesized poly(diethylaminoethyl methacrylate-*co*-hydroxyethyl methacrylate), poly(diethylaminoethyl acrylate-*co*-hydroxyethyl methacrylate), and poly(methacrylaminopropyl ammonium chloride-*co*-hydroxyethyl methacrylate) hydrogels. These polymers were unique from those mentioned above because they contained cationic pendant groups which resulted in pH sensitive behavior (see Fig. 6). These gels were highly swollen under acidic conditions and collapsed under basic conditions. Thus solutes could diffuse under acidic conditions. The swelling, solute release, and mechanical behavior of these copolymers were characterized as a function of pH and ionic strength. Many of the same trends seen with anionic networks were seen with these cationic networks. Similarly, Peppas and Foster [42] characterized the swelling behavior of cationic copolymers of poly(hydroxyethyl methacrylate-*co*-methacryloylaminopropyl trimethylammonium chloride). Hariharan and Peppas [41] derived an extensive mathematical model to describe the pH sensitive behavior of cationic networks.

Siegel and Firestone [43, 44] studied the pH sensitive swelling and transport behavior of hydrophobic polyelectrolyte gels. They studied copolymers of methyl methacrylate and *N*,*N*-dimethylaminoethyl methacrylate. These gels were collapsed at pH values above 6.6 and were expanded (swollen) at pH values below this point. Thus, their hydrophobicity was dominant at pH values above 6.6, and as the pH was lowered, the tertiary amine side chains on the *N*,*N*-dimethylaminoethyl methacrylate became protonated causing an increase in the charge density of the network. This resulted in electrostatic repulsion between these groups and swelling of the network ensued. The copolymer ratios in the gels also had an effect on the swelling of these materials. As the percentage of methyl methacrylate increased, the swelling transition shifted to lower pH values and the overall degree of swelling was decreased. It was also found in one study [45] that no caffeine was released at or above pH 7 while drug release

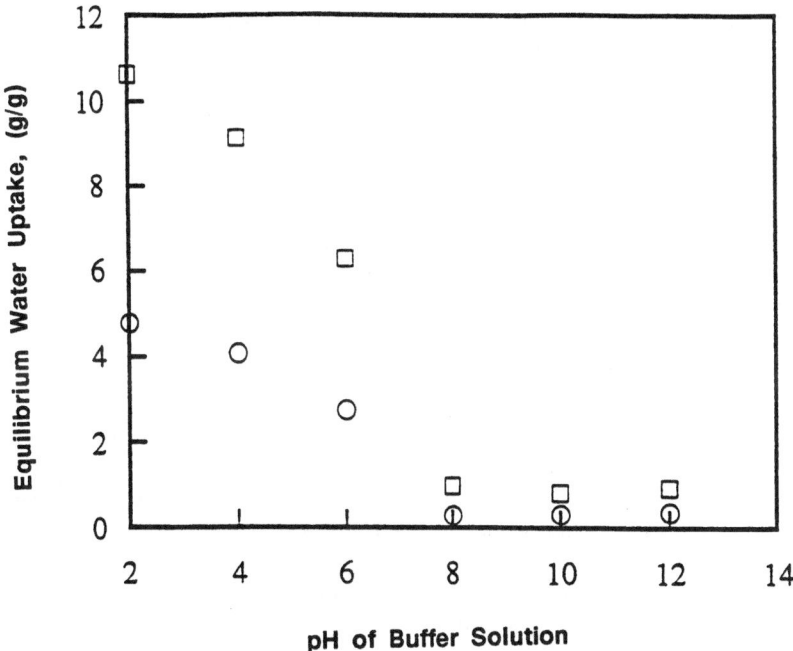

Fig. 6. Equilibrium water uptake of P(DEAEA-*co*-HEMA) samples with 30% (○) and 60% (□) DEAEA with a crosslinking ratio of 0.005 mol EGDMA/mol monomers in a citrate-borate buffer solution as a function of pH

occurred steadily for up to 250 h in pH of 3 and 5. The release in pH 3 was faster than that at pH 5. These results suggested that the release of caffeine from these gels was mainly controlled by the moving swelling front within the polymer gel as the gel swelled to a greater extent at lower pH.

A unique application for pH sensitive materials was studied by Tirrell [46]. He worked with phosphatidylcholine surfactant vesicles that formed synthetic vesicles in the center of which some bioactive agent could be solubilized. These materials could be triggered to release their contents through pH changes. They achieved this by anchoring poly(2-ethylacrylic acid) molecules to the hydrophilic heads of the surfactants such that they would be like a coating on the outside of the vesicles when the surfactants were aggregated. Because poly(2-ethylacrylic acid) is a polycarboxylic acid that exhibits a conformational change from an expanded to a contracted state upon transfer from base to acid, the vesicles produced were pH sensitive. Initially the surfactants formed large aggregates or vesicles in a basic solution; then, upon a decrease in pH, the poly(2-ethylacrylic acid) molecules collapsed causing a break-up of the large vesicles into many small micelles. When this occurred, any bioactive agent isolated within the vesicles would be released. In subsequent work, fluorescence and optical density techniques [47] were used to verify the transport mechanism.

Gehrke and Cussler [48] investigated mass transfer in gels that were sensitive to the pH of their surrounding environment. These gels were copolymers which contained acrylamide as the primary monomer with sodium methacrylate added to make the gel a polyelectrolyte. The swelling and collapse of the gels under basic and acidic conditions respectively were found to be reversible. Upon analysis, hydrochloric acid diffused into the gel much faster than the rate of gel collapse. This indicated that the rate-limiting factor during collapse was the rate of diffusion of water being exuded out of the gel and not the rate of ion exchange. On the other hand, the rate of swelling was equal to the rate of sodium hydroxide uptake indicating that the rate of swelling was limited by the rate of ion exchange. Ultimately, the rate of gel collapse was controlled by diffusion while the rate of swelling could be controlled by different mechanisms but was usually controlled by ion exchange.

Membranes that respond to the nature of the surrounding solution have potential applications in bioseparations. Bromberg [49] studied the pH sensitivity of composite membranes of nitrocellulose acetate ultrafilters impregnated with pH sensitive polyelectrolytes such as PAA. Below a pH of 6.5, the PAA was in its collapsed state. Thus it did not affect diffusion through the pores of the membranes. Above a pH of 6.5, the PAA in the pores swelled to a high degree and clogged the pores with dense hydrogel causing an abrupt decrease in the hydraulic permeability of the membrane. Bromberg [50] also found that at a pH above 6.5, when the carboxyl groups on the PAA were ionized, there was an excess of effective negative charges on the composite membrane. This effective charge resulted in a preferential permeability of potassium cations through the membrane as opposed to chloride anions.

In a system comprised of a straight pored polycarbonate membrane with PMAA grafts, Ito et al. [51] achieved similar pH sensitivity. The PMAA-grafted membranes showed a reproducible change in water permeability through a transition around pH 4. Above this pH, the PMAA grafts began to extend and interact sterically with one another in the pores of the membrane. This caused the pores to become blocked and permeability was decreased. The rate and pH response decreased with the amount of the grafts incorporated. The effect of the length of the PMAA chains was also investigated. Shorter chains were not able to block the pores of the membrane and thus did not affect the permeability of the membrane with pH. On the other hand, grafts of very long PMAA chains could cause blockage of the pores at high pH but, when the pH was lowered, the chains could not fully contract because they had become entangled during expansion.

Osada et al. [52] have also grafted PMAA to a porous substrate in order to achieve an environmentally sensitive membrane. This sensitivity was termed a "chemical valve function" because mechanochemical forces caused the pores to enlarge and contract. PMAA was grafted onto poly(vinyl alcohol) (PVA) films which had a mean pore radius of 4 μm. The water permeation of the membrane was strongly affected by the conformational state of the PMAA grafts. At low pH the chains were contracted and the water permeability was high while at high pH the chains expanded, covered the pores and lowered the water permeability. It was also shown that chelation of the PMAA with metal ions changed the permeability of the membrane. An increase in water permeability was seen in a solution containing a low concentration of Cu^{+2} and this effect was more pronounced than that seen with Na^+. Indeed, permeability was increased more by various divalent and trivalent metal ions than by monovalent metal ions. Addition of certain polymers that interacted with PMAA to form polymer- polymer complexes could also increase the permeability of the grafted membrane. Poly(ethylene glycol) (PEG) is an example of a polymer that forms a polymer-polymer complex with PMAA. Osada et al. [52] showed that addition of PEG caused an increase in the water permeability of these graft membranes due to this complexation. Albumin and γ-globulin were also able to form complexes with the PMAA grafts and thus affected the water permeability of the membrane.

2.2 Materials Sensitive to Temperature

Environmentally sensitive materials can also respond to the surrounding temperature. As before, materials that can respond to changes in the surrounding temperature have potential applications in a variety of fields, particularly those of controlled drug delivery and membrane applications. Tanaka [53] investigated the effect of temperature on polyacrylamide gels by fixing the acetone concentration in the swelling medium at about 40% and changing the temperature. The behavior of the gels was similar to that seen by changing acetone concentration. At temperatures above room temperature, the gels swelled. At room temperature the gels collapsed and remained collapsed while continuing to shrink below room temperature. These polyacrylamide gels were undergoing a coil-globule transition because of the reversible collapsing and expanding behavior they exhibited at specific acetone concentration and at specific temperature. The hydrodynamic radius of the polyacrylamide chains was determined as 500 Å at low acetone concentrations using laser light scattering spectroscopy. Around an acetone concentration of 39% the radius showed a sharp decrease to 200 Å and remained constant when the acetone concentration was increased further. Thus, a sharp coil-globule transition was seen.

Copolymer gels of N-isopropylacrylamide (NIPA) undergo phase transitions as a function of temperature in pure water solutions [54]. It was found that PNIPA gels exhibited a lower critical (or consulate) solution temperature or an LCST. When a crosslinked temperature sensitive gel is heated above its LCST, the gel shrinks dramatically. This behavior is reversible and occurs over a narrow temperature range. Pure PNIPA gels without ionic groups exhibited a continuous volume transition around 34.3 °C. In contrast, gels containing ionic groups showed a discontinuous transition. The transition temperature increased as the ionic concentration increased. Beltran et al. [55] observed similar results in their work with PNIPA gels containing a small amount of ionic comonomer. They modelled this behavior using an oriented-quasichemical model which had previously been applied to uncharged PNIPA gels [56].

Hydrogels of N-isopropylacrylamide and methacrylic acid copolymers (P(NIPA-co-MAA)) exhibited reversible shrinkage and swelling which is temperature and pH dependent [57]. Absorption and release of vitamin B12 was studied. Homopolymers deswelled sharply between 30 and 40 °C, and the release of vitamin B12 was controlled by the collapse and expansion of the gel. P(NIPA-co-MAA) copolymers showed a higher degree of swelling in the expanded state because of the hydrophilicity of the methacrylic acid groups. Dong and Hoffman [58] have also worked with copolymers of NIPA and acrylamide, which exhibit a LCST around 31–33 °C. Such gels were proposed for enzyme immobilization leading to thermally reversible shrinking and swelling. This could allow the enzyme activity to be switched on and off since the substrate diffusion would be determined by the gel pore size. Thus, a thermally reversible gel could be used to regulate reaction rates. Indeed, the immobilized enzyme activity increased with temperature until the LCST region was reached. Here the enzyme activity was effectively shut off by the gel collapse because the loss of pore water from the gel retarded or eliminated diffusion of the reactants and products in and out of the gel.

Copolymers of NIPA and N-acryloxysuccinimide [59] exhibit an LCST in the range of 30–35 °C and contain functional groups which readily react with the lysine residues of proteins. This characteristic allows for the conjugation of proteins such as immunoglobins like human IgG. Using this copolymer-Ig conjugate, these researchers developed a novel polymer based immunoassay. At room temperature, the polymer-Ig conjugate sample was placed in a solution of the appropriate flourescein isothiocyanate labeled antigen. After the temperature was raised and the network collapsed, the gel was analyzed using a fluorimeter to determine the concentration of the antigen.

Thermal and pH sensitive heterogeneous copolymer hydrogels which contain silicone rubber domains within a temperature and pH sensitive copolymer of NIPA and acrylic acid have been synthesized by Dong et al. [60]. These materials contained macropores when swollen and collapsed much faster than homopolymers of N-isopropylacrylamide. Biocatalyst immobilization using copolymers of NIPA and NN'- dimethylaminopropylmethacrylamide have also been studied [61].

A synthetic method to produce PNIPA gels that could respond to temperature changes much faster than conventional gels has been recently achieved [62]. Polymerization of the monomer with a crosslinking agent was initiated below the LCST and prior to gelation the solution was heated above the LCST causing microscale phase separation. The polymerization was then continued, and the phase-separated microstructure became permanent. These gels could expand 120 times faster and contract 3000 times faster than comparable homogenous PNIPA gels.

Kim and collaborators [63-67] have done extensive work on the use of temperature sensitive hydrogels as membranes. Copolymers of NIPA and butyl methacrylate (BMA) and interpenetrating networks of PNIPA with poly-(tetramethylene ether glycol) (PTMEG) have been studied [63] for use in controlled drug release and drug permeation through membranes. The effect of copolymerizing PNIPA with BMA on the thermal sensitivity of these materials was found to be a lowering of the temperature where gel shrinkage began and a change of the volume transition from a discontinuous shrinkage to a gradual deswelling upon heating. This effect was more prominent as the percentage of BMA in the copolymer was increased. Interpenetrating networks of PNIPA with PTMEG showed a different effect in that the transition temperature for the gels was the same for all compositions of PTMEG. For both the copolymers and the interpenetrating networks in this study, the outer surface of the gel was the first area affected when the temperature increased past the transition. This outer layer formed a dense skin on the surface of the gel and retarded the flux of water out of the gels. This presented the possibility for use of these materials as membranes with variable permeability depending on temperature or as thermosensitive drug delivery systems.

Okano et al. [63] performed permeation experiments with insulin and glucose for membranes of the P(NIPA-co-BMA) gels. They found that both insulin and glucose reached constant permeation rates fairly quickly and upon sudden temperature change above the transition, the permeation of both substances was completely stopped. This permeation control was attributed to the formation of the dense skin layer as the polymer swelled. Similar results were seen with indomethacin by Yoshida et al. [68] who studied copolymers of NIPA with alkyl methacrylates of different alkyl side chain lengths in order to see this effect on drug permeation. Longer side chain lengths were found to form a skin layer faster than samples containing shorter side chains. This resulted in lower permeation rates through membranes containing comonomers with longer side chain lengths. Drug release control due to this skin layer in P(NIPA-co-BMA) gels was also investigated by Kim and coworkers [63]. Indomethacin was loaded into the copolymers and the interpenetrating networks mentioned earlier, and release studies were performed. Both types of gels exhibited high release at low temperatures and a blockage of release as temperature was increased above the transition. This behavior was attributed to the formation of this skin layer as well as to the difference in drug solubility at different temperatures.

Heparin release from gels of P(NIPA-co-BMA) and P(NIPA-co-AA) has been investigated [64]. Since BMA is a hydrophobic comonomer and acrylic acid is a hydrophilic comonomer, incorporation of these materials into the NIPA gels produced copolymers with different swelling behavior. P(NIPA-co-BMA) gels showed a lower temperature of gel collapse than pure PNIPA gels while P(NIPA-co-AA) gels showed a higher transition than the homopolymer. In comparison of the swelling behavior of these gels, the formation of a skin layer upon swelling, as mentioned above, was seen with the P(NIPA-*co*-BMA) gels while the P(NIPA-*co*-AA) gels did not exhibit this phenomenon. The release of heparin was studied as a function of the loading temperature for both types of gels. For the P(NIPA-*co*-BMA) gels, a lower loading temperature resulted in a higher amount of drug released. Gels loaded at 1 °C also showed more rapid release rates than those loaded at 15 °C while those loaded at 20 °C showed no heparin release. On the other hand, the P(NIPA-*co*-AA) gels showed similar release kinetics for all loading temperatures with slight increases in release rates with loading temperature. Again, the differences between the two copolymers was attributed to the effect of the skin barrier that forms in the case of the copolymer containing BMA and does not form in the case of the AA copolymers.

Use of P(NIPA-*co*-BMA) gels as thermosensitive separation membranes has also been researched by Feil et al. [65]. The gels were prepared as membranes and separation studies were done using uranine and dextrans of various molecular weights. Permeability of the membranes for uranine, which has a molecular weight of 376, and dextran of molecular weight 4400 was seen to be highly temperature dependent with high permeability at low temperatures and a gradual decrease in permeability up to about 27 °C where permeation of solutes was completely blocked. Uranine showed the highest permeability at all temperatures. Dextran of molecular weight 150 000 could not permeate the membrane at any temperature. Separation studies revealed that at 25 °C after 100 h, 94.3% of the uranine had permeated the membrane while only 3.6% of the dextran-4400 had passed through the membrane. After 100 h the temperature was decreased to 20 °C and permeation of dextran-4400 was seen. These results were analyzed using free volume theory. Bae et al. [66] have also studied insulin permeation through hydrogels made from poly(*N*-acryloyl pyrrolidine-*co*-hydroxyethyl methacrylate) (P(AP-*co*-HEMA). These hydrogels showed thermal sensitivity similar to the P(NIPA-*co*-BMA) gels mentioned previously. Insulin permeation through these gels fell to less than half its original value between temperatures of 27 and 37 °C as the volume transition of the gel took place.

Terpolymer gels of poly(*N*-isopropylacrylamide-ter-butylmethacrylate-ter-diethylaminoethyl methacrylate) (poly(NIPAAm-*ter*-BMA-*ter*-DEAEMA) that exhibit temperature and pH sensitivity have also recently been studied by Feil et al. [67]. These gels showed pH and temperature sensitivity in their swelling behavior because of the presence of ionizable groups on the DEAEMA comonomer. At low pH the gels were highly swollen because the DEAEMA was ionized, while at high pH the swelling greatly decreased because the DEAEMA was not ionized. This pH sensitive behavior was sensitive to temperature in that as temperature was increased the transition pH was lowered.

Thermally reversible gels based on NIPA polymers and copolymers of *N,N*-diethylacrylamide and sodium methacrylate collapsed abruptly upon heating at 33 °C and 55 °C, respectively [69]. These materials could be used in separation processes. Collapsed gel samples were added to a solution which contained a protein or other macromolecule and various small solutes. The gel swelled, absorbing water and the small solutes but not the large protein. The excluded protein could be found in a concentrated solution. The swollen gel could then be recovered and warmed so as to shrink and release the absorbed solvent. The collapsed gel could then be recovered for re-use. This process was fairly efficient, and separation efficiency increased with crosslinking of the gel network. This system was modified [70] by producing flat sheets of the temperature sensitive gels. They bonded the crosslinked polyisopropylacrylamide to a polyester film such that they were able to absorb water but exclude large molecules on the hydrophilic side of the film and warm and cool the gel from the hydrophobic side. A molecular theory was presented [71] for the volume changes in temperature sensitive gels based on a compressible lattice model. The model included free volume as a third component in the system, and the phase behavior predicted upper and lower consulate boundaries. Saito and his coworkers [72] also worked on a theoretical model for the thermally induced volume changes seen in these gels. Their model considered hydrophobic interactions as an important part of the observed volume transitions. They divided the free energy of the gel into four parts which included terms describing the free energy of elasticity of the network, the free energy arising from the osmotic pressure of dissociated counterions, the free energy of interactions excluding the hydrophobic interaction, and the free energy of the hydrophobic interaction.

2.3 Materials Sensitive to the Presence of Glucose

Numerous investigators have studied environmentally sensitive gels for use in insulin release. This is a significant area of research because of its potential in the treatment of diabetes. For an effective insulin controlled release system, one needs to control the insulin release with respect to the amount of glucose in the bloodstream [73]. Ishihara et al. [74] reported on insulin permeation through 2-hydroxyethyl methacrylate-methyl methacrylate P(HEMA-co-MMA) copolymers. They found increased permeation with increased percentage of HEMA in the copolymer because of the tendency of HEMA to swell to a large extent in water. The permeability of insulin was slightly decreased when other blood proteins were in the surrounding environment, probably due to the complexation between insulin and the proteins.

Ishihara et al. [73, 75] produced membranes by combining a pH sensitive membrane of poly(diethylaminoethyl methacrylate-co-2-hydroxypropyl methacrylate) and a membrane containing immobilized glucose oxidase; the membranes could be induced to control the permeation of insulin by the amount of glucose present. The glucose oxidase acted as a glucose sensor and produced gluconic acid by an enzymatic reaction with glucose. The gluconic acid product lowered the pH of the medium resulting in a swelling of the membrane because of its pH sensitivity. Thus the glucose oxidase "sensed" the amount of glucose in the medium and as it reacted with the glucose it caused the pH sensitive portion of the membrane to control the overall permeability.

For the same purpose porous cellulose membranes have been tested with surface-grafted PAA chains [76] resulting in pH sensitive membranes. The enzyme glucose oxidase, which catalyzes the oxidation of glucose to gluconic acid, was immobilized onto the PAA-grafted cellulose membrane. Thus, at neutral pH in the absence of glucose, the PAA grafts were fully extended because of the ionization and electrostatic repulsion of the carboxyl groups. These extended chains blocked the pores of the cellulose membrane. Then, in the presence of glucose the enzymatic reaction occurred and the resulting gluconic acid protonated the PAA and caused the graft chains to coil because of the reduced electrostatic repulsion. These investigators measured insulin permeability through these membranes in the presence and absence of glucose. Insulin diffused through the membrane in the absence of glucose but the permeability increased 1.7 times in the presence of glucose. These results seem fairly substantial since the pH difference in this system in the presence and absence of glucose was only 4 pH units (from 7.2 in the absence of glucose to 6.8 in the presence of glucose).

Chung et al. [77] achieved a more efficient on-off insulin delivery response by immobilizing insulin on a poly(methyl methacrylate) (PMMA) membrane through a disulfide bond and glucose dehydrogenase (GDH), an enzyme which oxidizes glucose. In the presence of glucose, the GDH oxidized the glucose and produced electrons which in turn reduced the disulfide bonds and allow the insulin to be released from the membrane. This system showed a fast response to glucose and the insulin released was biologically active. Other interesting work with glucose responsive systems includes that of Heller et al. [78], Kim et al. [79], and Horbett et al. [80].

2.4 Materials Sensitive to Other Factors

Materials that are sensitive to environmental factors other than solvent composition and temperature have been developed. For example, Doi et al. [81] studied the deformation of ionic gels in buffer solutions when an electric field is applied. They reported the cyclic bending-straightening motion of a PVA-PAANa composite hydrogel in electrolyte solution under a sinusoidally varying electric field [82]. Shinkai et al. [83] have reported the conformational changes similar to the motion of a butterfly in photoresponsive crown ethers in alternating dark-light cycles. Irie and Kuwatchaku [84] have observed similar results with photoresponsive acrylamide gels containing leucocyanide groups. Upon ultraviolet irradiation these gels swelled 13 times their initial weight in water. In the absence of the ultraviolet light, the gels deswelled to their initial size. Irie [85] has also reported bending of acrylamide gels in the form of rods when irradiated with ultraviolet light. Environmental sensitivity of such gels has been achieved by cation/anion density [86] and by interaction with charged surfactants [87].

Thus there has been an abundance of work in recent years in the area of environmentally sensitive materials. Examples of this behavior have been shown to include the change in swelling of polymeric hydrogel or the change in the extension or coiling of a polymer molecule due to a change in the pH of the surrounding solution, the change in permeability of a thermosensitive membrane due to a change in temperature, or the change in drug release due to the presence of enzymatic reaction products. A wide range of behavior is possible with materials that are sensitive to some aspect of the surrounding environment and, because of this, the possibilities for unique applications of these materials are extremely diverse.

3 Interpolymer Complexation

Interpolymer complexes or polymer-polymer complexes are formed by the association of various macromolecules [88]. Complex formation is often sensitive to the surrounding environment. Such associations are often accompanied by a change in conformation, shape, or hydrophilicity and thus have potential use in the conversion of chemical energy into mechanical work. This could be applied in such areas as chemical engines, actuators, biosensors, separations, drug delivery, and underwater or space energy supply systems [89]. Interpolymer complexes are divided into four major categories depending on the dominant type of interaction. These categories are the following:

1. stereocomplexes which are due to van der Waals forces;
2. polyelectrolyte (or polyionic) complexes which form between macromolecular acids and bases or their salts and are stabilized by ionic bonds;
3. complexes formed by hydrogen bonding between polyacids and non-ionic polymers; and
4. coordination complexes.

There are complexes which exhibit a combination of these phenomena as well as complexes of one of these types that are stabilized by other forces such as hydrophobic interactions.

Interpolymer complexes often display physical properties very different from their individual constituent molecules. As a result, these systems have been studied using a variety of techniques such as potentiometry, conductimetry, turbidity, viscometry, calorimetry, sedimentation, light scattering, spectroscopy, and electron microscopy [88].

3.1 Polyelectrolyte Complexes

Polyelectrolytes can be used to investigate intermolecular interactions because parameters such as ionic strength and concentration can be changed to monitor reactions between oppositely charged macromolecules. Upon formation of a polyelectrolyte complex, there is also the formation of a low molecular weight acid or base. The polyelectrolyte complex is also insoluble in water because the charged groups responsible for solubility are involved in the complex. Also, these complexes form in a narrow pH range. These traits allow for investigation of the complexation behavior by a variety of techniques. To form a polyelectrolyte complex, interaction between sufficiently long chains is needed and the interaction occurs in a "zipping" fashion between the polyelectrolytes. With bulky pendant groups along the macromolecular chains, there exists a steric barrier for complexation and as a result, there may be loops and other imperfections along the complexed portions of the chains [88]. An understanding of synthetic polyelectrolyte complexes may provide useful insight into biological polymers, most of which are polyelectrolytes [90].

PAA and PMAA are polyelectrolytes which have been shown to form polyelectrolyte complexes with various polybases. Hydrogels based on the polyelectrolyte complex of PAA and poly(ethyleneimine) (PEI) can form polyelectrolyte complexes between the carboxylic acid groups on the PAA and the amine groups on the PEI [91]. Using IR spectroscopy and potentiometric titration, it was concluded that the number of interchain polyelectrolyte complexes could be varied by the pH and ionic strength of the solution in which the gel was swollen. Formation of the complexes acted to tighten the three dimensional network of the gel and thus decreased its degree of swelling. As a result, addition of salt, which interfered with the the complexation, brought about a higher degree of swelling in the gel. Indeed, addition of enough salt caused a complete disappearance of the complexation behavior. Minimum swelling capacity was observed in the region of pH between 4 and 7.8 because in this region the maximum number of complexes formed. Swelling was high in strongly acidic and strongly alkaline regions where the complexes were unstable.

Kono et al. [92] made use of this polyelectrolyte complexation between PAA and PEI in the formation of pH sensitive capsules that could release the contents contained in their aqueous centers as a function of pH. In this work, PAA-PEI complex capsule membranes were produced which contained 43.7% PAA and 56.3% PEI (confirmed by elementary analysis). These capsules showed a minimum degree of swelling between a pH of 3.5 and 7.0 due to complex formation. Consequently, the permeability of phenylethylene glycol through the capsule membrane was lowest between pH 3.5 and 7.0 and increased above 7.0 and below 3.5. These investigators also used turbidimetric and potentiometric techniques to confirm the presence of these polyelectrolyte complexes.

Polyelectrolyte complexation in aqueous solution between PEI and PMAA has been studied through viscometry, conductometry, potentiometry, and IR spectroscopy [90]. Upon addition of increasing concentrations of PMAA to an aqueous PEI solution, viscosity dropped suddenly around a 1 to 4 ratio of PMAA to PEI because of the complexation and subsequent coiling of the complexed chains. Reduced viscosity then rose past this ratio indicating that the stoichiometry of the complex occurs in a 1: 4 (PMAA groups : PEI groups) formation. Conductance and titration experiments agreed with this theory. The use of IR spectroscopy confirmed this complexation and provided evidence that the macromolecules complexed in a "ladder-like" configuration.

Membranes based on PAA and poly(ethylene piperidine) in equimolar fractions exhibit better mechanical stability in ultrafiltration than conventional membranes because of the presence of the ionic bonds. In ultrafiltration, these membranes showed from 4 to 20 times the permeability of conventional membranes. They have also been proven as effective as conventional membranes in dialysis [93]. Further investigation of the mechanical properties of these membranes under uniaxial tension has shown interesting trends. In order to further clarify the role of ionic complexation in the mechanical properties of the membrane, tensile tests were performed on membranes swollen in solutions of dioxane, a slightly polar solvent, and NaCl, a low molecular weight electrolyte. The idea was that if polyelectrolyte complexation was a key in strengthening the network, a slightly polar electrolyte should increase the electrostatic interaction between the polyelectrolytes and produce a stronger membrane while a low molecular weight electrolyte should weaken the electrostatic interactions because it would screen the functional groups on the polyelectrolytes thus weakening the membrane [94].

The behavior of polyelectrolyte gels in the presence of oppositely charged ionic surfactants has been investigated by Khokhlov and coworkers [95]. Ionic surfactants may concentrate in the polyelectrolyte network and form micelle-like aggregates. The charged heads of the surfactants which form the outer surface of the micelles formed complexes with the ionic groups along the polymer chains in the gel. This lowered the osmotic pressure of counterions in the gel and resulted in gel collapse. The complexation of the oppositely charged micelles with the polymer chains also induced additional effective crosslinking in the gel and contributed to the gel collapse. An increase in the charge density of the ionic network resulted in an increase in the amplitude of gel collapse. These systems were effective absorbers of organic compounds dispersed in water because of the organic centers of the micelles.

Complexation of polyelectrolytes with biological polymers such as trypsin and insulin is possible [96]. Insulin can form a complex with PMAA and in this complexed form, insulin can resist the enzymatic action of pepsin. When injected into the muscles of lab rabbits, the PMAA- insulin complex exhibited prolonged blood sugar level control. Trypsin in complex with polyvinylimidazole has shown an increased rate of thrombi dissolution. However, if trypsin interacted with PMAA, inactivation of the trypsin occurred and likewise for the interaction of insulin with polyvinylimidazole. This may be attributed to the energy of interaction between the interacting species. A high interaction energy between the two species can lead to deformation of the secondary and tertiary structures of the proteins. Trypsin has a high isoelectric point which causes strong interactions with PMAA resulting in deactivation while insulin has a low isoelectric point such that it is not deactivated but even stabilized by complexation with PMAA. Similarly, trypsin is not deactivated by polyvinylimidazole which is a weak base.

3.2 Interpolymer Complexes Stabilized by Hydrogen Bonding

Hydrogen bonding occurs between an electron-deficient hydrogen and a region of high electron density. Macromolecular complexation through hydrogen bonding occurs in many biological systems and this type of complexation behavior has also been seen in a few synthetic polymer systems between proton-accepting and proton-donating polymers [97]. Certain features are present in the majority of these systems. For example, the polymer nature of the complexing species is a key factor in complexation. Low molecular analogs of the interacting components do not exhibit complexation behavior. In other words, there exists a critical chain length below which complexation does not occur. Also, these complexes are generally observed in aqueous media and form within a narrow range of pH, ionic strength, and solvent composition [88]. The stability of the complexation is also affected by temperature, polymer concentration, polymer structure, and other interactions such as hydrophobic effects [97].

PAA and PMAA are known to form hydrogen bonded complexes with poly(N-vinyl-2-pyrrolidone) (PNVP), often considered among the most stable interpolymer complexes [88]. Ferguson and Shah [98] investigated polymerization of PAA in the presence of PNVP. During polymerization, the reaction mixture became turbid and these investigators used this as a method to follow the polymerization. Knowing the existence of this interaction between PNVP and PAA, they investigated its possible effects on the polymerization reaction since often when a polymerization process is studied, the results are analyzed assuming no interaction between monomers of different chemical structure. They found that PNVP greatly affected the polymerization of acrylic acid. The rate of polymerization exhibited a maximum at a little less than a 50:50 ratio of PNVP to AA. No acceleration was observed in the presence of a noncomplexing species like methyl pyrrolidone. Also, the maximum rate of polymerization was 163 times that of AA without PNVP. The complex formed by a mixture of PNVP and PAA was compared to the complex formed during the polymerization of PAA in the presence of PNVP. Upon examination using electrophoresis it was found that the mechanism of polymerization in the presence of PNVP proceeded as a type of replica or template polymerization. A mechanism was suggested such that the PNVP molecules adsorb the acrylic acid monomer creating a high local concentration. When initiated, the monomers polymerize in a "zipping" fashion and the resulting polymer would precipitate at a critical molecular weight of PAA. A similar phenomenon has been observed for the polymerization of methacrylic acid in the presence of poly(2-vinylpyridine) [99].

Template polymerization based on hydrogen bonded interpolymer complexes has also been studied by Van de Grampel et al. [100]. Crosslinked PMAA networks were used as templates for the polymerization of N-vinylimidazole (VIm) which is known to form complexes with PMAA. The effects of solvent nature on the stability of the PMAA-PNVP complex have been studied by comparing complex stability in aqueous and organic solvents [101]. In water, since the hydrophilic carboxylic acid groups of the PMAA were involved in the complexation, the hydrophobic interaction of the methyl groups on the PMAA chains caused the formation of globules in order to stabilize the complex. In DMSO, the PMAA-PNVP complex was not very stable due to the weakening of the hydrophobic interactions. The globules of complexed PMAA and PNVP seen in water broke up and caused a considerable increase in solution viscosity. Also, upon heating, the complexes in water were found to have excellent stability up to 100 °C.

Similar results were found for PMAA-PNVP in DMSO by other investigators [102]. NMR studies of PMAA in DMSO proved that there is a very strong hydrogen bonding interaction between PMAA and DMSO. Through similar studies with a variety of solvents such as water, methanol, ethanol, and DMF, it was determined that if the interaction between one of the component polymers and the solvent was stronger than the polymer-polymer interaction, no complexes were formed. Complexes between PMAA and PNVP were formed in DMF but became unstable at around 80 °C. In order to investigate the role of hydrophobic interactions in this case, PAA-PNVP was examined in DMF since PAA does not contain the methyl side group found on PMAA chains. Complexes of PAA-PNVP could not remain stable in DMF illustrating the effect of hydrophobic interactions.

The behavior of three component polymer complexes has been studied. The behavior of a system containing hydrogen bonding PMAA and PNVP in addition to poly(ethylene imine) (PEI), which was described in the previous section as forming a polyelectrolyte complex with PMAA, was studied by Chatterjee's group [103]. In this work, the investigators were looking at the stability of the two types of forces making up this three component complex as a function of temperature. Hydrogen bonding is known to break down beyond certain temperatures whereas electrostatic interactions are relatively unaffected with increasing temperature. Mixtures containing 80% PEI, 10% PMAA, and 20% PNVP showed greater stability at higher temperatures than samples containing less PEI because they contained a higher proportion of electrostatic interactions. Thermodynamic parameters for the complexing systems were also calculated and the samples containing the higher amounts of PEI showed higher ΔS values than those containing less PEI. This was attributed to the release of more solvated molecules during the formation of the polyelectrolyte complex as opposed to the hydrogen bonded complex.

Acrylamide units are also known to form hydrogen bonds with carboxylic acid groups on MAA and AA. Chatterjee et al. [104] used several experimental techniques to investigate hydrogen bonding complexation of copolymers of methacrylic acid- methacrylamide and copolymers of acrylic acid-methacrylamide with Cu^{+2} and polyvinylpyrrolidone. In this system a copolymer-copolymer-metal complex was formed by first adding Cu^{+2} to P(MAA-co-AAm) which resulted in a complexation between the carboxylic acid groups and the metal. Next, P(AA-co-AAm) was added causing complexation of the unreacted acid groups on the MAA with the acrylamide groups. Subsequent addition of metal ions caused further complexation with acid groups on the MAA or the

AA portions of the chains. This process could be repeated to produce a bulky three component complex network. Viscometry, potentiometry, and IR and UV spectroscopy confirmed the formation of this copolymer-copolymer-metal complex.

PMAA has also been found to form hydrogen bonded complexes with poly(L-proline) (PLP). Ree and associates [105, 106] investigated this complexation in regard to conformational changes upon complexation. In solution, PMAA takes on a random coil conformation while PLP has a helical structure. The conformational change which must take place in either or both polymers was of interest to these researchers. The formation of a 1:1 complex was confirmed through turbidity, pH, conductance, and polarimetric measurements. Scanning electron microscopy showed the complexes taking on the form of long entangled fibers. X-ray diffraction data elucidated the conformational changes in that no diffraction patterns which appear in pure PLP appeared in the data for complexed samples. This suggested the destruction of the helical structure of PLP during complexation.

Other examples of interpolymer complexation due to hydrogen bonding include complexes of poly(ethylene-co-methacrylic acid) with polyethers, poly(2-vinylpyridine), and polyethyloxazoline [107–109], and PAA with poly(vinyl alcohol-co-vinyl acetate) [110]. PAA has even been shown to form stable interchain hydrogen bonded association under high shear flow [111].

3.3 Interpolymer Complexes of PAA and PMAA with PEG

Of particular interest are two systems that exhibit interpolymer complexation due to hydrogen bonding: PMAA with PEG and PAA with PEG. These polycarboxylic acids are proton donors while PEG is a proton acceptor. As seen in both systems, the protons of the carboxylic acid groups form hydrogen bonds with the ether oxygens on the PEG chain. Since the acidic groups on the macromolecular chains are involved in this complexation, it is accompanied by an increase in the pH of the solution. Also, since complexation in these systems occurs between the hydrophilic acid groups of the PAA or PMAA with the oxygens on the PEG chain, the resulting interpolymer complex is more hydrophobic than the constituent polymers. An added hydrophobic group, as with PMAA, causes an added increase in hydrophobicity. Complexes between PMAA and PEG have proven more stable than those of PAA with PEG because the complex is further stabilized in aqueous solutions by the hydrophobicity of the CH_3 groups found on the PMAA chain [88].

Early work with these complexes made use of viscometry in order to investigate the effect of pH on the complexation formation. In working with solutions of PAA and PEG, Bailey et al. [112] found that the complex formed into a precipitate from water solution at around pH 3.8. The pH of the solution affects the ratio of acid to acrylate (neutralization) on the PAA chain thus affecting the number of complexes that may be formed. Because of this, these researchers saw varying degrees of association in this system. The viscosity of a PEG-PAA solution dropped sharply from pH 4 to 6 then remained constant between around 7 to 12. At higher pH values, there was polymer precipitation so the viscosity dropped again past pH 12. The complexation between PEG and PMAA was found to be even stronger than that of PEG and PAA indicating the importance of the hydrophobicity of the methyl groups on PMAA in complex stabilization.

As stated earlier, the polymeric nature of species that form interpolymer complexes is an important factor. Complexation can only occur if there are sufficiently long macromolecular chains. It has been shown that there is a critical chain length below which complexation cannot occur. Potentiometric titrations have proven useful in investigating this phenomenon. Since the complexation between PAA or PMAA with PEG occurs through the dissociated hydrogens on the carboxylic acid side groups, the formation of complexes results in an increase in solution pH. Thus a potentiometric method can measure the concentration of interpolymer complexes by measuring the concentrations of the hydrogen ions. As shown by Antipina et al. [113], in the PMAA-PEG system in water, when PEG of molecular weight 1000 was added to a dilute solution of PMAA there was no effect on the pH of the solution. In fact, up to a PEG molecular weight of 3000 there was no pH increase observed. After a molecular weight of 3000, the pH of the solution rose gradually. Addition of PEG samples of molecular weight of 6000, 15 000, and 40 000 caused a rapid rise in pH. Similar results were seen for the addition of PEG to a dilute PAA solution except that the critical chain length occurred around a molecular weight of 6000 and the effect was less drastic than with PMAA. Also, for PMAA-PEG complexes in methanol-water solution, the minimum chain length for complexation increased with concentration of methanol [88]. The above results agree with those obtained by Osada and Sato [114, 115] who studied the complexation in water and ethanol-water mixtures. These observations illustrate again the effect of the hydrophobic interactions on stabilization of the complex. These results were further confirmed for both systems using viscometry.

Baranovsky et al. [116] studied complex formation between PMAA and monosubstituted PEG in water and found that by introducing a hydrophobic group onto the PEG, the critical chain length for complexation was lowered. The authors looked at the effects of different substituted groups on the critical chain length for complexation. These groups included phenol, p-tertbutyl phenol, 1-naphthol, n-octanol, and hydroquinone. In complexation between PMAA with 1-naphthol-PEG (naph-PEG), the critical chain length was lowered to a molecular weight of around 500. This is considerably lower than the values reported above for PMAA-PEG complexation. The PMAA complexation with phenol-PEG (ph-PEG) also showed a decreased critical chain length but the effect in the PMAA-naph-PEG system was more dramatic; in other words these complexes were more stable because the 1-naphthol provided a larger hydrophobic "anchor" on the PEG. The p-tertbutyl phenol and n-octanol substituted

PEGs also showed enhanced stability similar to that of ph-PEG. PEG containing hydroquinone substitutions showed a decrease in critical chain length but
again not as drastically as the naph-PEG because the bulky hydrophobic group
on the hydroquinone is in the middle of the chain while that of the 1-naphthol is
on the end of the chain. Thus the effects of hydrophobic interactions have been
shown to be a universal stabilizing factor in interpolymer complexation.

In related work, Bokias et al. [117] made substitutions along PAA chains
and investigated the effects on complexation with PEG using viscometry and
potentiometry. These researchers substituted various groups in place of the
carboxylic acid groups along PAA chains. At constant acidic pH and ionic
strength, when ionic groups were substituted, the complexation was prevented
when the degree of substitution reached about 10%. On the other hand, the
substitution of neutral, relatively hydrophobic groups did not hinder complexation until the degree of substitution reached about 30%. The differences were
due to the fact that the ionic groups interrupted the sequence of complexing
carboxylic acid groups and also increased the rigidity of the chains due to
intrachain electrostatic repulsion. As a result the chains did not have the needed
flexibility to complex with the free PEG chains. In contrast, the neutral hydrophobic groups, while still interrupting the sequence of complexing acid groups,
strengthened the complexation through hydrophobic interactions and did not
alter the flexibility of the chains. Thus only at higher degrees of substitution did
these substituted groups interfere with the complexation behavior.

In considering the thermodynamics of PMAA-PEG complexes in aqueous
media, the heat of formation was found to be 0.33 kcal/mol while the entropy of
formation was 1.32 cal/mol °C. The heat of formation of the complex was found
to be independent of the molecular weight of PEG which is logical because the
stoichiometry of the actual complex is not a function of molecular weight. The
fact that complexation was found to be endothermic was attributed to dehydration of the PEG and PMAA molecules upon complexation. Because of the sign
of the heat of formation, the stability of the complex increased with increasing
temperature in water. In methanol-water solutions the heat of formation
changed signs causing the stability of the complex to decrease with increasing
temperature. This was attributed to the loss of the added stabilization of
hydrophobic interactions in methanol-water solutions [118].

Techniques other than viscometry and potentiometry have also been used to
study the complexation between PAA or PMAA with PEG. For example, Frank
and collaborators [119] have done extensive studies on these systems using
pyrene labeled excimer fluorescence by labeling the PEG chains on both ends
with pyrene groups. This technique proved more sensitive to complex formation
than conventional techniques used previously. In the PAA-PEG system, this
technique revealed a decrease in chain mobility upon addition of PAA. Also, the
local concentration of PEG was increased in the region near the PAA chains
because of the hydrogen bonding. These effects were more pronounced as
molecular weight of the PAA was increased. For the PMAA-PEG system,
PMAA was confirmed as a stronger proton donor than PAA. The PMAA-PEG

complex formed a more compact structure and addition of PMAA to PEG solution produced a more drastic decrease in chain mobility than PAA. The addition of methanol to the solution broke down the hydrophobic stabilization such that hydrogen bonding became the primary force in the complexation. In this situation, PMAA behaved like PAA in all respects [120, 121]. The nature of the complex structure mentioned above for the PMAA-PEG system was studied by Hemker and Frank [122] using photon correlation spectroscopy. The aggregation size, $\langle R \rangle$, was found to be a function of pH. A critical pH of 1.9 was observed such that, below this value, the aggregates formed more rapidly and were larger. They modelled this behavior using a cluster-cluster aggregation model and found good agreement with experimental data. Another model of the complexation of free and graft oligomers of complementary polymers such as PEG with PMAA or PAA was proposed by Peppas et al. [123]. This model was based on a statistical mechanical framework and irregularities in complex structure such as loops and tails were incorporated. Other techniques used in investigation of these PMAA- or PAA-PEG complexes include solution micro-calorimetry [124], differential scanning calorimetry, Fourier-transform infrared spectroscopy [125], and other fluorescence techniques [126, 127].

As seen, various aspects of the complexation of free chain polycarboxylic acids, namely PMAA and PAA, with PEG in solution have been studied. However, considerable work has recently been done in applying this complexation in the form of interpenetrating networks (IPNs), hydrogels, and membranes. Kotaka and coworkers [128-130] have created IPNs consisting of equimolar quantities of crosslinked PAA and crosslinked PEG. They characterized the swelling and mechanical properties of these IPNs and showed that the IPN underwent elongation and contraction as the pH of the swelling medium was raised and lowered. They also prepared the IPN in the form of a membrane with the idea of using it in ultrafiltration. For testing, the IPNs were preswollen unrestricted in a pH 7 and then fixed in a ultrafiltration cell. Thus, at the beginning of the experiment the IPNs were in a highly swollen state in which there was no complexation so the PAA and PEG chains interpenetrated each other throughout the network. Under these conditions, water flow rate through the membrane was low and solutes above a molecular weight of 2600 were rejected. Upon a change from basic to acidic conditions, the complexes formed between the PAA and PEG causing a contraction of the network but since the membrane was held in place in the cell and could not contract, pores opened up in the membrane allowing increased permeation and water flow rate. This behavior was shown to be completely reversible [131].

Osada and Takeuchi [132] have done pioneering work based on inter-polymer complexation and specifically with PMAA-PEG and PAA-PEG systems. In some interesting work, they looked at the effects of PEG treatment on a PMAA membrane. Drastic dilations and contractions because of complexation were seen as a function of pH when a PMAA membrane was treated with PEG. Upon treatment with a small amount of PEG, a PMAA membrane could lift and lower a weight by contraction and dilation to 90% of its length. Based on this, they produced environmentally sensitive membranes and coined the term "chemical valve" in reference to these membranes. Indeed, these membranes were found to act as chemical valves in that a PMAA membrane treated with PEG showed high water permeability and flow rate. When treated with basic solution, the membrane recovered its initially low water permeability. This occurred as discussed above; a MAA membrane was immobilized in an ultra-filtration cell and, when treated with PEG such that complexation occurred, the stress created by the contraction opened up pores in the membrane. When the complexes were broken by treatment of base, swelling of the membrane closed the pores and restored low permeability. Tests with hemoglobin and albumin solutions showed that these proteins could permeate the PEG-treated membrane without decreasing the water permeability or clogging the membrane which are common problems with untreated PMAA membranes. Protein separation was improved 240 times for albumin and 55 times for hemoglobin as compared to untreated membranes.

The chain length of the PEG used to treat the PMAA membrane was investigated by Osada [133]. PEGs with molecular weights 600 and 1000 produced quick contractions but the contractions were small. PEG of 2000 molecular weight produced rapid and considerable contraction while molecular weights of 7500, 20 000 and 83 000 also showed good contraction but the contraction took place over a long period of time. Due to this behavior, Osada proposed that the PEG not only adsorbed to the surface of the membrane but permeated the pores as well. A thermodynamic theory of this complexation behavior between linear polymers and polymer networks was proposed by Khokhlov and Kramarenko [134]. The theory was based on the Flory-Huggins treatment but incorporated a free energy contribution due to the formation of complexes. An excellent review of work done on chemical valves and the idea of conversion of chemical energy into mechanical work was also presented by Osada [89].

A unique material based on the complexation of PMAA and PEG was produced by Peppas and collaborators [135]. This material was a graft copolymer of P(MAA-g-EG). It was prepared by free radical polymerization of PMAA with PEG monomethacrylate to create a PMAA matrix with PEG grafts "branching" off the PMAA units. This copolymer was prepared in the form of hydrogels and microspheres. The idea behind these materials was to produce complexing, environmentally sensitive materials by actually bonding the complexing species together. Characterization of this copolymer was done in several ways. NMR spectroscopy confirmed the presence of a PMAA matrix with PEG grafts as indicated in Fig. 7. Swelling studies showed good pH sensitivity in that swelling was low in acidic pH where the PEG grafts interacted with the PMAA matrix to form a tight network and, swelling was high in basic pH where the complexation could not occur because of neutralization of the PMAA (see Fig. 8). Swelling in alcohol disrupted the hydrophobic stabilization of the complexes and caused the disruption of the complexes [136]. Recently, we have found [137] that variation in the copolymer composition affected the

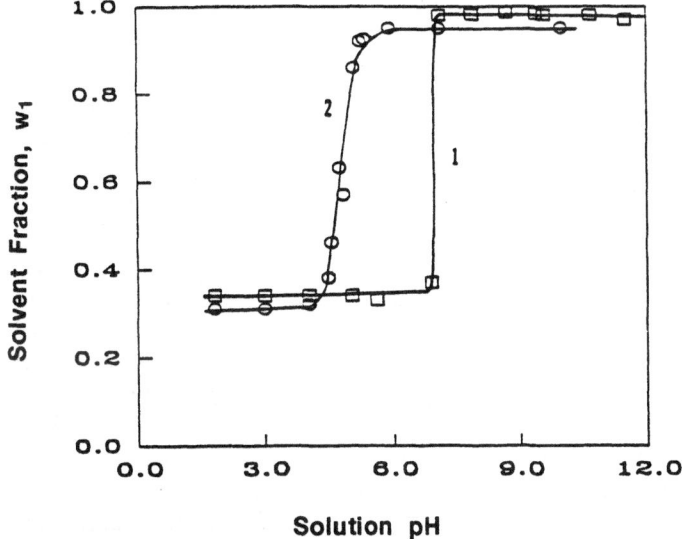

Fig. 7. Solvent weight fraction in P(MAA-*g*-EG) complexing gels vs swelling pH in distilled water (*curve 1*) or in 0.1 M NaCl solution (*curve 2*) at 37 °C. Gels prepared with PEG of $\bar{M}_n = 200$ and PEG:MAA = 60:40

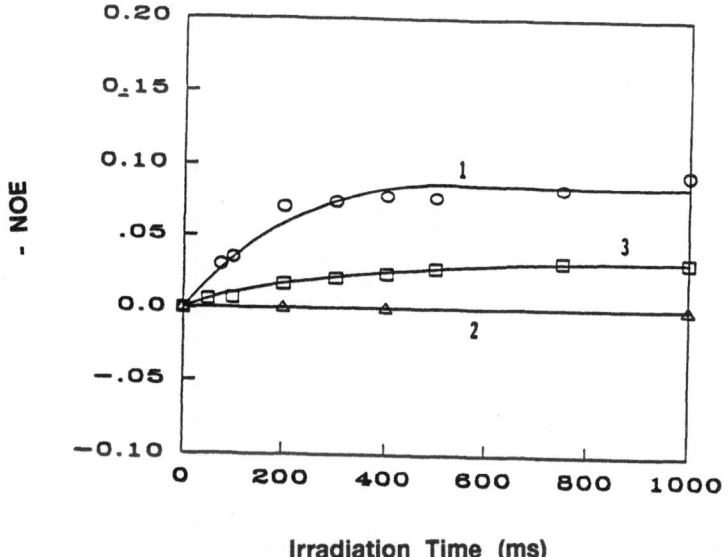

Fig. 8. Nuclear Overhauser enhancement of PEG-ethylene protons vs irradiation time in P(MAA-*g*-EG) gels exhibiting complexation. Proton enhancements of graft copolymer with PEG $\bar{M}_n = 400$ in D_2O (*curve 1*), graft copolymer in NaOD solution (*curve 2*), and polymer mixture with PEG $\bar{M}_n = 1000$ in D_2O (*curve 3*). The PEG concentration was 0.01 wt%, PMAA concentration was 0.09 wt%, copolymer concentration was 0.1 wt%, and temperature was 21 °C

swelling behavior of these materials. Copolymers with higher percentages of PMAA showed higher degrees of swelling in the uncomplexed states because of the hydrophilicity of the carboxylic acid groups (see Fig. 9). In oscillatory swelling studies in which the swelling agent was changed from acidic to basic over several cycles, this pH sensitive behavior was seen to be completely reversible, and the copolymers responded to pH changes in less than five minutes (see Fig. 10). Obviously, because of their environmentally sensitive behavior, these copolymers have potential uses in controlled drug delivery. Controlled release of guaiacol glyceryl ether was studied as a function of the network swelling. In general, release from disks in basic solution, where complexation cannot occur and swelling was high, was faster with a diffusion coefficient of 2.56×10^{-6} cm^2/s. Release from complexed gels, gels in acidic solution, showed diffusion coefficients of 3.0×10^{-9} cm^2/s [138]. We have also investigated these materials as environmentally sensitive membranes [137]. Uncomplexed membranes showed good permeation of solutes with the permeability of the membranes decreasing as solute size increased. Complexed or collapsed membranes only showed limited permeation of very small solutes. In the uncomplexed or expanded membranes, the permeability was also seen to decrease as the molecular weight of the PEG grafts in the copolymer increased. This was attributed to the fact that, in the uncomplexed state, the PEG grafts dangled out into the mesh space of the polymer network. Thus in the uncomplexed state the PEG grafts interfered with solute diffusion, and the longer grafts interfered to a higher degree. These researchers also showed that the permeation

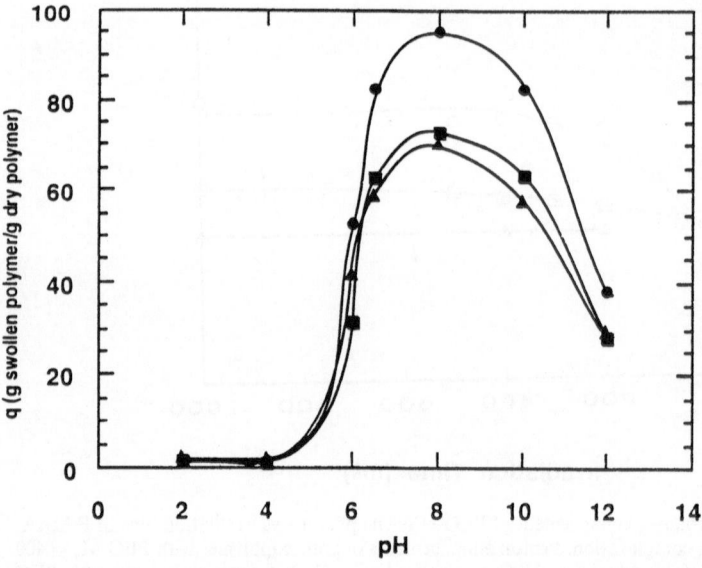

Fig. 9. Equilibrium swelling ratio as a function of pH for a samples containing PEG grafts of molecular weight 1000 with the ratio of MAA:PEG being (●) 60 : 40, (■) 50 : 50, (▲) 40 : 60

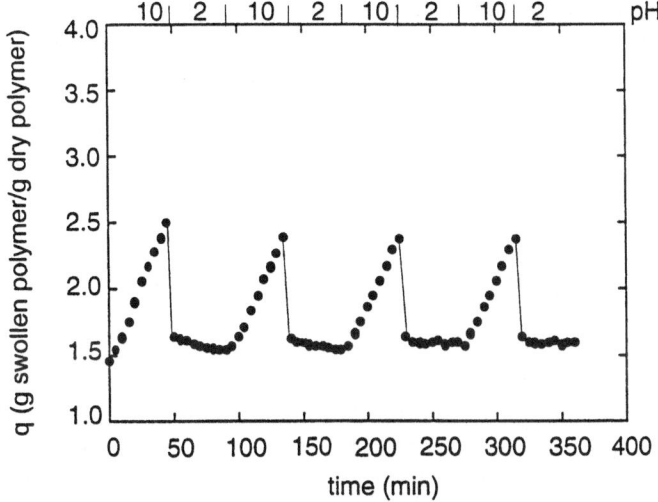

Fig. 10. Oscillatory swelling behavior as a function of time and pH for a sample containing 50% PMAA and 50% PEG with the molecular weight of the PEG grafts being 1000

of vitamin B12 through P(MAA-*g*-EG) membranes could be turned on and off by altering the pH of the surrounding solution.

4 Transport in Biomedical Membranes from Hydrogels or Interpolymer Complexes

Polymer network structure is important in describing the transport through biomedical membranes [139, 140]. The mechanism of diffusion in membranes may be that of pure diffusion or convective transport depending on the mesh size of the polymer network. With this in mind, polymer membranes are typically divided into three major types described below [141].

1. Macroporous membranes – these are membranes containing large pores. The pore size is usually between 0.1 and 1 µm. Convective transport through the pore space is the mechanism of diffusion in this case.

2. Microporous membranes – pore size in these membranes ranges from 50 to 200 Å. In this case, the pores are usually only slightly larger than the solutes and this results in hindered transport through the pores.

3. Nonporous gel membranes – these membranes do not contain a porous structure and thus diffusion occurs through the space between the polymer chains (the mesh). Obviously in this case, molecular diffusion rather than convective transport is the dominant mechanism of diffusion in these membranes.

Various theories have been proposed to describe the transport in all of these types of polymer membranes. Theories for macroporous and microporous membranes have been based on hydrodynamic and frictional considerations while those for nonporous gels have been based on Eyring's theory and use a free volume approach to describe the movement of solute through the mesh of the polymer.

4.1 Transport in Macroporous and Microporous Polymer Membranes

In the case of macroporous membranes, the pores are of large enough size and the mechanism of diffusion is such that the diffusion coefficient of a solute through the membrane may be described as the diffusion coefficient through the solvent-filled pores of the membrane. The macroporous membrane may be characterized by a porosity, ε, and a tortuosity, τ, as well as a partition coefficient, K_p, which describes how the solute distributes itself in the membrane. These parameters are usually included in the description of transport in macroporous membranes by incorporating them into the diffusion coefficient as

$$D_{eff} = D_{iw} \frac{\varepsilon K_p}{\tau} \tag{47}$$

where D_{eff} is the effective diffusion coefficient of the solute through the solvent in the pores, and D_{iw} is the diffusion of the solute in the pure solvent (usually water) [141].

The case of transport through microporous membranes is different from that of macroporous membranes in that the pore size approaches the size of the diffusing solute. Various theories have been proposed to account for this effect. As reviewed by Peppas and Meadows [141], the earliest treatment of transport in microporous membranes was given by Faxen in 1923. In this analysis, Faxen related a normalized diffusion coefficient to a parameter, λ, which was the ratio of the solute radius to the pore radius

$$\lambda = \frac{r_s}{r_p} \cdot \tag{48}$$

This relationship was given by

$$\frac{D_{im}}{D_{iw}} = (1 - \lambda)^2 (1 - 2.104\lambda + 2.09\lambda^3 - 0.95\lambda^5) \tag{49}$$

where D_{im} is the diffusion coefficient of the solute in the polymer membrane, and D_{iw} is the diffusion coefficient in the pure solvent.

Transport in a microporous biomedical membrane is described in Fig. 11. Membranes consist of cylindrical liquid-filled pores of length l and radius r_p with spherical solute molecules of radius r_s diffusing through the pores. The solute

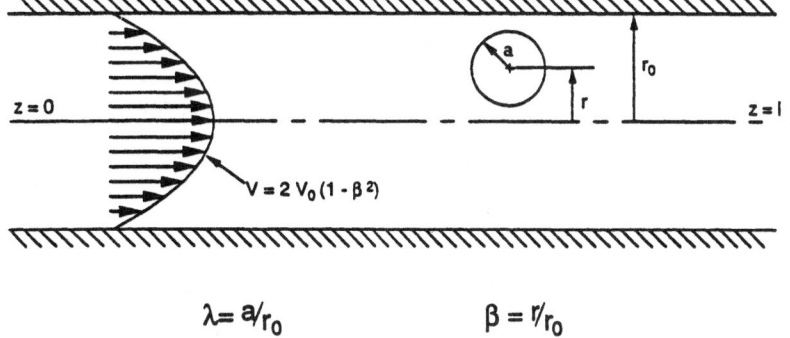

$$\lambda = a/r_0 \qquad\qquad \beta = r/r_0$$

Fig. 11. Spherical solute transport in a microporous membrane

diffusivity in a bulk solution (not in the pores), D_∞, possesses hydrodynamic and Brownian characteristics described by the Stokes-Einstein equation,

$$D_\infty = \frac{kT}{6\pi\eta r_s} \tag{50}$$

for a spherical solute. In this equation, k is the Boltzmann constant and T is the absolute temperature [142]. This equation has been found accurate for globular proteins [143] as well as smaller molecules such as glucose [144].

During solute transport a balance can be written between the driving force for solute flux and the drag between solute and solvent, F_s,

$$-kT\frac{d(\ln C)}{dz} = F_s \tag{51}$$

with the drag on each solute molecule being

$$F_s = f_\infty K[U_s - gV] \tag{52}$$

with U_s being the net solute velocity, K the ratio of pore to bulk friction coefficients and g the "lag coefficient" which accounts for the retarding effect of the pore wall on solute velocity. The resulting average solute flux through one pore is then

$$\bar{N}_s = -\zeta D_\infty \frac{dc}{dx} + \chi C V_0 \tag{53}$$

where

$$\zeta = \frac{D}{D_\infty} = \int\limits_0^{1-\lambda} [2\beta K^{-1}]\,d\beta \tag{54}$$

and

$$\chi = \int\limits_0^{1-\lambda} [4\beta(1-\beta^2)G]\,d\beta. \tag{55}$$

Here, ζ is the ratio of pore to bulk diffusivity while χ is fa sieving coefficient. The upper limit in the above integrals represents the fact that the spherical molecule cannot be any closer to the pore wall than its radius. Thus the values of K^{-1} and G must be determined. For λ less than 0.4, the following were found as good approximations for K^{-1} and G,

$$K^{-1} = 1 - 2.1044\lambda + 2.089\lambda^3 - 0.948\lambda^5 \tag{56}$$

$$G = 1 - (2/3)\lambda^2 - 0.163\lambda^3. \tag{57}$$

The final results, further derivation of which may be seen in the original work, were the following expressions for ζ and χ,

$$\zeta = \frac{D}{D_\infty} \approx (1 - \lambda)^2 K^{-1} \tag{58}$$

$$\chi \approx (1 - \lambda)^2 [2 - (1 - \lambda)^2] G. \tag{59}$$

Brenner and Gaydos [145] modified these expressions and obtained

$$\frac{D}{D_\infty} = 1 - (1.125)\lambda \ln \lambda^{-1} - 1.539\lambda + O(\lambda). \tag{60}$$

Other modifications to the theory of Anderson and Quinn [142] have been reviewed by Deen [146]. Malone and Quinn [147] modified the above theory to include the effect of electrostatic interactions on transport in microporous membranes. Smith and Deen [148] have also looked at these electrostatic or double layer interactions. More recently, Kim and Anderson [149] investigated the hindrance of solute transport in polymer lined micropores. Also, as briefly mentioned above, an excellent review of the theories presented for transport in microporous membranes has been given by Deen [146].

4.2 Transport in Nonporous Polymer Gel Membranes

The theories developed for transport in microporous membranes cannot be applied to nonporous gel membranes. The pore structure in microporous membranes is not analogous to the mesh of the nonporous gels. Thus a different set of theories had to be developed for the treatment of nonporous polymer gel membranes. These theories are based on the idea of the existence of free volume in the macromolecular mesh. As a result, diffusion through nonporous membranes is said to occur through the space in the polymer gel not occupied by polymer chains.

The first theory of transport through nonporous gels was presented by Yasuda et al. [150] and was proposed as a result of previous experimental results [151, 152]. This theory relates the ratio of diffusion coefficient in the polymer membrane and diffusion coefficient in the pure solvent to the volume fraction of solvent in the gel membrane or in Yasuda's terminology, the degree of hydration of the membrane, H (g water/g swollen polymer). Yasuda et al. use the subscripts 1,2, and 3 to denote the water (solvent), solute, and polymer, respectively. Thus, $D_{2,13}$ represents the diffusion coefficient of the solute in the water swollen polymer, and $D_{2,1}$ represents the solute diffusion coefficient in pure water. Assuming that the free volume available for solute diffusion in the membrane is the free volume of the water in the polymer/water system, the theory of Yasuda et al. states that

$$\frac{D_{2,13}}{D_{2,1}} = \varphi(q_s) \exp[- B(q_s/V_{f,1})(1/H - 1)] \tag{61}$$

where $V_{f,1}$ is the free volume of the water in the swollen membrane, q_s is the cross sectional area of the diffusing molecule, and $\varphi(q_s)$ describes a sieving mechanism which allows small molecules that may be accommodated in the network to pass through while rejecting larger molecules that cannot fit into the "spaces" in the macromolecular network. If the solute molecules are sufficiently small with respect to the network, this sieve coefficient is equal to one and a plot of log $(D_{2,13} / D_{2,1})$ vs $1/H$ yields a straight line. Also, log $(D_{2,13} / D_{2,1})$ decreases as solute size increases and this effect is weakened as H is increased. From this theory, the permeability coefficient is given by

$$P_{2,13} = K'D_{2,13} \tag{62}$$

where K' is the partition coefficient which describes how the solute is distributed in the polymer with respect to the solution (g solute per cm^3 polymer / g solute per cm^3 solution). In experiments with urea, creatinine, and secobarbital, Yasuda and Lamaze [153] confirmed that log $P_{3,21}$ decreased linearly with membrane hydration and solute size with the effect of solute size decreasing with increased membrane hydration.

In considering the solute permeation through PHEMA hydrogel membranes, Kim et al. [154] found good agreement with Yasuda's dependence of diffusion coefficient on membrane hydration and solute size when considering hydrophilic solutes. Also, with hydrophilic solutes, these investigators found that as membrane hydration increased, the diffusion coefficient became less dependent on solute size. From these results, they concluded that hydrophilic solutes diffuse through the solvent filled mesh space of the polymer network as proposed by Yasuda et al. In other words, the diffusion of hydrophilic solutes was concluded to occur through a "pore" mechanism. On the other hand, studies with hydrophobic solutes revealed substantial deviation from the proposed dependencies. Also, the diffusion coefficients of hydrophobic solutes were, in general, two orders of magnitude lower than that of hydrophilic solutes, and the partition coefficients for hydrophobic solutes were found to be about two orders of magnitude higher than those of hydrophilic solutes. These results indicated strong interactions between the hydrophobic solutes and the macromolecular segments in the hydrogels. In later work, Kim and other collaborators [65] again found good agreement with Yasuda's theory when investigating the diffusion of dextran and uranine through temperature sensitive hydrogel membranes.

Peppas and Reinhart have also proposed a model to describe the transport of solutes through highly swollen nonporous polymer membranes [155]. In highly swollen networks, one may assume that the diffusional jump length of a solute molecule in the membrane is approximately the same as that in pure solvent. Their model relates the diffusion coefficient in the membrane to solute size as well as to structural parameters such as the degree of swelling and the molecular weight between crosslinks. The final form of the equation by Peppas and Reinhart is

$$\frac{D_{im}}{D_{iw}} = k_1 \left(\frac{\bar{M}_c - \bar{M}_c^*}{\bar{M}_n - \bar{M}_c^*} \right) \exp \left(- \frac{k_2 r_s^2}{Q - 1} \right) \tag{63}$$

where D_{im} is the solute diffusion coefficient in the swollen membrane, D_{iw} is the solute diffusion coefficient in pure water, k_1 and k_2 are structural parameters of the polymer/water system, Q is the degree of swelling of the membrane, r_s is the radius of the diffusing solute, $\bar{M}_c$ is the molecular weight between crosslinks, $\bar{M}_n$ is the molecular weight of the uncrosslinked polymer (molecular weight before crosslinking occurred), and $\bar{M}_c^*$ is the theoretical molecular weight between crosslinks below which diffusion of a solute of size r_s could not occur.

Thus the factor $(\bar{M}_c - \bar{M}_c^*)/(\bar{M}_n - \bar{M}_c^*)$ may be thought of as the sieving term mentioned in the theory of Yasuda et al. [150]. In the Peppas-Reinhart theory, the sieving mechanism takes an understandable form which is a function of the structure of the network. It must be noted that the presence of semi-crystalline regions in the polymer membrane leads to deviations from the predicted dependencies in this theory. These researchers found that as the crosslinking density in the polymer membrane increased, the solute diffusion coefficient decreased, further illustrating the importance of structural parameters of the polymer network in predicting the solute diffusion coefficient [156].

Burczak et al. [157] investigated the permeation of various proteins through PVA hydrogel membranes. These researchers interpreted their results in terms of the Peppas-Reinhart theory. Good agreement with the theory was found in the experimental trends. The solute diffusion coefficient was seen to increase with increased membrane swelling as well as with increased values of the molecular weight between crosslinks. They also found that the dependence of the diffusion coefficient on the membrane degree of swelling became less pronounced as solute size increased.

As mentioned, the Peppas-Reinhart theory is valid in the case of highly swollen membranes. Additional work by Peppas and Moynihan [158] resulted in a theory for the case of moderately swollen networks. This theory was derived much like the Peppas-Reinhart theory with the exceptions that in a moderately swollen network, one may not assume that the diffusional jump length of the solute in the membrane, $\lambda_{2,13}$, is equal to the diffusional jump length of the solute in pure solvent, $\lambda_{2,1}$ and, also, one may not assume that the free volume of the polymer/solvent system is equal to the free volume of the solvent. The initial form of the Peppas-Moynihan theory is

$$\frac{D_{2,13}}{D_{2,1}} = \frac{\lambda_{2,13}^2}{\lambda_{2,1}^2} B(\upsilon_{13}^*) \exp\left[-\upsilon_s\left(\frac{1}{V_{13}} - \frac{1}{V_1'}\right)\right] \tag{64}$$

where $\lambda_{2,13}$ and $\lambda_{2,1}$ are the diffusional jump lengths in the polymer/water system and pure water, respectively, $B(\upsilon_{13}^*)$ is a term representing the characteristic size of the space available for diffusion in the membrane, υ_s is the size of the diffusing solute, and V_{13} and V_1' are the free volumes in the swollen membrane and pure water, respectively. Based on thermodynamic considerations and scaling laws, the forms of the diffusional jump lengths, free volumes, solute size, and $B(\upsilon_{13}^*)$ were postulated. The form of the diffusional jump lengths was taken as

$$\frac{\lambda_{2,13}^2}{\lambda_{2,1}^2} = \exp[k_3(\bar{M}_c - \bar{M}_n)]. \tag{65}$$

The contributions of the free volumes were expressed as

$$\Phi(V) = \frac{V_1 - V_3}{(Q-1)V_1^2 + V_1V_3}. \tag{66}$$

The solute size was written as the cross-sectional area, πr_s^2 times the length l_s and $B(\upsilon_{13}^*)$ was taken as $f(\upsilon_{3,s}^{-3/4})$. Thus the final form of the Peppas-Moynihan model for moderately swollen membranes is

$$\frac{D_{2,13}}{D_{2,1}} = f(\upsilon_{3,s}^{-3/4}) \exp[k_3(\bar{M}_c - \bar{M}_n) - \pi r_s^2 l_s \Phi(V)]. \tag{67}$$

In another model, Harland and Peppas [159] considered the diffusion of solutes through semicrystalline hydrogel membranes. These types of membranes were assumed to consist of a crosslinked, swollen (amorphous) phase through which solute diffusion occurred and an impermeable, crystalline phase. A simplified form of the model assumes uniform amorphous regions. With this assumption, the diffusion coefficient through a semi-crystalline membrane, D_c, was written as

$$D_c = \frac{(1-\upsilon_c)D_a}{\tau}. \tag{68}$$

Here, D_a is diffusion coefficient in the amorphous phase alone, υ_c is the volume fraction of crystalline polymer, and τ is a scalar quantity that denotes the tortuosity of diffusional path of the solute. The value of D_a may be estimated by the Peppas-Reinhart model if the amorphous regions of the polymer are highly swollen. This substitution yields

$$D_c = \frac{(1-\upsilon_c)D_w}{\tau} f\left[\frac{\bar{M}_c - \bar{M}_c^*}{\bar{M}_n - \bar{M}_c^*}\right] \exp\left[\frac{-\pi r_s^2 l_s \upsilon_a}{V_w \upsilon_s}\right]. \tag{69}$$

Or in the case of a moderately or poorly swollen amorphous polymer region, the

Peppas-Moynihan model may be used to estimate D_a. This yields

$$D_c = \frac{(1 - \upsilon_c)D_w}{\tau} f[\upsilon_a^{-3/4}] \exp[k_3(\bar{M}_c - \bar{M}_c - \pi r_s^2 l_s \Phi(V))]. \tag{70}$$

In both equations, D_w is the solute diffusion coefficient in pure water, r_s is the molecular radius of the solute, l_s is its characteristic size, V_w is the water free volume, $\bar{M}_c$ is the molecular weight between crosslinks in the amorphous phase, $\bar{M}_n$ is the number average molecular weight of the polymer before crosslinking, $\bar{M}_c^*$ is the minimum value of $\bar{M}_c$ below which the solute cannot diffuse, $\Phi(V)$ is the free volume function mentioned earlier, and k_3 is a constant.

As seen, diffusion in nonporous gel membranes differs from that in macroporous or microporous membranes. Various theories based on solute diffusion through the macromolecular free volume in the membrane have been proposed. It is clear from these theories that structural parameters of the polymer network such as degree of swelling, molecular weight between crosslinks, and crystallinity in addition to factors such as solute size and solvent free volume play important roles in this type of transport.

5 References

1. Kudela V (1985). In: Mark HF, Kroschwitz JI (eds) Encyclopedia of polymer science and technology. Wiley-Interscience, New York pp 783–807
2. Peppas NA (1991) J Bioact and Compat Polym 6: 241–246
3. Peppas NA, Barr-Howell BD (1986). In: Peppas NA (ed) Hydrogels in medicine and pharmacy. CRC, Boca Raton pp 28–55
4. Brannon-Peppas L (1990). In: Brannon-Peppas L, Harland RS (eds) Absorbent polymer technology. Elsevier, Amsterdam pp 45–66
5. Flory PJ, Rehner R Jr (1943) J Chem Phys 11: 521
6. Flory PJ (1953) Principles of Polymer Chemistry. Cornell University Press, Ithaca
7. Peppas NA, Lucht LM (1984) Chem Eng Comm 30: 291
8. Kovac J (1978) Macromolecules 11: 362
9. Galli A and Brummage WH (1983) J Chem Phys 79: 2411
10. Oppermann W (1992). In: Harland RS, Prud'homme RK (eds) Polyelectrolyte gels: Properties, Preparation, and Applications, American Chemical Society, Washington, DC pp 159–170
11. Siegel RA (1990). In: Kost J (ed) Pulsed and self-regulated drug delivery. CRC Press, Boca Raton, Florida pp 129–155
12. Peppas NA, Khare AR (1993) Adv Drug Del Rev 11: 1–35
13. Brannon-Peppas L, Peppas NA (1991) Chem Eng Sci 46: 715–722
14. Brannon-Peppas L, Peppas NA (1988) Polym Bull 20: 285–289
15. Brannon-Peppas L, Peppas N (1990). In: Brannon-Peppas L, Harland RS (eds) Absorbent Polymer Technology. Elsevier pp 67–75
16. Hooper HH, Baker JP, Blanch HW, Prausnitz JP (1990) Macromolecules 23: 1096–1104
17. Konak C, Bansil R (1989) Polymer 30: 677–680
18. Nielsen LE (1974) Mechanical properties of polymers and composites. Marcel Dekker, Inc, New York vol I p 39
19. Treloar LRG (1975) The physics of rubber elasticity. Oxford University Press, Oxford
20. Dubrovskii SA, Ilavsky M, Arkhipovich GN (1992). Polym Bull 29: 587–594
21. Zrinyi M, Horkay F (1987) Polymer 28: 1139–1143
22. Kuhn W, Hagarty B, Katchalsky A, Eisenberg H (1950) Nature 165(4196): 514–516

23. Liu G, Guillet JE, Al-Takrity ETB, Jenkins AD, Walton DRM (1991) Macromolecules 24: 68–74
24. Horsky J (1987) Polym Comm 29: 110–111
25. Katchalsky A (1949) Experientia pp 319–320
26. Katchalsky A, Kunzle O, Kuhn W (1950) J Polym Sci 5: 283–300
27. Katchalsky A, Michaeli I (1955) J Polym Sci 15: 68–86
28. Katchalsky A (1949) J Polym Sci 4: 393–412
29. Steinberg IZ, Oplatka A, Katchalsky A (1966) Nature 210: 568–571
30. Ohmine I, Tanaka T (1992) J Chem Phys 77: 5725–5729
31. Nicoli D, Young C, Tanaka T, Pollak A, Whitsides G (1983) Macromolecules 16: 887–890
32. Kokufuta E, Tanaka T (1991) Macromolecules 24: 1605–1607
33. Sato E, Tanaka T (1992) Nature 358: 482–485
34. Annaka M, Tanaka T, Osada Y (1992) Macromolecules 25: 4826–4827
35. Khare AR, Peppas NA (1991) Polym News 16: 230–236
36. Brannon-Peppas L, Peppas NA (1991) Inter J Pharm 70: 53–57
37. Brannon-Peppas L, Peppas NA (1991) Chem Eng Sci 46: 715–722
38. Brannon-Peppas L, Peppas NA (1991) J Contr Rel 16: 319–330
39. Brannon-Peppas L, Peppas NA (1989) J Contr Rel 8: 267–274
40. Khare AR, Peppas NA (1993) Polymer 34: 4736–4739
41. Hariharan D (1993) PhD thesis. Purdue University, West Lafayette, IN, USA
42. Peppas NA, Foster LK (1994) J Appl Polym Sci 52: 763–768
43. Siegel RA, Firestone BA (1988) Macromolecules 21: 3254–3259
44. Firestone BA, Siegel RA (1988) Polym Comm 29: 204–208
45. Siegel RA, Falamarzian M, Firestone BA, Moxley BC (1988) J Contr Rel 8: 79–182
46. Tirrell DA (1987) J Contr Rel 6: 15–21
47. Borden KA, Eum KM, Langley KH, Tirrell DA (1987) Macromolecules 20: 454–456
48. Gehrke SH, Cussler EL (1989) Chem Eng Sci 44: 559–566
49. Bromberg LE (1991) J Membr Sci 62: 117–130
50. Bromberg LE (1991) J Membr Sci 62: 131–143
51. Ito Y, Inaba M, Chung D-J, Imanishi Y (1992) Macromolecules 25: 7313–7316
52. Osada Y, Honda K, Ohta M (1986) J Membr Sci 27: 327–338
53. Tanaka T (1979) Polymer 20: 1404–1412
54. Hirotsu S, Hirokawa Y, Tanaka T (1987) J Chem Phys 87: 1392–1395
55. Sagrario Beltran, John P Baker, Herbert H Hooper, Harvey W Blanch, John M Prausnitz (1991) Macromolecules 24: 549–551
56. Beltran S, Hooper HH, Blanch HW, Prausnitz JM (1990) J Chem Phys 92: 2061–2066
57. Hoffman AS, Afrassiabi A, Dong LC (1986) J Contr Rel 4: 213–222
58. Dong LC, Hoffman AS (1982). In: Paul S Russo (ed) ACS Symposium Series: Reversible Polymeric Gels and Related Systems. American Chemical Society pp 237–244
59. Cole C-A, Schreiner SM, Priest JH, Monji N, Hoffman AS (1987). In: Paul S Russo (ed) ACS Symposium Series: Reversible Polymeric Gels and Related Systems. American Chemical Society pp 245–254
60. Dong LC, Yan Q, Hoffman AS (1992) J Contr Rel 19: 171–178
61. Park TG, Hoffman AS (1992) J Appl Polym Sci 46: 659–671
62. Kabra BG, Gehrke SH (1991) Polym Comm 32: 322–323
63. Okano T, Bae YH, Jacobs H, Kim SW (1990) J Contr Rel 11: 255–265
64. Gutowska A, Bae YH, Feijen J, Kim SW (1992) J Contr Rel 22: 95–104
65. Feil H, Bae YH, Feijen J, Kim SW (1991) J Membr Sci 64: 283–294
66. Bae TH, Okano T, Kim SW (1989) J Contr Rel 9: 271–279
67. Feil H, Bae YH, Feijen J, Kim SW (1992) Macromolecules 25: 5528–5530
68. Yoshida R, Sakai K, Okano T, Sakurai Y (1992) J Biomater Sci Polym (ed) 3: 243–252
69. Freitas RFS, Cussler EL (1987) Chem Eng Sci 42: 97–103
70. Trank SJ, Cussler EL (1987) Chem Eng Sci 42: 381
71. Marchetti M, Prager S, Cussler EL (1990) Macromolecules 23: 1760–1765
72. Otake K, Inomata H, Konno M, Saito S (1989) J Chem Phys 91: 1345–1350
73. Ishihara K, Matsui K (1986) J Polym Sci: Polym Lett (ed) 24: 413–417
74. Ishihara K, Kobayashi M, Shinohara I (1984) Polymer J 16: 647–651
75. Ishihara K, Kobayashi M, Ishimaru N, Shinohara I (1984) Polymer J 16: 625–631
76. Ito Y, Casolaro M, Kono K, Imanishi Y (1989) J Contr Rel 10: 195–203
77. Chung DJ, Ito Y, Imanishi Y (1992) J Contr Rel 18: 45–54

78. Heller J, Chang AC, Rodd G, Grodsky GM (1990) J Contr Rel 13: 295–302
79. Kim SW, Pai CM, Makino K, Seminoff LA, Holmberg DL, Gleeson JM, Wilson DE, Mack EJ (1990) J Contr Rel 11: 193–201
80. Horbett TA, Ratner BD, Kost J, Singh M (1984). In: Anderson JM, Kim SW (eds) Recent advances in drug delivery systems. Plenum Press, New York, pp 209–220
81. Doi M, Mitsumoto M, Hirose Y (1992) Macromolecules 25: 5504–5511
82. Shiga T, Hirose Y, Okada A, Kurauchi T (1993) J Appl Polym Sci 47: 113–119
83. Shinkai S, Nakaji T, Ogawa T, Shigematsu K, Manabe O (1981) J Am Chem Soc 103: 111–115
84. Irie M, Kuwatchaku D (1986) Macromolecules 19: 2476–2480
85. Irie M (1986) Macromolecules 19: 2890–2892
86. Peiffer DG, Lundberg RD (1985) Polymer 26: 1058–1068
87. Khoklov AR, Kramarenko EYu, Makhaeva EE, Starodubtzev SG (1992) Macromolecules 25: 4779–4783
88. Bekturov E, Bimendina L (1981) Adv Polym Sci 41: 99–147
89. Osada Y (1987) Adv Polym Sci 82: 2–46
90. Chatterjee SK, Yadav D, Sudipta Ghosh, Khan AM (1989) J Polym Sci: Part A: Polym Chem 27: 3855–3863
91. Kopylova YeM, Valuyeva SP, El'tsefon BS, Rogacheva VB, Zezin AB (1987) Vysokomolekul Soed A29: 517–524
92. Kono K, Tabata F, Takagishi T (1993) J Membr Sci 76: 233–243
93. Kalyuzhnaya RI, Rudman AR, Vengerova NA, Razvodovskii YeF, El'tsefon BS, Zezin AB (1975) Vysokomolekul Soed A17: 2786–2792
94. Kalyuzhnaya RI, Volynskii AL, Rudman AR, Vengerova NA, Razvodovskii YeF, El'tsefon BS, Zezin AB (1976) Vysokomolekul Soed A18: 71–76
95. Khokhlov AR, Kramarenko EYu, Makhaeva EE, Starodubtzev SG (1992) Macromolecules 25: 4779–4783
96. Samsonov GV (1979) Vysokomolekul Soed A21: 723–733
97. Tsuchida E, Abe K (1982) Adv Polym Sci 45: 1–119
98. Ferguson J, Shah SAO (1968) European Polym J 4: 343–354
99. Matuszewska-Czerwik J, Polowinski S (1988) Polym Bull 19: 149–154
100. Van de Grampel HT, Santing AGM, Tan YY, Challa G (1992) J Polym Sci: Part A, Polym Chem (ed) 30: 787–796
101. Bimendina LA, Roganov VV, Ye A Bekturov (1974) Vysokomolekul Soed A16: 2810–2814
102. Ohno H, Abe K, Tsuchida E (1978) Makromol Chem 179: 755–763
103. Chatterjee SK, Rajabi FH, Farahani BV, Chatterjee N (1991) Polym Comm 32: 473–476
104. Chatterjee SK, Chatterjee N, Amarendra M Khan, Sudipta Ghosh (1991) Polym Comm 32: 220–224
105. Jeun SH, Park SM, Ree T (1989) J Polym Sci: Part C: Polym Lett 27: 161–165
106. Park SM, Jeun SH, Ree T (1989), J Polym Sci: Part A: Polym Chem 27: 4109–4117
107. Lee JY, Painter PC, Coleman MM (1988) Macromolecules 21: 346–354
108. Lee JY, Painter PC, Coleman MM (1988) Macromolecules 21: 954–960
109. Lichkus AM, Painter PC, Coleman MM (1988) Macromolecules 21: 2636–2641
110. Staikos G, Bokias G (1991) Makromol Chem 192: 2649–2657
111. Kim O-K, Choi LS, Long T, McGrath K, Armistead JP, Yoon TH (1993) Macromolecules 26: 379–384
112. Bailey FE Jr, Lundberg RD, Callard RW (1964) J Polym Sci: Part A 2: 845–851
113. Antipina AD, Baranovskii WYu, Papisov IM, Kabanov VA (1972) Vysokomol Soyed A14: 941–949
114. Osada Y, Sato M (1976) J Polym Sci: Polym Lett (ed) 14: 129–134
115. Osada Y (1979) J Polym Sci: Polym Chem 17: 3485–3498
116. Baranovsky V, Shenkov S, Rashkov I, Borisov G (1991) European Polym J 27: 643–647
117. Bokias G, Staikos G, Iliopoulos I, Audebert R (1994) Macromolecules 27: 427–431
118. Papisov IM, Baranovskii VYu, Sergieva YeI, Antipina AD, Kabanov VA (1974) Vysokomol Soyed A16: 1133–1141
119. Oyama HT, Tang WT, Frank CW (1987) Macromolecules 20: 474–480
120. Oyama HT, Tang WT, Frank CW (1987) Macromolecules 20: 1839–1847
121. Hemker DJ, Garza V, Frank CW (1987) Macromolecules 23: 4411–4418
122. Hemker DJ, Frank CW (1987) Macromolecules 23: 4404–4410
123. Scranton AB, Klier J, Peppas NA (1991) J Polym Sci: Part B: Polymer Physics
124. Daoust H, Darveau R, Laberge F (1990) Polymer 31: 1946–1949

125. Jeon SH, Ree T (1988) J Polym Sci: Part A: Polym Chem 26: 1419–1428
126. Chen H-L, Morawetz H (1982) Macromolecules 15: 1445–1447
127. Soutar I, Swanson L (1990) Macromolecules 23: 5170–5172
128. Adachi H, Nishi S, Kotaka T (1982) Polym, J 14: 985–992
129. Nishi S, Kotaka T (1985) Macromolecules 18: 1519–1525
130. Nishi S, Kotaka T (1989) Polym J 21: 393–402
131. Nishi S, Kotaka T (1986) Polym J 21: 393–402
132. Osada Y, Takeuchi Y (1981) J Polym Sci: Polym Lett (ed) 19: 303–308
133. Osada Y (1980) J Polym Sci: Polym Lett (ed) 18: 281–286
134. Khokhlov and Kramarenko EY (1993) Makromol Chem, Theory Simul 2: 169–177
135. Drummond RK, Klier J, Alameda JA, Peppas NA (1989) Macromolecules 22: 3816–3818
136. Klier J, Scranton AB, Peppas NA (1990) Macromolecules 23: 4944–4949
137. Bell CL (1994) PhD thesis. Purdue University
138. Peppas NA, Klier J (1991) J Contr Rel 16: 203–214
139. Flynn GL, Yalkowsky SH, Roseman TJ (1974) J Pharm Sci 63: 479–510
140. Crank J (1975) The mathematics of diffusion. Clarendon Press, Oxford, 2 edition
141. Peppas NA, Meadows DL (1983) J Membr Sci 16: 361–377
142. Anderson JL, Quinn JA (1974) Biophys J 14: 130–150
143. Dubin SB, Lunacek JH, Benedek GB (1967) Proc Natl Acad Sci USA 57: 1164–1171
144. Gladden JK, Dole M (1953) J Am Chem Soc 75: 3900–3904
145. Brenner H, Gaydos LJ (1977) J Coll Interfac Sci 58: 312–356
146. Deen WM (1987) AIChE J 33: 1409–1425
147. Malone DM, Quinn JL (1978) Chem Eng Sci 33: 1429–1440
148. Smith III FG, Deen WM (1980) J Coll Interfac Sci 78: 444–465
149. Kim JT, Anderson JL (1991) Ind Eng Chem Res 30: 1008–1016
150. Yasuda H, Peterlin A, Colton CK, Smith KA, Merrill EW (1969) Die Makromol Chem 126: 177–186
151. Yasuda H, Lamaze CE, Ikenberry LD (1968) Die Makromol Chem 118: 19–35
152. Yasuda H, Ikenberry LD, Lamaze CE (1969) Die Makromol Chem 125: 108–118
153. Yasuda H, Lamaze CE (1971) J Macromol Sci Phys B5: 111–134
154. Kim SW, Cardinal JR, Wisnewski S, Zentner GM (1990). In: Rowland SP (ed) Water in Polymers. American Chemical Society, Washington, DC pp 347–359
155. Peppas NA, Reinhart CT (1983) J Membr Sci 15: 275–287
156. Reinhart CT, Peppas NA (1984) J Membr Sci 18: 227–239
157. Burzak K, Fujisato T, Hatada M, Ikada Y (1994) Biomaterials, 15: 231–238
158. Peppas NA, Moynihan HJ (1985) J Appl Polym Sci 30: 2589–2606
159. Harland RS, Peppas NA (1987) Polym Bull 18: 553–556

Polymeric Dental Composites: Properties and Reaction Behavior of Multimethacrylate Dental Restorations

K. S. Anseth[1], S. M. Newman[2], and C. N. Bowman[1, *]

[1] Department of Chemical Engineering, University of Colorado, Boulder, Colorado 80309-0424

[2] Department of Restorative Dentistry, Health Sciences Center, University of Colorado, Denver, Colorado 80262-0284

With over 200 million dental restorations performed each year, the importance of developing a restorative material with tooth-like appearance and properties cannot be underestimated. In this article, the use of poly (multimethacrylates) as dental composites is summarized from both fundamental and practical sides. Detail is provided regarding the utilization, procedures, and problems with polymeric composite restoratives, and a complete discussion of the polymerization kinetics and the polymer structural evolution is presented. In the final sections, properties of current composite materials and suggestions for what areas of research would prove most promising are presented.

---

* Author to whom correspondence should be addressed

1 Introduction and Scope of This Article

With more than 200 million dental restorations performed each year, the importance of using a restorative material which is both safe and durable should not be underestimated. Currently, dental amalgam is used in the vast majority of these restorations; however, recent scrutiny of mercury levels in dental amalgam and the desire for tooth colored restorations have led to increasing demand for polymeric dental composites. Polymeric composites, generally composed of a multimethacrylate and a ceramic glass filler, have primarily been used for anterior tooth restorations in which color matching is imperative for aesthetic purposes.

Composite resins allow for color matching, conservative cavity preparation, and simple preparation through intraoral photopolymerization. These advantages have made composites an increasingly popular substitute for amalgam in dental restorations, especially when aesthetics are of concern. In this article, we will focus on the actual process of forming dental composites, the properties of the composites that are formed, and a complete description of the photopolymerization of the multimethacrylates that produce the dental composite. We will only be focusing on the use of polymers as dental restorations. Other dental applications of polymers, e.g. dentures and ionomer cements (reviewed elsewhere by Scranton and Klier) will not be addressed.

In the first portion of this section, we will focus on the materials and processes used to form polymer dental composites. This section will be followed by a discussion of the problems associated with polymer composite materials. An overview of the photopolymerization behavior and the polymer structure of these highly crosslinked materials is presented in Sects. 3 and 4. Lastly, some of the properties of current composite resin formulations are presented.

2 Materials

2.1 Motivation for Polymeric Composites

Dental amalgams have been used as dental restorative materials in the United States for the past 160 years and are currently used in approximately 75% of all direct restorations [1]. Of the direct posterior permanent fillings placed in the United States by private practice dentists, 88% are dental amalgam [2]. This material has been considered the preferred material for several reasons. The first is amalgam's history of long term clinical success [3]. The dental amalgam is also easy to place with relatively little technique sensitivity. Dental amalgam is inexpensive to the patient when compared with currently competitive materials

[4]. Finally, amalgam approximates many of the properties required for replacement of tooth structure.

Despite these advantages, amalgam has been subject to repeated controversies surrounding the safety of the mercury levels in the material. There is a release of mercury vapor during the placement and removal of amalgam restorations [5]. The release of mercury during the functional life of the restoration is a major point of controversy [6–11]. Correlations between the total surface of amalgam restorations and the levels of mercury in the blood [12–14] or urine [15] are still being debated, but reported tissue levels are all well below any values of toxic concern. The Swedish government has even proposed the progressive elimination of amalgam as a dental restorative material by 1997 [16], but the motivation is that of environmental concerns about waste management of removed dental amalgam [17, 18].

While the risks associated with mercury in dental amalgam are debatable, other disadvantages are also apparent. The preparation of the tooth for amalgam restorations incorporates mechanical undercuts which inherently weaken the structure of the remaining tooth, whereas newer materials have the potential to be bonded into place with a good prognosis. Since amalgam is not strong in a thin layer, it must be placed in a layer thick enough to withstand mechanical biting forces. These tooth preparation requirements often necessitate the removal of healthy tooth structure beyond the carious tissue.

Amalgam is aesthetically unattractive compared to new materials. It has a metallic color that does not reproduce the natural appearance of the ceramo-organic tooth structure. The release of metallic ions from the amalgam restoration also can discolor the neighboring tooth structure [19].

The demand for aesthetic dental restorative materials continues to increase and may be the most important criterion for the promising future of the aesthetic polymeric composite resins. As the physical, mechanical, and wear properties of these materials improve, their use in dentistry will expand. The acid-etching of dental enamel [20] and dentin bonding procedures [21] will allow for conservative cavity preparation and the preservation of healthy tooth structure.

2.2 Composite Resins for Dental Restorations

Polymer resins were first introduced in the early 1940s as an aesthetic alternative to repair defects in anterior teeth. Some of the first resins were unfilled polymers of methyl methacrylate. Presently, these unfilled resins have been replaced by filled composite materials that limit the problems associated with polymerization volume shrinkage, abrasion or wear resistance, mechanical properties, water sorption, solubility, and thermal expansion. Polymeric composite materials generally consist of a monomer resin, a ceramic filler, a polymerization initiator or initiating system, and a coupling agent which binds the polymer phase to the filler phase. The composite material is then polymerized in situ to restore and render a functional tooth.

The majority of resins are composed of two dimethacrylate monomers, 2,2′-bis [4(2-hydroxy-3-methacryloyloxypropyloxy)phenyl] propane (Bis-GMA) and triethylene glycol dimethacrylate (TEGDMA) [22–28]. Typically, TEGDMA or other methacrylate monomers are added as viscosity modifiers to Bis-GMA to make the solution less viscous and more appropriate for clinical use. These diluents also allow for better distribution of the components during manufacture of these composite systems. Another common monomer used to make dental composites, especially those manufactured in Europe, is urethane dimethacrylate [24, 29, 30]. Ethoxy bisphenol A dimethacrylate is another modification of the Bis-GMA monomer that can be used to make a more hydrophobic polymer that would better withstand the wet oral environment. Other diluents include low viscosity diacrylates and dimethacrylates. Table 1 lists some of these monomers [31–37].

The ceramic filler phase generally consists of irregularly shaped pieces of fused silica or special glasses which possess extremely low or negative thermal expansion coefficients [38]. These fillers must also have refractive indices that match, as closely as possible, that of the final polymer in order to produce as translucent a base material as possible to pigment to the appropriate tooth matching shades needed. Filler amounts range anywhere from 45–85 mass % [22, 39–43] and are incorporated to produce stronger, more wear resistant composites that have lower polymerization volume shrinkage and lower thermal expansion.

Critical to the final properties of the composite is the size of the filler particle. The sizes used vary over several orders of magnitude, which greatly affects the ability to load the composite due to the surface area that needs to be wetted by the monomer. The largest particle sizes of conventional composites have evolved from an average size of 30 microns down to 1–3 microns and in some cases

Table 1. Some monomers used in dental composites [31–37]

BisGMA	2,2 bis[4-(2-hydroxy-3-methacryloyloxypropoxy)phenyl]propane
BisEMA	2,2 bis[4-(2-methacryloyloxyethoxy)phenyl]propane
BisMA	2,2 bis[4-methacryloyloxyphenyl]propane
UDMA	dimethacryloxyethyl [trimethyl hexamethylene diurethane]
TeEGDMA	tetraethylene glycol dimethacrylate
TEGDMA	triethylene glycol dimethacrylate
EGDMA	ethylene glycol dimethacrylate
HXGDMA	1,6 hexamethylene glycol dimethacrylate
TMCMA	3,3,5-trimethylcyclohexyl methacrylate
BMA	benzyl methacrylate
HEMA	hydroxyethyl methacrylate
MMA	methyl methacrylate
MAA	methacrylic acid
THFMA	tetrahydrofurfuryl methacrylate

as small as 0.7 microns. These ceramic particles may contain barium or other large atoms to impart radiopacity to the system [44]. The radiopacity is an important aid in dental diagnoses [45, 46]. Composites based on some of these micron and submicron sizes may be called "small particle" composites. As opposed to these systems, there are composites called "microfilled" which contain a pyrollitic silica in the range of 0.01 to 0.04 microns. The microfilled composites contain significantly less ceramic filler thus decreasing the strength, increasing the water sorption, and increasing the polymerization shrinkage. These systems do offer the dentist the ability to produce a smoother polished surface.

There are composite systems that contain both small particle size distributions and microfill distributions, called "hybrids", that are attempting to achieve the best of both kinds of systems. These systems may provide, in addition, some degree of dispersion strengthening and increased wear resistance. A more complex system of nomenclature for describing dental composites based on the filler loading system was proposed by Lutz and Phillips [39], but has not been widely used due to the increasing diversity of loading systems. Common coupling agents for bonding the ceramic filler to the resin are functionalized silanes which possess chemical groups that facilitate attachment to the filler surface (e.g., $-SiX_3$ groups where X normally is a chloro, alkoxy, or acetoxy group) [47–50] along with functional groups for incorporation in the polymer structure (e.g., methacrylate groups) [47–50]. The most common agent used is γ-methacryloxypropyltrimethoxysilane [51, 52]. Silanation of the ceramic can produce layers of resinous silane that would significantly alter the mass of the filler particles. The effect would lead to weight percent additions of filler that do not accurately predict the actual mass % of ceramic filler.

The commercially available dental composite systems are provided with varying curing mechanisms from light cured to chemically cured to dual-cured systems. The light curing systems are generally a single paste. This paste contains an $\alpha - 1, 2$ diketone and an amine reducing agent [22, 30, 53] that initiates the reaction when exposed to a visible light curing system of high intensity blue light (near 470–490 nm). These light cured systems are provided as a single paste in opaque tubes or single dose tips. The light curing mechanism allows the dentist to place and manipulate the material with essentially an unrestricted working time, yet the material will set within a minute using the appropriate intense light source. In contrast, chemically cured systems are two paste systems that react when they are mixed. The initiator is an amine-peroxide that induces free radical polymerization upon mixing. Typically, these polymerizations are slower than their light cured counterpart. The stability of the peroxide/monomer combination in the paste is achieved by allowing air to permeate the packaging to provide a sufficient shelf life to the product [54]. An alternative delivery of the chemically cured mechanism is to provide the ceramic filler particle coated with the peroxide and a fluid resin containing the amine. Such systems have been used but are currently out of favor in the market. Finally, dual-cured systems are supplied as two pastes that will chemically cure when mixed, but if exposed to curing light will react on a much shorter time scale. Certain unique dental applications are enhanced by having both mechanisms available.

2.3 Problems Associated with Dental Composites

Despite many years of research and attempts to develop the ideal polymeric composite restorative material, significant problems still exist with present day composite resins. These problems are associated with the polymer itself, the polymerization reaction, the durability of the restoration, and the aesthetics of the material.

One of the most significant problem with polymeric composites is their lack of durability that leads to rapid wear, discoloration, and eventually, failure [31, 47, 55–60]. Unfortunately, current methods of evaluating wear in vitro do not accurately predict the in vivo wear behavior, but they are useful in comparing materials within the same type or class.

The mechanism by which clinical degradation/wear occurs in polymer composites is not well understood [59, 60]. Several factors which are known to play a role are the strength of the interfacial bonding between the polymer and the filler, the hardness of the surface, and the magnitude of the induced stresses within the restoration [61]. Stresses occur because of interactions with food and opposing or proximal teeth, differences in thermal expansion [22, 62, 63], swelling from moisture uptake [64–67], and volume shrinkage during polymerization [68–74].

Another significant problem is the potential for recurrent decay of tooth structure around the restorative material. This decay may have been due to an undetected residual bacterial contamination left behind during the preparation of the tooth or to the leakage of new bacteria along the interface of the composite restorative material with the tooth. In either case, a better sealing of the composite restorative material to the tooth would decrease recurrent decay due to the decrease of food sources and new bacteria from the oral fluids leaking along the restoration [75–77]. Basic composite restorative materials have no antibacterial activity after polymerization. Bonding procedures to both enamel and dentin have been developed that greatly decrease the leakage of material at the interface. The decrease in hydraulic flow also may decrease the tooth's postoperative sensitivity [78–80]. The longevity of the sealing ability of these newer bonding systems has not been established. Increased understanding of the resin tooth interaction will increase the quality of the bonding procedures. Decreased polymerization shrinkage and decreased thermal expansion differences would decrease the stresses on these bonds.

The setting reaction of the composite system still needs major improvement due to the incomplete double bond conversion that occurs during these polymerization reactions. The final double bond conversion ranges anywhere from 55–75% [22, 81–87] which implies that a minimum of 6.25% of the monomer is left as extractable. All unreacted double bonds and monomer which remain in the system act as plasticizers, diminishing the strength of the polymer and leading to increased swelling. Also, only 55–75% of the double bonds which could participate in bonding the filler to the polymer are reacted. Finally, the most undesirable effect of the unreacted monomer is the possibility that it may leach into the body where it may exhibit cytotoxic effects. The effects of unreacted monomer in the body have not been completely evaluated [88] despite preliminary studies [1, 89]. By increasing the double bond conversion, the crosslinking density is increased, the residual monomer is decreased, and the swelling is decreased.

For systems where the polymerization is not restricted to the mouth (i.e., inlays, outlays, and veneers of composite), increasing the temperature may lead to higher conversions [90] and degrees of crosslinking with expected improvements in physical properties of the material [91]. For the cases of crown and bridge prostheses, increasing the temperature might be a viable alternative to reach higher conversions. For the dental restorations that must be polymerized intraorally and the reaction temperature is fixed, the conversion is determined by the light intensity, initiator concentration, and the monomer resin system.

One alternative to reaching higher monomer conversions is to copolymerize Bis-GMA with a linear methacrylate or acrylate based monomer. One might expect lower crosslinking densities in such systems; and therefore, the mechanical and physical properties required of the restoration might be compromised. Alternatively, if the comonomer composition is chosen such that significantly higher levels of conversion are reached than in standard homopolymerizations of dimethacrylates, the resulting polymer networks might actually possess the desired and even higher degree of crosslinking. A second method of attaining higher conversions in these systems is to increase the rate of polymerization by increasing either the light intensity or the initiator concentration. As will be discussed later, higher rates of polymerization tend to produce excess free volume which enhances the mobility of the system and leads to higher conversions.

3 Reaction Properties

As described above, the majority of materials used in dental applications contain multifunctional monomers that polymerize to form highly crosslinked polymer networks. In addition, many of the applications, such as tooth restorations, require that the crosslinked polymer is polymerized intraorally. This restriction can often complicate the cure of the monomers since the material is exposed to oxygen and moisture in the oral environment. Also, depending on the thickness of the restoration, the material might not be uniformly cured because of variations in light intensity with depth in the sample. These problems can lead to unreacted functional groups in the material which may significantly alter the properties and wear of the restoration.

Thus, to improve upon the physical and mechanical properties of the polymer material, one must not only consider the materials used, but also the conditions under which the polymer was formed. These reaction conditions, along with the type of monomer system chosen, will completely control the conversion of functional groups in the system. More importantly, the conversion will ultimately determine the mechanical, physical, and wear properties of the material. Since most dental materials are crosslinked polymers, characterizing the polymerization reaction becomes even more important since the physical nature of a crosslinked polymer is fixed upon completion of the polymerization. For example, not only is the microstructure (i.e. the degree of crosslinking) largely unalterable after polymerization, but the system is insoluble and fixed macroscopically. Clearly, to produce crosslinked networks with the desired material properties, one must ascertain the appropriate reaction conditions and the effects of the reaction conditions on the network structure.

Hence, the ability to understand and characterize the polymerization reaction behavior is extremely important and fundamental in attempts to develop improved dental materials. The following sections will discuss the general characteristics of polymerization reactions with particular emphasis on the complexities occurring in the high crosslinking regime. The effects of the reactions conditions, like oxygen and sample thickness, on the polymerization rate will also be included.

3.1 Polymerization Rate

In classical free radical polymerizations, the rate of polymerization, R_p, is often expressed as [92]

$$R_p = \frac{k_p}{k_t^{1/2}} [M] \left(\frac{R_i}{2}\right)^{1/2} \tag{1}$$

where $[M]$ is the concentration of monomer or functional groups, R_i is the rate of initiation, and k_p and k_t are the kinetic constants for propagation and termination, respectively. In the development of Eq (1), the reactivity of radicals was assumed to be equal and independent of chain length and the rate of radical initiation and consumption was assumed to be equal (i.e., the pseudo-steady state assumption was made). The validity of these assumptions for the monomer systems used in dental materials, as well as their validity in the high crosslinking regime where the reactions are occurring, will be addressed later.

Since most of the dental applications employ a photoinitiated process (visible or ultraviolet light), the rate of initiation may be expressed as [92]

$$R_i = 2\phi I_a \tag{2}$$

where ϕ is the quantum yield and I_a is the absorbed light in terms of moles of light quanta per liter-second. Equation (2) also assumes that each initiator molecule decays to form two radicals. Physically, the quantum yield for initiation is the number of propagating chains initiated per light photon absorbed and represents a measure of the initiator efficiency. Because the light intensity decays exponentially through the sample, the intensity at a distance x into the sample is given by [92]

$$I = I_0 e^{-\varepsilon[A]x} \tag{3}$$

where I_0 is the incident light intensity in terms of moles of light quanta per liter-second, ε is the extinction coefficient, and $[A]$ is the photoinitiator concentration. The energy of one mole of light quanta (also referred to as an Einstein) is equal to Nhv where N is Avogadro's number, h is Planck's constant, and v is the frequency of the incident light. Integrating over the thickness, b, of the reacting sample, the expression for the total amount of light absorbed becomes [92]

$$I_a = I_0[1 - e^{-\varepsilon[A]b}]. \tag{4}$$

For thin systems with low absorbance, the exponential can be expanded and truncated at two terms so that the absorbed light intensity becomes [92]

$$I_a = \varepsilon I_0[A]b. \tag{5}$$

The rate of polymerization then becomes [92]

$$R_p = \frac{k_p}{k_t^{1/2}}[M](\phi \varepsilon I_0[A]b)^{1/2}. \tag{6}$$

The above expression indicates that the rate of polymerization is proportional to the square root of both the incident light intensity and the photoinitiator concentration. This expression is accurate to within 2.5% for thin films where $\varepsilon[A]b < 0.2$. As the thickness of the sample is increased, the light intensity can decrease appreciably across the sample, and Eq. (4) must be used.

3.2 Effects of Light Penetration and Oxygen

Figure 1 shows a simulated profile of conversion of functional groups in a multifunctional monomer vs reaction time at different depths in the sample, assuming an isothermal polymerization. At the surface of the restoration, the sample is exposed to the highest intensity of light and the conversions are correspondingly greatest. Depending on the thickness of the sample, the differences between conversions at the surface and in the bulk of the sample can be appreciably different. These variations in conversion with depth are dependent on the molar absorptivity and concentration of the photoinitiator. Considering the illustrated case where $\varepsilon[A] = 5$ cm^{-1}, differences in the extent of cure are seen 1.4 mm from the surface, and at a distance of 4.6 mm from the surface the average conversion can vary up to 15% from the surface conversion depending on the polymerization time.

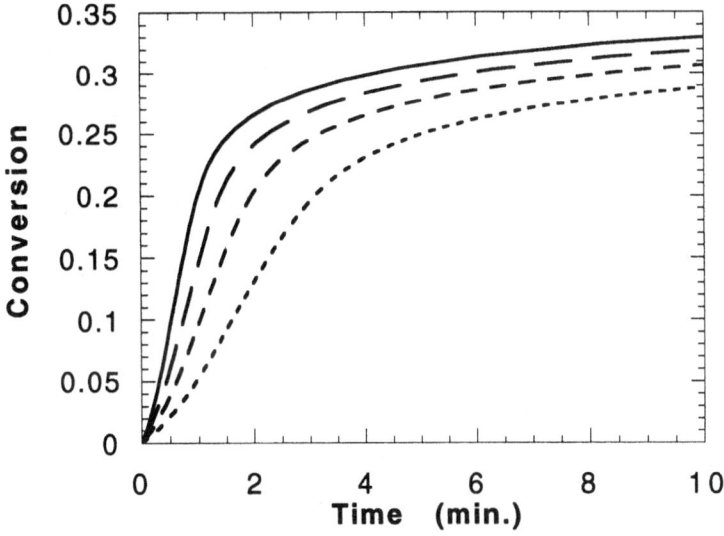

Fig. 1. Simulated profile of conversion of double bonds in a multifunctional monomer vs polymerization time at different depths in the polymer: (——, surface), (— — —, 1.4 mm), (— — —, 2.8 mm), and (- - -, 4.6 mm)

Hirose et al. [93] studied the differences between several visible-light curing units operating at intensity levels ranging from ca. 100 mW/cm² to 300 mW/cm² on the depth of cure in dental composite resins. The Knoop hardness was measured as a function of sample depth for the dental composite resin Occlusin. The system was polymerized with two different visible light curing units (146.6 mW/cm² and 222.7 mW/cm²) for 20, 40, and 60 s, and the Knoop hardness was reported at depths ranging from 0.5 to 9 mm for each of these conditions. Indicative of the variation in conversion with depth in the sample, the Knoop hardness decreased by more than 50% from 0.5 mm (67) to 5 mm (25) in a sample that was polymerized 20 s. As the curing time was increased to 60 s, the Knoop hardness increased to approximately 70 at 0.5 mm and 34 at 8 mm. In addition, as the intensity of the light source was decreased, the Knoop hardness was correspondingly lower at all curing times and depths. Again, these lower values for the Knoop hardness with lower light intensities were further evidence for the lower conversions reached with decreasing rates of polymerization.

While variations in light intensity across a sample may be controlled to some extent by molar absorptivity and concentration of the initiator chosen, as well as polymerizing larger restorations in layers rather than in bulk, the effects of oxygen are somewhat more difficult to overcome. In most practical applications of monomer resins in dental materials, the polymerization is carried out in an atmosphere where oxygen is present. The effects of oxygen on the polymerization are typically adverse [94–99]. They include [95] the presence of an induction period between the initial irradiation and the beginning of significant polymerization, a reduced rate of polymerization and crosslinking, and incomplete functional group conversion. Phan [97], for example, has shown a five-fold decrease in polymerization efficiency when oxygen was present in the ultraviolet polymerization of methyl methacrylate initiated by 2,2-dimethoxy-2-phenyl-acetophenone.

Oxygen has two possible interactions during the polymerization process [94], and these reactions are illustrated in Fig. 2. The first of these is a quenching of the excited triplet state of the initiator. When this quenching occurs the initiator will absorb the light and move to its excited state, but it will not form the radical or radicals that initiate the polymerization. A reduction in the quantum yield of the photoinitiator will be observed. The second interaction is the reaction with carbon based polymerizing radicals to form less reactive peroxy radicals. The rate constant for the formation of peroxy radicals has been found to be of the order of 10^9 l/mol-s [94]. Peroxy radicals are known to have rate constants for reaction with methyl methacrylate of 0.24 l/mol-s [100], while polymer radicals react with monomeric methyl methacrylate with a rate constant of 515 l/mol-s [100]. This difference implies that peroxy radicals are nearly 2000 time less reactive. Obviously, this indicates that even a small concentration of oxygen in the system can severely reduce the polymerization rate.

To overcome quenching, some photoinitiators decay quite rapidly via unimolecular fragmentation from the excited triplet state and leave little time for the

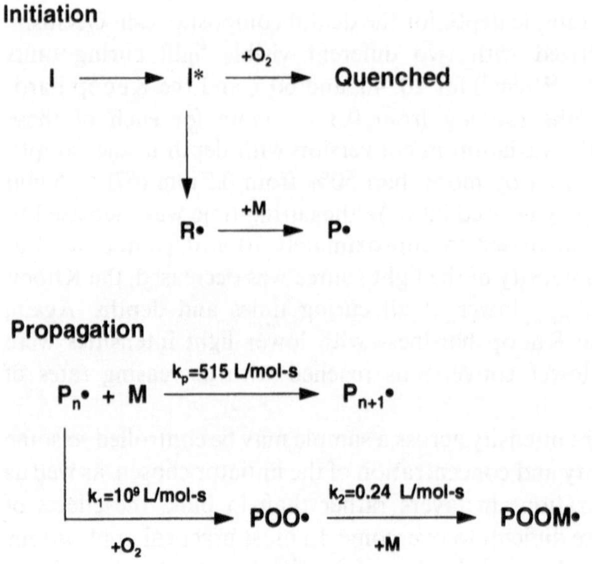

Initiation

$$I \longrightarrow I^* \xrightarrow{+O_2} \text{Quenched}$$

$$R\bullet \xrightarrow{+M} P\bullet$$

Propagation

$$P_n\bullet + M \xrightarrow{k_p=515 \text{ L/mol-s}} P_{n+1}\bullet$$

$$\xrightarrow[+O_2]{k_1=10^9 \text{ L/mol-s}} POO\bullet \xrightarrow[+M]{k_2=0.24 \text{ L/mol-s}} POOM\bullet$$

Fig. 2. Polymerization reactions in the presence of radical scavenging oxygen. Kinetic constants are approximate values for methyl methacrylate

oxygen to quench the initiator in the triplet state [95]. This, however, does not prevent the oxygen from reacting with free radicals to form the less reactive peroxy radicals. Clearly, this reaction can and will occur as long as oxygen is present in the system.

After curing is completed, oxygen can react with radicals that are trapped in the polymer network. Residual, trapped radicals have been experimentally detected by several researchers using electron spin resonance spectroscopy (ESR) [101–106]. The nature of these studies will be discussed later in the heterogeneity section of the modeling of network structure, and here we will restrict our discussion to the reactions of these trapped radicals. Trapped radicals can propagate with unreacted functional groups, terminate with other residual radicals in the network, or undergo side reaction with molecules such as oxygen.

Li et al. [107] characterized the reaction of oxygen with trapped radicals and related this reaction to the permeation rate of oxygen in crosslinked copolymer networks of methyl methacrylate (MMA) and ethylene glycol dimethacrylate (EGDMA). They used ESR to determine the concentration of residual radicals and monitored the radical decay as a function of oxygen permeation time. Interestingly, as the wt% of crosslinker, EGMDA, was increased the permeability coefficient of oxygen increased until reaching a maximum around 70 wt% EGDMA and then decreased. The authors related these permeation coefficients to network structure and concluded that the 70 wt% EGDMA network had the most porous structure.

Other possible explanations include the changes in overall conversion of functional groups with increasing crosslinker content. The crosslinking density of the network is a balance between increasing the relative amount of crosslinker in the system and the total functional group conversion. In general, as the amount of multifunctional monomer is increased, the polymerization becomes diffusion controlled at lower conversions and the maximum functional group conversion decreases. Even though the conversions were not reported, it is reasonable to assume that the networks with the highest percent EGDMA probably had the lowest functional group conversion. Thus, to produce networks with the highest crosslinking density, the amount of multifunctional monomer present must be balanced by the limiting functional group conversion. Another consideration is that the oxygen and radical reactions were characterized above the glass transition temperature of the network (110–180 °C). At these high temperatures, it is possible that depolymerization (i.e. the reverse propagation reaction) could be occurring.

Li et al. [107] also reported the change in residual radical concentration vs oxygen permeation time for a homopolymerization of EGDMA. At 110 °C, the radical concentration decreased by approximately 12.5% in 90 min. While at 180 °C, the radical concentration dropped nearly 62.5% over the same time period. While these reactions occurred over a relatively short time period, the temperatures were quite high and certainly one would expect significantly reduced rates at lower temperatures below the glass transition temperature of the polymer. For example in a dental restoration the polymer is maintained near body temperature (37 °C), below the glass transition temperature of the network, where the oxygen permeability would be considerably lower as compared to 110–180 °C. As for the stability of these radicals, Li et al. [107] have reported that radicals trapped in a glassy copolymer network of MMA/EGDMA remain stable with no decay in their concentration over a 2 year period when the system was kept at room temperature and in the absence of oxygen.

Kloosterboer et al. [106] also studied the reaction of oxygen with trapped radicals in photopolymerized networks of 1,6-hexanediol diacrylate (HDDA). Unlike the methacrylate networks, they found the radical concentration decayed to zero at room temperature in the presence of air after only 15 min. Considering that the accumulation of radicals in highly crosslinked photopolymers is relatively prominent, the reactions of these radicals with oxygen and the resulting peroxides that are formed might be worthy of further attention. In particular, Kloosterboer et al. [106] have suggested that these peroxides might well be related to the often claimed inferior long-term mechanical properties of photopolymers compared with thermally polymerized materials.

3.3 Reaction Behavior

Polymerization reactions of multifunctional monomers such as those used in dental restorations occur in the high crosslinking regime where anomalous behavior is often observed, especially with respect to reaction kinetics. This behavior includes autoacceleration and autodeceleration [108–112], incomplete functional group conversion [108, 109, 113–116], a delay in volume shrinkage with respect to equilibrium [108, 117, 118], and unequal functional group reactivity [119–121]. Figures 3 and 4 show a typical rate of polymerization for a multifunctional monomer as a function of time and conversion, respectively. Several distinctive features of the polymerization are apparent in the rate profiles.

Autoacceleration, i.e., the increase in rate of polymerization despite the consumption of monomer, is nearly always observed in these systems because of the extremely restricted diffusion of radicals in the highly crosslinked polymer. In these polymers, bulk mobility of the radicals is severely hindered leading to termination only when two radicals are able to diffuse together. Then, as the radical diffusivity continues to decrease, the termination rate is reduced and the radical concentration increases, often by several orders of magnitude [103, 104, 122]. The increased radical concentration leads to higher rates of polymerization, hence autoacceleration. In reexamining Eq. (1), one notes that autoacceleration occurs because k_t decreases by several orders of magnitude [123–126], thus increasing the polymerization rate.

Autodeceleration of the polymerization kinetics has also been frequently observed during polymerizations of this type, especially during polymerizations of small, high functionality monomers. Autodeceleration occurs because the

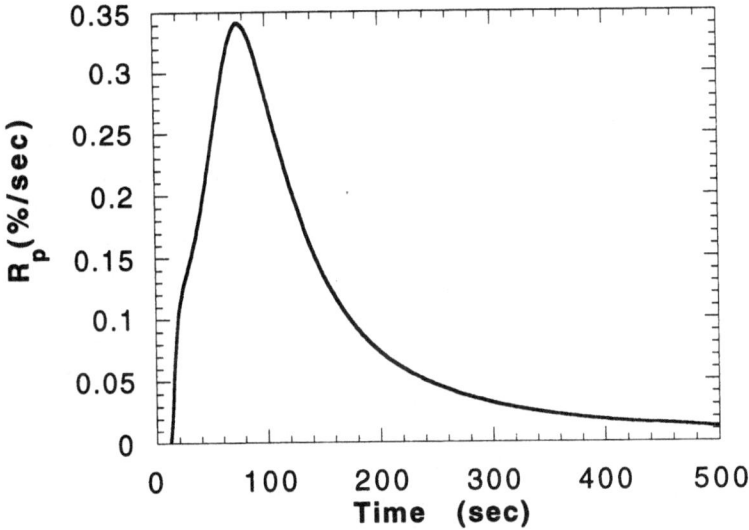

Fig. 3. A typical rate of polymerization, R_p, for a multifunctional monomer vs time

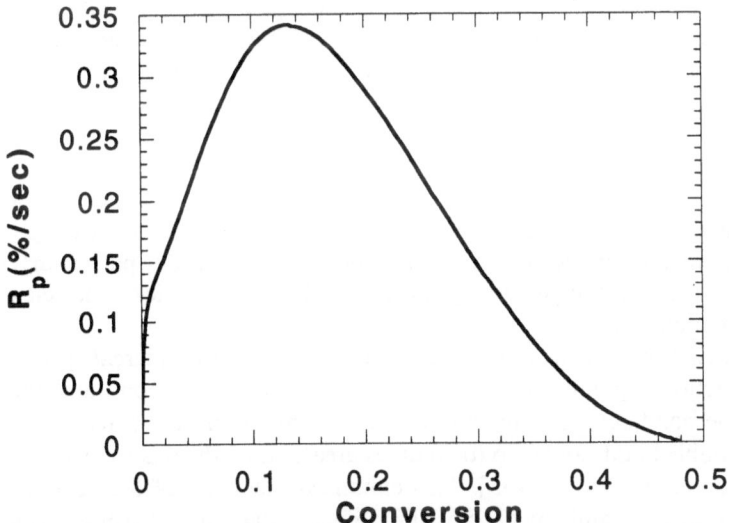

Fig. 4. A typical rate of polymerization, R_p, for a multifunctional monomer vs conversion of double bonds

propagation reaction eventually becomes diffusion controlled along with the termination reaction. As the mobility of the reactive functional groups is further reduced, the functional groups become less and less reactive until the reaction stops from vitrification. Again, looking at Eq. (1), one finds that k_p decreases much more rapidly than k_t during autoacceleration, thus causing the observed rate decrease.

Depending on the monomer system and the reaction conditions by which the system was polymerized, the polymerization typically ceases before complete conversion of the functional groups. This incomplete functional group conversion is a primary concern with dental materials especially with respect to the residual, unreacted monomer in the system. Unreacted pendant double bonds can affect the stability of the restoration while residual monomer poses the threat of leachability. Several researchers have studied monomer and functional group conversion in various resin systems under different curing conditions [85–87], resin compositions [22, 81–87], and filler content [84]. In general, the degree of conversion increased with increasing light intensity and decreasing filler content. A further discussion on determining the amount of residual, unreacted monomer from the methacrylate group conversion is included in a later section on kinetic gelation modeling of these systems.

To explain the increase in functional group conversion with increasing rates of polymerization, the physical effects of volume shrinkage/physical aging on the polymerization kinetics and reaction behavior must be considered. Several researchers have experimentally observed increased conversions with higher rates of polymerization [108, 113, 116, 127]. Effects of this have been seen in the properties of dental composites by several researchers [23, 85, 93, 113, 128–131]. As the rate of polymerization increases, the polymerizing system cannot shrink rapidly enough to maintain its equilibrium volume (i.e. volume shrinkage is occurring on a much slower time scale than reaction). The slow relaxation leads to a volume excess beyond equilibrium. This excess volume is a free volume and it enhances the mobility in the system. Higher conversions are reached in these systems as compared to equilibrium volume systems because of the higher mobility. Compared to linear polymers, volume shrinkage plays an increasingly important role in the kinetics of crosslinked network formation, particularly since these networks tend to gel at very low double bond conversions and relax very slowly to equilibrium.

Finally, when polymerizing multifunctional monomers, two different species of reactive functional groups can exist, the monomeric double bond and the pendant double bond(s). At low functional group conversions, the reactivity of the pendant double bond can be up to 50 times greater than the reactivity of the monomeric double bond [132, 133]. This enhanced reactivity of the pendant double bond is caused mainly by the close proximity of the pendant group to the active radical. As the pendant double bonds react, microgel regions are formed in the polymer and lead to structural heterogeneity. As will be discussed later, the fate of these pendant double bonds (i.e. whether they form cycles or crosslinks) is extremely important in determining the strength and properties of the restoration.

As the reaction proceeds to higher conversions, the unreacted pendant double bonds eventually become shielded in the microgel regions, and the reactivity of the monomeric double bond approaches, and in some cases surpasses, that of the pendant double bond. This changing reactivity, especially at the end of the reaction, can significantly affect the amount of unreacted monomer in the network. Unfortunately, most on-line experimental techniques such as differential scanning calorimetry (DSC) and infrared spectroscopy (IR) can only provide information about the average rate of reaction and total conversion of the functional groups. In addition, since the systems are highly crosslinked, techniques which extract the unreacted monomer to determine the fraction of unreacted monomer vs unreacted pendant functional groups are often impossible or prohibitively slow. Lacking a definitive experimental technique to measure the extent of monomer conversion vs pendant functional group conversion, one is limited either to the homogeneous assumption of equal reactivity or to models for the relative reactivity of the functional groups.

For completeness in the discussion of reactivity of functional groups, many of the polymeric dental materials are produced from a copolymerization of two different type monomer molecules like Bis-GMA and TEGDMA. In general, most of the systems employ comonomer systems that are equivalent in chemical reactivity (e.g. copolymerizing a methacrylate based monomer with another methacrylate based monomer). As long as the propagation reaction that consumes the monomer molecules remains chemically controlled, the two methacrylate monomers should possess approximately the same reactivity. However, later in the reaction when propagation becomes diffusion controlled, the monomers might have significantly different reactivities based on the diffusivity of the molecule rather than its chemical reactivity. This diffusion controlled regime can lead to compositional drifts in the network as smaller, more mobile molecules will have a relatively higher reactivity as compared to bulkier, stiffer molecules like Bis-GMA.

3.4 Kinetic Modeling of Crosslinked Network Reaction Behavior

3.4.1 Models

In multifunctional monomer polymerizations, the primary reaction characteristics that must be modeled are diffusion limited reactions, the volume shrinkage rate, and the attainment of a maximum conversion. As stated previously, autoacceleration is the phenomenon observed when the kinetic constant for termination decreases significantly faster than the kinetic constant for propagation and the rate actually has a maximum greater than the initial rate. Autoacceleration is observed when the termination reaction between radicals, which is always diffusion limited, is slowed as the diffusion of the macroradicals is slowed. Autodeceleration is observed in reactions following autoacceleration, and despite the presence of unreacted functional groups and radicals, the polymerization reaches a limiting conversion where the diffusion controlled propagation reaction occurs only at vanishingly slow rates.

Because of the strong dependence of composite properties on this final conversion, it is imperative that models of polymerizing systems be used to predict the dependence of the rate of polymerization and, hence, conversion on reaction conditions. The complexities of modeling such systems with autoacceleration, autodeceleration, and reaction diffusion all coupled with volume relaxation are enormous. However, several preliminary models for these systems have been developed [177, 125, 126, 134–138]. These models are nearly all based on the coupled cycles illustrated in Fig. 5.

The lower cycle represents the chemical changes occurring during polymerization and relates them to the free volume of the system. In general, free volume of a polymer system is the total volume minus the volume occupied by the atoms and molecules. The occupied volume might be a calculated van der Waals excluded volume [139] or the fluctuation volume swept by the center of gravity of the molecules as a result of thermal motion [140, 141]. Despite the obscurity in an exact definition for the occupied volume, many of the molecular motions in polymer systems, such as diffusion and volume relaxation, can be related to the free volume in the polymer, and therefore many free volume based models are used in predicting polymerization behavior [117,126, 138].

Since autoacceleration and autodeceleration arise from mass transfer limitations in the polymerizing system, it is important to relate the kinetic constants to diffusivity of the monomer and polymer. Again, these diffusion coefficients for the monomer and polymer can be developed in terms of the free volume in the polymerizing system. Likewise, k_p and k_t will depend on the diffusion coefficients of the monomer and polymer respectively.

Since the free volume of the system changes with functional group conversion, the dependence of the kinetic constants on conversion is determined. Thus, in the lower cycle, the free volume of the system affects the diffusion coefficients

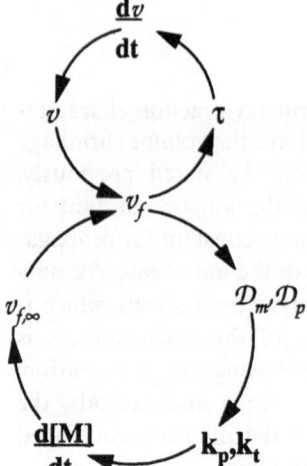

Fig. 5. Coupled polymerization (*lower*) and volume relaxation (*upper*) cycles that describe the changes occurring during polymerization

of the monomer and polymer. These diffusion coefficients directly control the kinetic constants for propagation and termination, which determine the rate of functional group conversion, $d[M]/dt$. As functional groups are consumed, the equilibrium free volume, $v_{f,\infty}$, changes based on the conversion in the system. When polymerizing multifunctional monomers, the volume relaxation is often much slower, and the system is not always in equilibrium with respect to the specific volume or free volume.

The volume relaxation (represented by the top cycle in Fig. 5) that occurs has often been referred to as physical aging. Because of significant volume relaxation during crosslinking polymerization reactions, it is necessary to account for the physical aging in addition to change in equilibrium conditions. This coupling of physical aging and polymerization kinetics is essential in order to describe systems that react and relax on similar time scales.

In the aging cycle, the characteristics relaxation time, τ, is a function of the free volume of the system. At low conversions and high free volumes, this relaxation can be nearly instantaneous, while it may take several days or years at higher conversions where the mixture is a glassy polymer. This characteristics relaxation time determines the rate at which the specific volume of the system changes, dv/dt [142]. Thus, the characteristic relaxation time along with the deviation of the specific volume from the equilibrium volume determines the rate at which the specific volume of the system changes. The actual free volume of the system can then be in excess of the equilibrium free volume of the system when relaxation is significantly slow.

For crosslinking reactions, this coupling of the physical aging cycle with the polymerization kinetic cycle is essential in modeling the reaction behavior. In particular, models which ignore the effects of physical aging are incapable of predicting the observed increases in conversion with increasing rate of polymerization. For dental restorations, this behavior becomes increasingly important since the total functional group conversion can be controlled to some extent by the rate of the polymerization. Thus, polymerization kinetic models which incorporate physical aging are extremely useful in elucidating the effects of reaction conditions on the extent of reaction in the composite material.

3.4.2 Kinetic Behavior

Another unique attribute of polymerizations of multifunctional monomers is the dominance of reaction diffusion as a termination mechanism [134, 136, 143–146]. Reaction diffusion involves the mobility of radicals by propagation through unreacted functional groups. This termination mechanism is physically different from translation and segmental diffusion termination mechanisms which involve the diffusion of polymer macroradicals and chain segments to bring radicals within a reaction zone before terminating. Whereas normal termination mechanisms are related to the diffusion coefficient of the polymer, reaction diffusion must be considered differently. In essence, reaction diffusion is a propagating termination mechanism and, therefore, is related to the propagation kinetic constant.

In multifunctional monomer polymerizations, the mobility of radicals through segmental diffusion falls well before their mobility through reaction diffusion at very low functional group conversions (as compared to linear polymerizations). From this point in the reaction, the termination and propagation kinetic constants are found to be related, and the termination kinetic constant as a function of conversion may actually exhibit a plateau region. Figure 6 illustrates the typical behavior of k_p and k_t vs conversion as predicted by a kinetic based model.

The propagation kinetic constant remains relatively constant at low to moderate conversions where diffusion of the smaller monomer molecules is unhindered. As the critical free volume for the diffusion control of propagation is reached, k_p begins to decrease. Diffusion of monomer molecules is now the rate controlling step for propagation. As polymerization continues and conversion increases (free volume decreases), the diffusion of the monomer and k_p drastically decrease.

The termination kinetic constant exhibits a somewhat more complex behavior. From the onset of reaction, termination is diffusion controlled (segmental diffusion controlled). The diffusion of the macroradicals is the controlling step and the primary means of free radical termination. At some later conversion, the termination mechanism changes from segmental to reaction diffusion control. In this region, a plateau in k_t occurs. Reaction diffusion is a propagation controlled

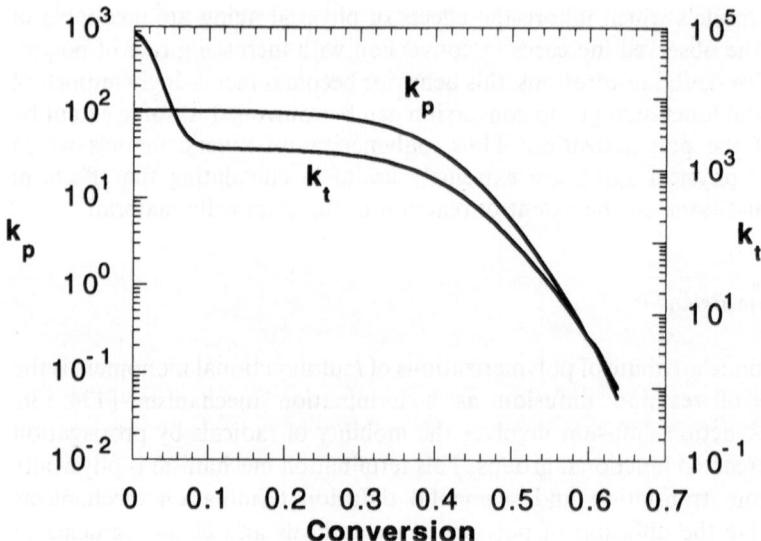

Fig. 6. The characteristic behavior of the propagation kinetic constant, k_p, and the termination kinetic constant, k_t, as a function of double bond conversion for a multifunctional monomer polymerization

mechanism, and therefore, when propagation remains chemically controlled and constant, so does termination. Finally, as propagation becomes diffusion controlled and decreases, the reaction diffusion controlled termination mechanism is monomer diffusion controlled and correspondingly decreases.

Anseth et al. [146, 147] have experimentally characterized the kinetic constant for a series of multifunctional methacrylate and acrylate monomers. In particular, they explored the kinetic evidence for the importance of reaction diffusion for polymerizations occurring in the high crosslinking regime. When reaction diffusion is the controlling termination mechanism, it was hypothesized that k_t would be proportional to $k_p[m]$ where $[m]$ is the concentration of double bonds. The works of Anseth et al. [146] then characterized the proportional constant between k_t and $k_p[m]$ for the methacrylates and acrylates studied.

When reaction diffusion controlled termination, the ratio of $k_t/k_p[m]$ was found to be nearly the same for all monomers of the same type of functionality (methacrylate or acrylate) and appeared to be independent of the reaction conditions (i.e., temperature and light intensity). The reported values for this ratio was approximately 3 for the methacrylates and between 6 and 8 for the acrylates.

4 Kinetic Gelation Modeling of Crosslinked Network Structure

4.1 Heterogeneity

While the properties of polymer based dental restoratives are strongly dependent on the network microstructure, multifunctional monomer polymerizations and the resulting polymer structures are often extremely difficult to characterize either experimentally or theoretically. This difficulty arises primarily from the structural heterogeneity that develops in the polymer during the polymerization of multifunctional monomers [108]. As discussed previously, since these monomers have more than one functional group, each group may possess a different reactivity. In general, not only is the reactivity of the functional groups different, but the relative reactivity changes with conversion in the system. Any polymerization involving multifunctional monomers has this added complexity of varying reactivity of functional groups on the same molecule. The primary result is the formation of microgel regions early in the reaction (because of high pendant double bond reactivity) [127, 148, 149], which can lead to significant reduction in the mechanical properties of the restoration when compared with a homogeneous polymer network.

When characterizing polymer networks, the following definitions are typically applied [150] and are illustrated in Fig. 7. When a radical on a polymer chain propagates through a pendant double (i.e. a double bond from a monomer with one double bond already reacted), a crosslink, secondary cycle, or primary cycle can be formed. A crosslink forms when the radical reacts with a pendant

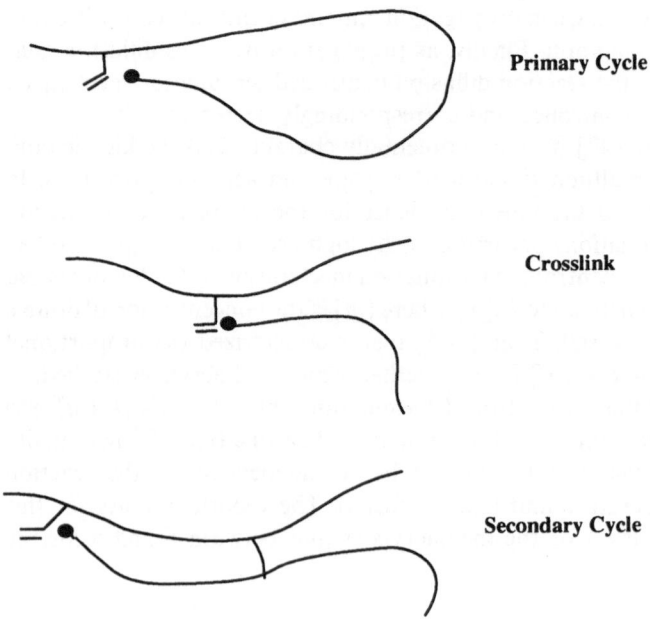

Primary Cycle

Crosslink

Secondary Cycle

Fig. 7. Classification of the reactions of pendant double bonds during free radical crosslinking polymerizations

double bond on a different kinetic chain. When the radical reacts with a pendant double bond on its own kinetic chain, a primary cycle results. Finally, if the radical reacts with a pendant double bond on a different kinetic chain with which it is already crosslinked, a secondary cycle forms. Whether the pendant double bond reacts to form a crosslink or cycle can significantly alter the mechanical and physical properties of the dental restoration.

Several researchers [151–156] have studied the effect of the chain length and flexibility between double bonds in divinyl monomers on the extent of cyclization in the system. In general, as the chain flexibility was increased, the reactivity of the pendant functional groups was enhanced which led to extensive cyclization at low conversion. Considering the significant differences between chain flexibility in Bis-GMA and TEGDMA, the relative amount of each monomer in the copolymerizing system might significantly affect the heterogeneity of the polymer microstructure.

Other researchers have experimentally observed heterogeneity in crosslinked polymers by studying radical concentrations and environment with ESR [101–106]. Knowledge of the structure and reactivity of trapped radicals is especially important when considering the long term physical and mechanical properties of dental polymers. Kloosterboer et al. [106] has studied the structure of trapped acrylate radicals while Hamielec and coworkers [102–105] have extensively studied the structure of methacrylate radicals in crosslinking polymerizations. In addition, the shape of the ESR spectrum was found to be indicative of the local environment of the radicals. Radical concentrations were divided into two distinct species depending on their local environent, and this classification further supports the hypothesized heterogeneous nature of highly crosslinked polymer networks. ESR allows active radicals in a mobile environment to be distinguished from inactive radicals that are trapped in densely crosslinked regions.

Ottaviani et al. [157] conducted ESR studies on ten commercially available dental composites. They investigated the effects of the different structure characteristics of the composites, composition of the resin and filler materials, and the dimensions of the filler particles on the radical concentration and stability. It was reported that the presence of inorganic filling material slowed the polymerization process, while it accelerated the decay of radicals. The structural stability and the resistance of the composites were confirmed by both the long period of decay and the high temperatures (100–150 °C) needed to overcome the potential barrier for starting the decay process. Finally, the composites were shattered to understand how the radicals might react if a portion of the composite were to come loose over time (e.g. from abrasion as the result of chewing). These investigations indicated that the particles removed by surface abrasion experience rapid decay, thus reducing the possibility of harmful effects on internal organs if ingested.

4.2 Models

In general, three different approaches have been used to model free radicals polymerizations and crosslinking reactions. The first method considers a statistical approach [158–164] whereby polymer structures evolve according to certain probabilistic rules for the formation of bonds between smaller monomer molecules. The second approach is kinetic based [165–171] and involves solving the differential equations that describe the concentration of each reacting specie. Finally, the third approach involves simulation of the structure in space using a percolation type simulation called a kinetic gelation model [132, 133, 172–177]. While statistical and kinetic methods are useful in modeling lightly crosslinked networks of relatively homogeneous nature, they are limited by their mean-field nature and are incapable of predicting the heterogeneities that arise in the high crosslinking regime. In contrast, kinetic gelation models can be used to predict and model spatial heterogeneities that arise during polymerization of multifunctional monomers. While limited by their fixed lattice structure and the difficulty in introducing realistic mobility of reacting species, kinetic gelation models are presently the method of choice for modeling reactions of high crosslinking density such as those in dental restorations.

4.2.1 Development of Kinetic Gelation Models

Kinetic gelation simulations are percolation-type simulations for the free radical copolymerization of difunctional and tetrafunctional monomers in any relative proportions. Manneville and de Seze [172] first proposed the kinetic gelation method for analyzing the simulating free radical polymerizations of multifunctional monomers. It has since been used and modified by various researchers over the course of the last decade [132, 133, 172–177]. The primary advantage of kinetic gelation simulations for large amounts of multifunctional monomers is its ability to predict the heterogeneity that is so prominent in reactions of this type.

Kinetic gelation models are based on lattice-type structures where each site represents a molecular structure in the system (e.g. monomer molecules or functional groups, initiator molecules, solvent). As radicals are generated in the system, propagation occurs through reaction of the radicals with one of its randomly chosen nearest neighbors, that, of course, must be an unreacted functional group for reaction to occur. Termination occurs when two radicals that are nearest neighbors are selected to react. The progress of the reaction and the structural evolution of the polymer can then be determined at any step in the computer algorithm. Essentially, a snapshot of the system is available at any instant from which the conversions, molecular weights, and polymer structure, including relative amounts to cycles and crosslinks as well as the amount of unreacted monomer, can be determined. A sample lattice slice (two dimensional) from the model of Anseth and Bowman [133] is presented in Fig. 8 showing the distribution of monomer, polymer, initiator, and free volume in the system during polymerization at different double bonds conversions. The system modeled is a homopolymerization of a tetrafunctional monomer (i.e., a monomer molecule with two double bonds). The inhomogeneity of the system is apparent as polymer regions are concentrated around initiation sites, and unreacted double bonds primarily exist in monomer pools.

Kinetic gelation models [178] have been used to determine, within experimental error, the fraction of constrained and unconstrained double bonds over a wide range of conversions in the polymerization of ethylene glycol dimethacrylate. The amount of unconstrained and constrained functional groups was determined experimentally by solid state nuclear magnetic resonance spectroscopy. The rules for determining constraint in the model were that all pendant double bonds and all monomers in pools of six or less are constrained. Monomers in pools of seven or more are assumed to be unconstrained. Whether a site is constrained or not does not affect the reactivity; only the analysis of the model is affected by the rules defining constraint.

Most recently, Bowman and Peppas [132] developed a simulation with several modifications. Their improvements included the incorporation of monomer molecules which occupy multiple lattice sites, distinct initiator molecules which exponentially decay into two radicals per initiator, and a crankshaft type motion of all species on a face center cubic lattice. They found the results from

Fig. 8a, b. Snapshots of a homopolymerization of a tetrafunctional monomer in two dimensions on a 40 × 40 lattice **a** at 10%; **b** at 25% conversion of double bonds. Initial species concentrations were 1.0 mol % initiator and 15% free volume. The initiator molecules are represented by , reacted double bonds by, free volume by, and unreacted double bonds by

the simulation in agreement with those found experimentally as the simulation predicts the attainment of a maximum conversion due to shielding and trapping of unreacted functional groups within the polymer. In addition, results for the relative reactivity of the pendant functional group of the monomeric functional group indicate that, early in the reaction, pendant functional groups are up to 50 times more reactive than monomeric functional groups. At later stages of the reaction, monomeric functional groups are actually more reactive than pendant functional groups.

4.2.2 Prediction Capabilities

The present state of kinetic gelation models allows one to elucidate much of the information regarding development of polymer networks that is often extremely difficult or impossible to determine either experimentally or theoretically. Depending on the nature of the system being studied and the kinetic gelation model used, this information might range from qualitatively showing trends to quantiative agreement with experimental data. For the dental material researcher, the information of primary interest that might be gleaned from kinetic gelation models includes the amount of residual monomer in a restoration, the relative reactivity of the functional groups, and the microstructure of the polymer which is inherently related to the mechanical and wear characteristics of the material.

Figure 9 shows a typical kinetic gelation model prediction of monomer conversion vs double bond conversion for the polymerization of a tetrafunctional monomer as compared to the homogeneous prediction of equal reactivity of functional groups. Because of the enhanced reactivity of pendant functional groups in these systems, the kinetic gelation model predicts a lower monomer conversion than the homogeneous assumption. For example, at 50% conversion of the functional groups, the homogeneous assumption predicts that 75% of the monomer has at least one double bond reacted and is unextractable. In contrast, the kinetic gelation model, which does not assume equal reactivity of functional groups, predicts only 69% of the monomer is reacted at 50% functional group conversion. Since the amount of residual monomer is a primary concern in dental restorations, this difference is quite important.

Another predictive capability of the kinetic gelation model is shown in Fig. 10. Figure 10 shows the relative fraction of crosslinks, primary cycles, and secondary cycles as a function of double bond conversion for a homopolymerization of a tetrafunctional monomer. Initially, primary cyclization dominates crosslinking and secondary cyclization, as the pendant double bonds have an increased reactivity in the localized region of the free radical. This behavior accounts for the formation of microgel regions and the heterogeneity of the network. As conversion and polymer concentration increase, crosslinking and secondary cyclization begin to increase and a network forms. Finally, a transition region is reached where secondary cyclization and primary cyclization cross over. The trend is now towards a more homogeneous network

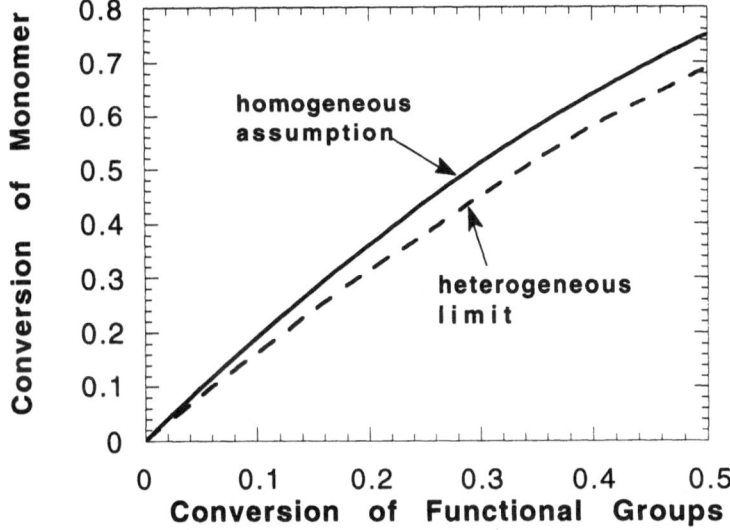

Fig. 9. Kinetic gelation model prediction (heterogeneous limit), – – –, of monomer conversion vs double bond conversion as compared to the homogeneous prediction, ———, of equal reactivity of functional groups during the polymerization of a multifunctional monomer

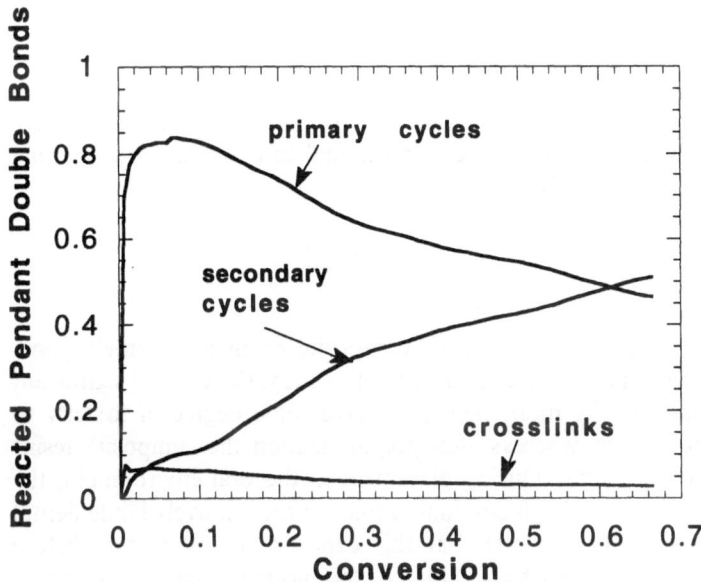

Fig. 10. Kinetic gelation model prediction of the relative fraction of crosslinks, primary cycles, and secondary cycles as a function of double bond conversion for the polymerization of a multifunctional monomer

structure with a very high degree of secondary cyclization. Since the elastic activity of a primary cycle, secondary cycle, and crosslink are vastly different, this information about the polymer microstructure is extremely important in determining and predicting the mechanical properties of the dental restoration.

Despite its limitations, kinetic gelation modeling is still a very useful tool in simulating network structure in highly crosslinked systems. While kinetic gelation models have gained widespread use in the polymer science field, the application of these models to dental materials and their development appears to be an area appropriate for further exploration.

5 Properties of Dental Composite Resins

The dental industry is currently accepting the International Standards Organization (ISO) standard for "Polymer-based filling and restorative materials." This standard will require testing of the following properties:

   a. biocompatibility
   b. working time, setting time, depth of cure
   c. flexure strength, flexural modulus
   d. water sorption and solubility
   e. color stability.

Though these may provide a standard for screening production quality, they are merely representative. The flexural properties will be a consistent test of the many possible mechanical property testing modalities. Other areas of physical properties that are important to the success of a composite dental restorative are radiopacity, polymerization shrinkage and thermal interactions, e.g., thermal expansion and thermal diffusivity.

5.1 Biocompatibility

In general, the biocompatibility of dental composite resins is extremely good. Though the materials are not implanted within the body, there is still significant potential for exposure. The monomers used have some degree of toxicity to bacterial cells and tissues, whereas after polymerization the composite resins have minimal toxic potential. During placement in the oral environment, the monomers can penetrate through dentinal tubules to the relatively labile dental pulp tissue. Current data suggests that this exposure to monomers before polymerization poses little toxic hazard [179]. Once polymerized, the composite restorative passes all tests for toxicity. The current testing guideline is American Dental Association Council of Dental Materials, Instruments, and Equipment Specification #41. Testing required includes several in vitro tests for cellular toxicity of the composite or extractable agents and mutagenicity. On occasion, a potential variation in initiating system or monomer test positive on the Ames Test for mutagenicity and never reaches the market [180].

There is also a possibility that someone can develop a type IV delayed hypersensitivity to some monomers. Of the monomers used, TEGDMA and HEMA may have the highest incident of sensitivity, but that incidence is extremely low. Though there are no literature reports of toxicity to composite resins over the 30 years of dental use, there are investigations into the potential for at least an allergic response to degradation products or extractable components [181].

5.2 Setting Times, Depth of Cure

When a restoration is light cured, there is a relatively long time for the dentist to place the material and adapt the composite to the preparation. The dentist can then use a light source (470 nm, with an output of at least $300 \, \text{mW/cm}^2$) to obtain complete cure of the material within 60 s [182].

With the light curing mechanism, there is a limitation to the penetration of the light. The dentist may have to place a restoration that is 6 + mm thick, whereas the light may penetrate only 2 mm [182]. Factors that affect this penetration are the translucence of the material, the color or shade used to match the tooth, the ability to place the light source close to the material being polymerized, and the intensity of the source. Under relatively ideal conditions, the mean depth of cure is approximately 4–5 mm. Thus, the dental application requires that the material be placed in layers. Due to the oxygen inhibition of the outside surface of the resin layers, additional layers can be laminated and cured with the appearance of uniformity of the final restoration.

5.3 Mechanical Properties

A number of mechanical properties have been studied that may affect the clinical success of dental composite restorative materials. Among these are diametral tensile strength (DTS), flexural strength, fracture toughness, elastic modulus, hardness, and fatigue resistance. The mechanical properties should approximate those of tooth structure [183], but correlation of clinical success to any of these properties is limited.

The primary failure modality identified clinically for restorations in posterior teeth is loss of material through abrasion. The complex nature of this failure mode in composite materials makes it difficult to correlate this phenomenon with any one mechanical property. A number of studies have suggested improvements in the system by using various mechanical properties as evidence. These studies have identified major factors such as ceramic filler loading and type of filler [186–191]. Some effects have been identified related to the comonomer system, polymerization methodology, and degree of polymerization [192–197]. The ability to post-cure or heat cure a composite resin after an initial light polymerization procedure has been an intriguing concept for dental application; however, the enhancement in mechanical properties has not been of sufficient magnitude to have clinical effect [198].

The DTS has been used as a strength criterion for meeting American Dental Association specification #27. Reported DTS values for microfilled composites range from 30–50 MPa; and for the conventional and hybrid composites, from 45–75 MPa [199–204]. The wide variation in the ranges of properties reported may relate not only to differences in materials, but also to test methodologies such as state of hydration, surface roughness of the samples, and strain rates [205, 206].

Flexural strength and flexural modulus are being used as the screening criteria for a new ADA specification #27 which adopts the new ISO standards. The minimum flexural strength required will be 50 MPa. In addition, the criterion makes it mandatory that the flexural strength must exceed a value, N, based on the flexural modulus, where N = (flexural modulus *0.0025) + 40. This stipulation will require higher flexural moduli from the stiffer composites like the conventional, small particle and hybrid systems.

There may be an effect of the layering necessitated by light curing mechanisms. It has been noted that if there is crack initiation along the laminate interface, there may be a slight reduction in fracture toughness [207]. This reduction in fracture toughness would occur only in highly filled systems where the oxygen inhibited layer could have a decreased concentration of filler.

Clinical abrasion is the primary mode of failure of posterior composite restorations [55, 208, 209]. Three- to five-year clinical studies are required to determine if the material is suitable for use as a posterior restorative. The general consensus is that these materials do not yet have the ability to replace dental amalgam as a posterior restorative material. Advanced systems are being investigated using ceramic inlays for the restoration and a low viscosity composite resin as a cementing agent. These systems still expose the low viscosity resin to wear at the cementing line [210], and the composite resin still wears significantly.

In anterior teeth, where aesthetics is more critical to the patient ad where the restoration can be shielded from the strongest biting forces, the composite can serve as an excellent, long lasting restorative. The current mechanical properties of the material allow excellent clinical success in these applications.

5.4 Water Sorption and Solubility

The new specification for dental composite resin systems will establish criteria for both water sorption and solubility. According to the new ISO standard, these measurement are to be made on a 15 × 1 mm disc immersed in water for one week. The criteria allows up to 40 µg/mm^3 of water sorption into composite and 7.5 µg/mm^3 solubility of the composite system in water and is easily satisfied by the highly filled conventional and hybrid systems. In the microfilled composites, the high volume fraction of resin will make this criterion difficult to satisfy because the absorption occurs primarily in the resin phase [211, 212]. The older specification for sorption allowed up to 0.7 mg/cm^2 for the conventional and hybrid composites and 1.7 mg/cm^2 for the microfilled composites in a 20 × 1 mm sample disc. It has been suggested that the hygroscopic expansion of the microfilled resins could compensate over time for the polymerization shrinkage [213], but such a system sacrifices mechanical properties and may not efficiently fulfill the desired expectations [214].

Increasing the hydrophobic nature of the resin should decrease the water sorption and the resulting undermining of the mechanical properties [195, 215, 216]. The solubility of these systems is extremely small (< 0.01 mg/cm^2 in a 1 mm thick sample) [204], but some of the water sorption and solubility problems could be attributed to variations in the silanation process [217, 218] and to the different solubilities of the quartz and barium-silica glass [219–220].

5.5 Color Stability

The tooth structure has a highly translucent, colorless to light gray-blue enamel coating, with a supporting dentin that is more opaque and has a tendency toward yellow or orange coloring. Other factors contribute to characterize a highly complex, healthy looking tooth. ADA specification #27 provides a color stability criterion for composites restorative materials based on exposure to intense light for 24 h and 37 °C for one week. There should be no change in color during these exposures.

To produce a translucent restorative, composite formulation requires matching the refractive indices of the polymer and the ceramic filler. When one considers that such materials must meet stringent requirements for appropriate physical and mechanical properties, it is difficult to satisfy the aesthetic requirements of shade and translucence. The light curing catalyst, camphoroquinone, can impart a yellow color to the composite system that biases any shading, and there can be a significant color shift after the polymerization. The composite shading is achieved by the addition of colorants to produce tooth colors that are in the range of very low chroma yellows and oranges, and very low value grays. Thus, the greater the thickness of the composite the closer it approximates the desired tooth shade [221]. The refraction of light through the tooth and the restorative can provide some degree of blending of the composite to the tooth coloring.

Studies have been done to identify several mechanisms of color change in a composite restoration. Exposure to ultraviolet light can significantly alter the shading over time [222, 223]. The composite resin system can also be stained during the clinical life of the restoration by products such as coffee, tea, grape juice, and smoking [224–227]. Even though the material may not fail due to physical breakdown, the restoration may have to be replaced due to aesthetic failure.

5.6 Radiopacity

Because of the need to use X-rays to help in future diagnosis of dental disease, the composite resin should provide in 1 mm thickness the radiopacity of 2 mm of aluminum. This radiopacity is usually obtained by the addition of a ceramic containing a large atom such as barium or strontium. These glasses can be ground to small particle and submicron sizes but have not been made as a fused silica small enough for a microfilled composite. Therefore, the microfilled composites are completely radiolucent. One brand of composite uses yttrium trifluoride to produce radiopacity in what would otherwise be a microfilled composite.

5.7 Polymerization Shrinkage

During the polymerization reaction, a certain amount of volume ($22.5 \, \text{cm}^3/\text{mol}$) will be consumed for each methacrylate group that reacts [228]. Because of this relationship, minimizing the total number of methacrylate groups that react will minimize the volume shrinkage. The method of curing may also contribute to the degree of shrinkage. Light curing mechanisms have been shown to slightly increase the contraction over chemical curing of the same resin/ceramic system [229]. One of the major problems associated with polymerization shrinkage is marginal leakage [230]. Marginal leakage results when the resin does not adhere to the tooth structure, or stresses are induced between the polymer and filler interface. Bowen and Marjenhoff [231] have reported that a 2% polymerization volume shrinkage can lead to stresses that are sufficient to cause either adhesive failure resulting in marginal leakage or cohesive failure resulting in microcracking of the composite. Immediately after placement, the forces are not sufficient to lead to fracture of the tooth [232], but the induced strain [233] on the tooth could lead to a post-operative sensitivity in the dental pulp. Post operative sensitivity may be alleviated by cutting of a slot in the new composite restoration to reduce the strain and re-restoring the small slot with additional composite.

Unfortunately the minimization of volume shrinkage would lead to a lower degree of composite and a lower degree of crosslinking. In addition, the swelling behavior of polymers is controlled by the amount of unreacted monomer as well as the crosslinking density and compatibility of the polymer and solvent. In order to minimize swelling, the polymer should be highly crosslinked, have a high double bond conversion (with little unreacted monomer or double bonds), and be as hydrophobic as possible.

Bailey [234] has reported the results of some ring opening polymerizations based on spiroothocarbonates (SOCs). These monomers tend to expand upon polymerization. Ideally, this polymerization expansion might be used to eliminate or reduce the polymerization volume shrinkage associated with methacrylate (or vinyl) monomers by copolymerizing SOCs and methacrylate systems. Bailey [235] demonstrated reduced volume shrinkage in some high-strength industrial composites by introducing SOCs into the system.

The further use of SOCs in dental composites has been investigated by several researchers [236–241]. In general, the compatible blends of SOC/Bis-GMA resins that were developed did show some reduction in the composite shrinkage. Unfortunately, the shrinkage was still significant because of the unsaturated SOCs that were used to make copolymers of the SOC and Bis-GMA through a free radical initiated mechanism. Others [241–251] have begun work with SOCs and epoxy resins to minimize further the shrinkage and to develop new dental composite resins that expand during polymerization. Presently, no satisfactory expanding monomer system has been developed for dental composite resins. The current technology has relied on larger monomers and the inorganic fillers to reduce the polymerization shrinkage of the composite system.

5.8 Thermal Properties

Thermal expansion differences exist between the tooth and the polymer as well as between the polymer and the filler. The tooth has a thermal expansion coefficient of $11 \times 10^{-6}/°C$ while conventional filled composites are 2–4 times greater [63, 252]. Stresses arise as a result of these differences, and a breakdown between the junction of the restoration and the cavity margin may result. The breakdown leads to subsequent leakage of oral fluids down the resulting marginal gap and the potential for further decay. Ideal materials would have nearly identical thermal expansion of resin, filler, and tooth structure. Presently, the coefficients of thermal expansion in dental restorative resins are controlled and reduced by the amount and size of the ceramic filler particles in the resin. The microfilled composites with the lower filler loading have greater coefficient of thermal expansions that can be 5–7 times that of tooth structure. Acrylic resin systems without ceramic filler have coefficients of thermal expansion that are 9 times that of tooth structure [202–204, 253].

An additional thermal property of interest is thermal diffusivity. The dental pulp sensory system is extremely sensitive to changes in temperature. These sensory inputs are interpreted only as pain. Metallic restorations of deep carious lesions of the tooth frequently need to have a low thermal conductor placed beneath them to avoid causing pulpal pain. The thermal diffusivity of composite varies from approximately that of tooth structure ($0.183 \text{ mm}^2/\text{s}$) to twice that value [204, 254]. Metallic restorations of concern have diffusivities at least an order of magnitude greater than that of the composite restorations. The composite restoration materials have diffusivities in the range of the bases which are used to insulate the metallic restorations.

Thermal changes within the range to which restoratives may be exposed in the mount, 0–60 °C, may have other effects on the composite restorative. It is possible that these changes can cause decreases in the mechanical properties of the composites [255–257]. These temperatures may approach the glass transition temperatures of some of the less densely crosslinked resin systems used, increasing the negative impact on the mechanical properties.

6 Bonding to Tooth Structure

There are two radically different hard structures in the tooth to which bonding can be obtained. The enamel is about 96% inorganic hydroxyapatite crystals, 1% organic material and 3% water by weight [258]. The hard tissue of dentin is about 70% inorganic hydroxyapatite, 20% collagen fibrillar network and 10% water by weight [259]. The dentin is interlaced with tubules containing cell processes from the odontoblastic layer in the pulp. Bonding techniques must be refined to achieve strong, consistent, and long lasting results to both surfaces. There are two goals to bonding: 1) to obtain strong retention onto the tooth structure; and 2) to maintain an impervious seal of the restoration against the tooth. Failure to maintain the seal can lead to hypersensitivity of the tooth and to recurrent decay. The retention gained by the bonding procedures may allow the dentist to retain more complex fillings with less cutting of undecayed tooth structure.

6.1 Bonding to Enamel

The system for bonding to enamel was developed by Buonocore in the 1950s [260]. This acid etch procedure requires the preparation of the enamel surface with an acidic solution, usually about 37% phosphoric acid. The surface then has altered surface tension and altered topography with enamel prismatic "tags" approximately 25 microns long and 5 microns apart [258, 261]. An unfilled, low viscosity resin can be allowed to flow between these tags and then polymerize to form a tight junction with the tooth enamel [262].

The monomers used in the commercial enamel bonding systems have usually been that of the composite. Primarily the systems have been based on BisGMA with significantly larger amounts of diluents such as triethyleneglycol dimethacrylate [263]. Other agents could be added to further reduce the surface tension of the liquid and increase the efficiency of the flow into the enamel tags. This bonding technique offers strength of retention to the enamel [264] and significant resistance to leakage around the restoration [265].

6.2 Bonding to Dentin

Technology for bonding to dentin is currently going through significant changes and critical controversy. The current systems rely on a multistep procedure. Generalization of these procedures follow.

Step 1: etch the dentin with a mild acidic solution or agent to remove some inorganic material. Such procedures produce a brush network of collagen fibers on the surface.
Step 2: a primer containing a monomer solvent system is applied and allowed to dry.
Step 3: an unfilled resin is applied to penetrate into the intricacies that have been impregnated with primer. This unfilled resin is polymerized.

The major developments in this area have been led by Bowen [266]. Mild demineralization treatments of the dentin include 10% phosphoric acid, 2% nitric acid, 17% ethylene diamine tetraacetic acid, 10% citric acid and maleic acid. The primers have contained such monomers as N- glycidyl methacrylate (NPG-GMA), and pyromellitic acid dianhydride-2-hydroxyethyl methacrylate (PMDM). Other systems include 4-methacryloxyethyltrimellitic anhydride (4-META) [267]. The theory is that these primers carried in the solvent will penetrate wet dentin and produce a hybrid layer containing collagen and resin [268]. The acetone/alcohol solvent carries the monomer into the wet dentin removing the water and evaporating. This leaves a monomer impregnated collagen network that allows the resin to penetrate into the partially de-mineralized structure below the exposed collagen [269, 270].

The resin adhesive applied in the third step is primarily the enamel bonding resin systems with further modifications by such monomers as 2-hydroxyethyl methacrylate. Though there are considerable variations in the results obtained for bond strengths to dentin by many investigators, the newest systems are capable of producing bond strengths to dentin comparable to that previously obtained for enamel [271]. The ability to seal has been improved with these dentin bonding systems to the point that some are claiming complete elimination of leakage. Studies on the theory that elimination of any bacterial contamination removes any pulpal response suggest that these bioactive agents are biocompatible [272]. The stability of these bonds have as yet not been established. Earlier dentinal bonding mechanisms revealed significant failure within one year [273].

Should these bonds prove to have a long prognosis clinically then improvements in the properties of the composite resin restorative system could make dental composite a superior direct filling material for any application in the dentition.

7 Conclusions

The development of an ideal material for dental restorations has proved to be an elusive and perplexing problem. The traditional material for dental restorations is an amalgam composed of mercury, silver, tin, and copper; however, recent concerns over the relatively large amount of mercury in the formulation, as well as the desire for an aesthetically suitable material, has led to considerable effort in the development of polymeric composite materials suitable for dental applications. The advantages of composite resins are that they allow for color matching, conservative cavity preparation, and simple preparation through intraoral photopolymerization. Unfortunately, the polymeric dental materials developed to date have several deficiencies, and there is considerable need for the development of improved materials and greater understanding of the polymerizations used to produce these materials.

With these deficiencies in mind, future directions in composite dental materials research are likely to include the development of materials with lower volume shrinkage, higher degrees of conversion, lower soluble/unreacted monomer fractions, lower toxicity, lower thermal expansion, greater wear resistance, and improved mechanical properties. Techniques to probe the polymer microstructure and its relationship to reaction conditions should be developed along with correlations for how the wear resistance depends on the microstructure. In general, a better understanding of the polymerization mechanism and the polymer microstructure as well as advanced materials development is necessary before polymer composite materials replace dental amalgam in the majority of restorations.

8 References

1. Mongkolnam P (1992) Australian Dent J 37: 360
2. Bentley J (to be published 1994) Survey of dental services rendered, ADA Bureau of Economic and Behavioral Research
3. Newman S (1991) J Am Dent Assoc 122: 67
4. Anderson PE (1994) Dental Economics 84: 37
5. Powell LV, Johnson GH, Yashar M, Bales DJ (1994) Operative dentistry 19: 70
6. Eley BM, Cox SW (1993) Brit Dent J 175: 355
7. Vimy MJ, Lorscheider FL (1985) J Dent Res 64: 1069
8. Newman SM (1987) (Abstract #41) J Dent Res 66: 112
9. Mackert JR (1991) J Am Dent Assoc 122: 54
10. Newman SM (1984) Ontario Dentist 61: 31
11. Marek M (1994) (Abstract #26) J Dent Res 73 (special issue): 105
12. Abraham JE, Svare CW, Frank CW (1984) J Dent Res 63: 71
13. Ott KHR, Loh F, Kroncke A, et al., (1984) Dtsch Zahnarztl Z 39: 199
14. Newman S (1986) International Dental Journal 36: 35
15. Olstad ML, Holland RI, Wandel N, Hensten Pettersen A (1987) J Dent Res 66: 1179
16. Anusavice KJ, Soderholm K, Grossman DG (Sept. 1993) MRS Bulletin 64

17. Arenholt-Bindslev D (1992) Advances in Detal Research 6: 125
18. Naleway CA, Ovsey B, Mihailova C, et al. (1994) (Abstract #25) J Dent Res 73(special issue): 105
19. Halse A (1975) Arch Oral Biol 20(1): 87
20. Buonocore MG, Wileman W, Brudevold W (1955) J Dent Res 34: 849
21. Suh BI (1991 Jul–Aug) J Esthet Dent 3(4): 139
22. Ruyter IE, Oysaed H (1988) CRC Critical Reviews in Biocompatibility 4: 247
23. Urabe H, Wakasa K, Yamaki M (1990) J Mateirals Sci.: Materials in Medicine 1: 163
24. Ruyter IE, Sjovik IJ (1981) Acta Odont Scand 39: 133
25. Bowen RL (Nov. 27, 1962) U.S. Pat 3,066,112
26. Bowen RL (Apr. 20, 1965) U.S. Pat 3,179,623
27. Smith LT, Powers JM (1991) Int J Prosthodont 4: 445
28. Bowen RL (1970) J Dent Res 49: 810
29. Asmussen E (1975) Acta Odont Scand 33: 129
30. Taira M, Urabe H, Hirose T, Wakasa K, Yamaki M (1988) J Dent Res 67: 24
31. Roulet JF (1987) Degradation of Dental Polymers. Karger, New York
32. Miyazaki K, T. Horibe T (1988) J Biomed Mat Res 22: 1011
33. Ruyter IE, Svendsen SA (1977) Acta Odont Scand 36: 75
34. Ruyter IE, Oysaed H (1987) J Biomed Mat Res 21: 11
35. Asmussen E (1975) Acta Odont Scand 33: 129
36. Suzuki S, Nakabayashi N, Masuhara E (1982) J Biomed Mat Res 16: 275
37. Bunker JE, Fields RP, U.S. Pat 4,544,467
38. Bowen RL (1963) J Am Dent Assoc 66: 57
39. Lutz F, Phililps RW (1983) J Prosthet Dent 50: 480
40. Watts DC (1992) in: Cahn RW, Haasen P, Kramer EF (eds) Materials Science and Technology 14: 209
41. Albers HF (1985) Tooth colored restoratives, 7th edn Cotati, Alto Books
42. Leinfelder KF (1985) Dent Clin North Am 29: 359
43. Phillips RW (1991) Science of Dental Materials. WB Saunders Co., Philadelphia
44. Randklev RM, U.S. Pat 4,503,169
45. Drinnan AJ (1967) J Am Dent Asoc 74: 446
46. Espelid I, Tvet AB, Erickson RL, et al. (1991) Dental Materials 7: 114
47. Clark HA, Plueddemann EP (1963) Modern Plastics 40: 133
48. Plueddemann EP (1978) Additives for Plastics 1: 123
49. Ranney MW, Berger SE, Marsden JG (1974) in: Plueddemann EP (ed) Interfaces in Polymer Matrix Composites, Composite Materials 6
50. Bowen RL (1979) J Dent Res 58: 1101
51. Chen TM, Brauer GM (1982) J Dent Res 61: 1439
52. Kass RL, Kordos JL (1971) Polymer Eng Sci 11: 11
53. Swartz ML, Phillips RW, Rhodes BF (1982) J Dent Res 61: 270
54. Lee HL, Smith FF, Swartz ML (1975) US Patent 3,926,906
55. Lutz F, Phillips RW, Roulet JF, Sectos JC (1984) J Dent Res 63: 914
56. Roulet JF, Roulet-Mehrens TK (1982) J Periodont 53: 257
57. Powers JM, Fan PL (1980) J Dent Res 59: 815
58. Powers JM, Fan PL, Marcotti M (1980) J Dent Res 59 special issue B: 936
59. Reid CN, Fisher J, Jacobsen PH (1990) J Dent 18: 209
60. Larsen IB, Freund M, Munksgaard EC (1992) J Dent Res 71: 1851
61. Kusy RP, Leinfelder KE (1977) J Dent Res 56: 544
62. Watts DC, McAndrew R, Lloyd CH (1987) J Dent Res 66: 1576
63. Kandil SH, Kamar AA, Shabaan SA, Taymour N, Morsi SE (1988) Eur Polym J 24: 1181
64. Pearson GJ (1979) J Dent 7: 64
65. Swartz ML, Moore BK, Phillips RW, Rhodes BF (1982) J Prosthet Dent 47: 163
66. Cowperthwaite GF, Foy JJ, Malloy MA (1979) J Dent Res 58 special issue A: 242
67. Venz S, Dickens B (1991) J of Biomedical Mat Res 25: 1231
68. Davidson CL, de Gee AJ (1984) J Dent Res 63: 146
69. Feilzer AJ, de Gee AJ, Davidson CL (1988) J Prosthet Dent 59: 297
70. Penn RW (1986) Dent Mater 2: 78
71. Rees JS, Jacobsen PH (1989) Dent Mater 5: 41
72. Smith J, Fitchie J, Puckett A, Hembree J (1992) J Dent Res 71: 209, Abstr No 825
73. de Gee AJ, Feilzer AJ, Davidson CL (1993) Dent Mater 9: 11

74. Lai JH, Johnson AE (1993) Dent Mater 9: 139
75. Brannstrom M (1987) Oper Dent 12: 158
76. Felton D, Bergenholtz G, Cox CF (1989) J Dent Res 68: 491
77. Cox CF, Bergenholtz G, Heys DR, Syed SA, Fitzgerald M, Heys RJ (1985) J Oral Pathol 14: 156
78. Pashley DH (1990) Dent Clin North Am 34: 449
79. Pashley DH, Matthews WG (1992) Arch Oral Biol 38: 577
80. Watanabe T, Sano M, Itoh K, Wakumoto S (1991) Dental Materials 7: 148
81. Ferracane JL, Greener EH (1984) J Dent Res 63: 1093
82. Ferracane JL, Greener EH (1986) J Biomed Mater Res 20: 121
83. Eliades GC, Voughiouklakis GJ, Caputo AA (1987) Dent Mater 3: 19
84. Barron DJ, Rueggeberg FA, Schuster GS (1992) Dent Mater 8: 274
85. Vaidyanathan J, Vaidyanathan TK (1992) J Mat Sci.: Mat in Med 3: 19
86. Kalipcilar B, Karaagaclioglu L, Hasanreisoglu U (1991), J of Oral Rehabilitation 18: 399
87. Tanaka K, Taira M, Shintani H, Wakasa K, Yamaki M (1991), J of Oral Rehabilitation, 18: 353
88. Effects & Side Effects of Dental Restorative Materials. National Institutes of Health, Technology Assessment Conference Statement. August 26–28, 1991.
89. Cox CF, Keall CL, Keall HJ, Ostro E, Bergenholtz G (1987) J Prosthet Dent 57: 1
90. Wendt SL (1991) American Journal of Dentistry 4: 10
91. Ferracane JL, Condon JR (1992) Dent Mat 8: 290
92. Odian G (1982), Prinicples of Polymerization. McGraw-Hill, New York.
93. Hirose T, Wakasa K, Yamaki M (1990) J of Mat Sci 25: 1209
94. Kloosterboer JG, Lijten GFCM, Zegers CPG (1989) Polym Mater Sci Eg Proceed 60: 122
95. Hageman HJ (1989) Polym Mater Sci Eng Proceed 60: 558 and references therein
96. Baxter JE, Davidson RS (1988) Makromol Chem 189: 2769
97. Phan XT (1986) J Rad Cur 13: 18
98. Kishore K, Bhanu VA (1988) J Polym Sci Polym Chem Ed 26: 2831
99. Kloosterboer JG, Lijten GFCM, Greidanus FJAM (1986) Polym Comm 27: 268
100. Kerber V, Serini V (1970) Makromol Chem 140: 1
101. Best ME, Kasai PH (1989) Macromolecules 22: 2622
102. Tian Y, Zhu S, Hamielec AE, Fulton DB, Eaton DR (1992) Polymer 33: 385
103. Zhu S, Tian Y, Hamielec AE, Eaton DR (1990) Macromolecules 23: 1144
104. Zhu S, Tian Y, Hamielec AE, Eaton DR (1990) Polymer 31: 1726
105. Zhu S, Tian Y, Hamielec AE, Eaton DR (1990) Polymer 31: 154
106. Kloosterboer JG, Lijten GFCM, Greidanus FJAM (1986) Polymer Reports 27: 268
107. Li D, Zhu S, Hamielec AE (1993) Polymer 34: 1383
108. Kloosterboer JG (1988) Adv Polym Sci 84: 1
109. Kloosterboer JG, Litjen GFCM (1988), in: Dickie R, Labana S, Bauer R (eds) Cross-Linked Polymer Chemistry. ACS Symposium Series, Vol 367, 409, ACS, Washington, D.C.
110. Miyazaki K, Horibe TJ (1988) J Biomed Mater Res 22: 1011
111. Allen P, Simon G, Williams D, Williams E (1989) Macromolecules 22: 809
112. Cook WD (1992) Polymer 33: 2152
113. Turner D, Haque Z, Kalachandra S, Wilson T (1987) Polym Mater Sci Eng 56: 769
114. Siimon G, Allen P, Bennett D, Williams D, Williams E (1989) Macromolecules 22: 3555
115. Allen P, Bennett D, Hagias S, Hounslow A, Ross G, Simon G, Williams D, Williams E (1989) Eur Polym J 25: 785
116. Kloosterboer J, Lijten G, Boots H (1989) Makromol Chem Macromol Symp 24: 223
117. Bowman CN, Peppas NA (1991) Macromolecules 24: 1914
118. Greiner R, Schwarzl FR (1984) Rheologica Acta 23: 378
119. Kloosterboer JG, van de Hei GMM, Boots HM (1984) Polym Commun 25: 354
120. Landin DT, Macosko CW (1988) Macromolecules 21: 846
121. Matsumoto A, Oiwa M (1990) Polym Prepr Japan 39: 928
122. Garrett RW, Hill DJT, O'Donnell JH, Pomery PJ, Winzor CL (1989) Polym Bull 22: 611
123. Sack R, Schulz GV, Meyerhoff G (1988) Macromolecules 21: 3345
124. Ballard MJ, Napper DH, Gilbert RG (1984) J of Polym Sci Polym Chem 22: 3225
125. Soh SK, Sundberg DC (1982) J of Polym Sci Polym Chem 20: 1315
126. Marten FL, Hamielec AE (1979) in: Henderson J, Bouton T (eds) Polymerization reactors and processes. ACS Symposium Series Vol. 104, 43
127. Matsumoto A, Ando H, Oiwa M (1989) Eur Polym J 25: 185

128. Wison TW, Turner DT (1987) J Dent Res 66: 1032
129. Turner DT, Haque ZU, Kalachandra S, Wilson TW (1984) in: Labana S, Dickie R (eds) Characterization of highly crosslinked polymers. ACS symposium series Vol 243, ACS Washington, D.C., p. 185
130. Cook WD (1991) J of Appl Poly Sci 42: 1259
131. Yoshida K, Matsumura H, Tanaka T, Atsuta M (1992) Dent Mater 8: 137
132. Bowman CN, Peppas NA (1992) Chem Engng Sci 47: 1411
133. Anseth KS, Bowman CN (1994) Chem Engng Sci 49: 2207
134. Cook WD (1993) J of Poly Sci Polym Chem 31: 1053
135. Russell GT, Napper DH, Gilbert RG (1988) Macromolecules 21: 2133
136. Stickler M (1983) Makromol Chem 184: 2563
137. Buback M (1990) Makromol Chem 191: 1575
138. Anseth KS, Bowman CN (1992–93) Polym React Eng 1: 499
139. Haward RN (1970) J Macromol Sci, Rev Macromol Chem C4: 191
140. Robertson RE, Simha R, Curro JG (1985) Macromolecules 18: 2239
141. Robertson RE, Simha R, Curro JG (1988) Macromolecules 21: 3216
142. Kovacs A, Aklonis J, Hutchinson J, Ramos AJ (1979) J Polym Sci Polym Phys 17: 1097
143. Soh SK, Sundberg DC (1982) J Polym Sci Polym Chem 20: 1345
144. Ballard MJ, Napper DH, Gilbert RG (1986) J Polym Sci Polym Chem 24: 1027
145. Sack R, Schulz GV, Meyerhoff G (1988) Macromolecules 21: 3345
146. Anseth KS, Wang CM, Bowman CN (1994) Macromolecules 27: 650
147. Anseth KS, Wang CM, Bowman CN (1994) Polymer 35: 3243
148. Bastide J, Leibler L (1988) Macromolecules 21: 2647
149. Funke W (1989) Brit Polym J 21: 107
150. Dotson NA, Macosko CW, Tirrell M (1992) in: Aharoni SM (ed) Synthesis, Characterization and theory of polymeric networks and gels. Pleunum Press, New York, p 319
151. Soper , Haward RN, White EFT (1972) J Polym Sci A-1 10: 2545
152. Dusek K, Ilavsky M (1975) J Polym Sci Symp 53: 47
153. Dusek K, Ilavsky M (1975) J Polym Sci Symp 53: 75
154. Landin DT, Macosko CW (1988) Macromolecules 21: 846
155. Shah AC, Holdaway I, Parson IW, Haward RN (1978) Polymer 19: 1067
156. Matsumoto A, Matsuo H, Oiwa M (1987) Makromol Chem Rapid Commun 8: 373
157. Ottaviani MF, Fiorini A, Mason PN, Corvaja C (1992) Dent Mater 8: 118
158. Macosko CW, Miller DR (1976) Macromolecules 9: 199
159. Miller DR, Macosko CW (1976) Macromolecules 9: 206
160. Williams RJJ (1988) Macromolecules 21: 2568
161. Williams RJJ, Vallo CI (1988) Macromolecules 211: 2571
162. Gordon M (1962) Proc Roy Sco London A268: 240
163. Dotson NA, Galvan R, Macosko CW (1988) Macromolecules 21: 2560
164. Scranton AB, Klier J, Peppas NA (1991) Macromolecules 24: 1412
165. Mikos AG, Takoudis CG, Peppas NA (1986) Macromolecules 19: 2174
166. Mikos AG, Peppas NA (1986) J Controlled Release 5: 53
167. Tobita H, Hamielec AE (1989) Macromolecules 22: 3098
168. Tobita H, Hamielec AE (1988) Makromol Chem Makromol Symp 20/21: 501
169. Zhu S, Hamielec AE (1993) Macromolecules 26: 3131
170. Zhu S, Hamielec AE, Pelton RH (1993) Makromol Chem, Theory Simul 2: 587
171. Zhu S, Hamielec AE (1992) Macromolecules 25: 5457
172. Manneville P, de Seze L (1981) Numerical methods in the study of critical phenomena. Springer, Berlin
173. Herrmann HJ, Landau DP, Stauffer D (1982) Phys Rev Lett 49: 412
174. Bansil R, Herrmann HJ, Stauffer D (1984) Macromolecules 17: 998
175. Boots HMJ, Pandey RB (1984) Polym Bull 11: 415
176. Boots HMJ, Kloosterboer JG, van de Hei GMM (1985) Brit Polym J 17: 219
177. Boots HMJ, Dotson NA (1988) Polym Comm 29: 346
178. Simon GP, Allen PEM, Bennett DJ, Wiliams DRG, Williams EH (1989) Macromolecules 22: 3555
179. Heys RJ, Heys DR, Cox CF, Avery JK (1977) J Oral Pathol 6: 63
180. Fredericks HE (1981) Am J Orthod 3: 316
181. Spahl W, Budzikiewicz H, Geurtsen W (1994) (Abstract #1548) J Dent Res 73: 295
182. Newman SM, Murray GA, Yates JL (1983) J Prosth Dent 50: 31

183. Hannah CM, Combe EC (1976) Brit Dent J 140: 167
184. Tyas MJ (1990) Austr Dent J 35: 46
185. Ferracane JL (1993), "What are the appropriate uses of posterior composites?" in symposium on esthetic restorative materials, ADA/NIDR, pp. 24–30
186. Cross M, Douglas WH, Fields RP (1983) J Dent Res 62: 850
187. Roberts JC, Powers JM, Craig RG (1977) J Dent Res 56: 748
188. Chung KH, Greener EH (1990) J Oral Rehab 17: 478
189. Chung KH (1990) J Dent Res 69: 852
190. Draughn RA (1985) in: Vanherle G, Smith DC (eds) International symposium on posterior composite resin dental restorative materials, 3M Corporation, 299
191. Soderhold KJ (1982) Acta Odont Scand 40: 145
192. McCabe JF, Kagi S (1991) Br Dent J 171: 246
193. Stansbury JW (1990) J Dent Res 69: 844
194. Stannard JG, Sornkul E, Collier R (1993) (Abstract #254) J Dent Res 72(special issue): 135
195. Ferracance JL, Hopkin JK, Condon JR (1993) (Abstract #256) J Dent Res 72(special issue): 135
196. Asmussen E (1982) Scand J Dent Res 90: 484
197. Oysaed H, Ruyter IE (1989) J Biomed Mat Res 23: 719
198. Wendt SL, Leinfelder KF (1990) J Am Dent Assoc 120: 177
199. Coury TL, Miranda FJ, Duncanson MG (1981) J Prosth Dent 45: 296
200. Berrong JM, Dodge WW, Glenn VC (1982) J Prosth Dent 47: 275
201. Cullen DR, Nelson JA, Sandrik JL (1993) J Prosth Dent 69: 247
202. Vanherle G, Labrachts P, Braem M (1985) in: Vanherle G, Smith DC (eds) International symposium on posterior composite resin dental restorative mateirals, 3M Corporation, 21
203. Phillips RW (1991), Skinner's science of dental materials, 9th edition. Sounders
204. O'Brien WJ (1989), Dental materials, properties and selection. Quintessence
205. Oysaed H, Ruyter IE (1986) J Biomed Mat Res 20: 261
206. Peutzfeldt A, Junggreen L (1992) Scand J Dent Res 100: 181
207. Kovarik RE, Ergle JW (1993) J Prosth Dent 69: 557
208. Leinfelder KF (1991) J Am Dent Assocc 122: 65
209. Leinfelder KF, Taylor DF, Barkmeier WW, Goldberg AJ (1986) Dent Mat 2: 198
210. O'Neal SJ, Miracle RL, Leinfelder KF (1993) J Am Dent Assoc 124: 48
211. Braden M, Causton EE, Clarke RL (1976) J Dent Res 55: 730
212. Soderholm KJ (1984) J Biomed Mater Res 18: 271
213. Feilzer AJ, De Gee AJ, Davidson CL (1990) J Dent Res 69: 36
214. Bowen RL, Rapson JE, Dickson G (1982) J Dent Res 61: 654
215. Ruyter IE, Nilsen J (1993) (Abstract #588) J Dent Res 72(special issue): 177
216. Soderholm KJ, Roberts MJ (1990) J Dent Res 69: 1812
217. Misra DN, Bowen RL (1977) J Dent Res 56: 603
218. Soderholm KJ (1981) J Dent Res 60: 1867
219. Soderholm KJ, Zigan M, Ragan M, Fischlschweiger W, Bergman M (1984) J Dent Res 63: 1248
220. Soderholm KJ (1983) J Dent Res 62: 126
221. Yeh CL, Powers JM, Miyagawa Y (1982) J Dent Res 61: 1176
222. Powers JM, Fan PL (1981) J Dent Res 60: 1692
223. Ferracane JL, Moser JB, Greener EH (1985) J Prosth Dent 54: 483
224. Cooley RL, Barkmeier WW (1988) Gen Dent 36: 517
225. Khokhar ZA, Razzoog ME, Yaman P (1991) Quintessence International 22: 733
226. Gross MD, Moser JB (1977) J Oral Rehab 4: 311
227. Nunn WR, Hembree JH, McKnight JP (1979) Journal of dentistry for children May/June: 210
228. Patel MP, Braden M, Davy KWM (1987) Biomaterials 8: 53
229. Hannsen EK (1982) Scand J of Dent Res 90: 329
230. Jensen ME, Chan DCN (1985) in: Vanherle G, Smith DC (eds) Posterior composite resin dental restorative materials. Utrecht, Peter Szulc, 243
231. Bowen RL, Marjenhoff WA (1992) Advances in Dent Res 6: 44
232. Hegdahl T, Gjerdet NR (1977) Acta Odont Scand 35: 191
233. Lopes LM, Leitao JG, Douglas WH (1991) Quintessence Int 22: 641
234. Bailey WJ (1975) J Macromolecular Science-Chem A9: 849
235. Bailey WJ (1990) Mater Sci and Eng A126: 271
236. Stansbury JW (1991) J Dent Res 70: 527, Abstr No. 2088
237. Stansbury JW, Bailey WJ (1988) ACS Symposium on Progress in Biomedical Polymers 133

238. Thompson VP, Williams EF, Bialey WJ (1979) J Dent Res 58: 1522
239. Stansbury JW (1992) J Dent Res 71: 434
240. Byerley TJ, Eick JD, Chen GP, Chappelow CC, Millich F (1992) Dent Mater 8: 345
241. Eick JD, Byerley TJ, Chappell RP, Chen GR, Bowles CQ, Chappelow CC (1993), Dent Mat 9: 123
242. Eick JD, Byerley TJ, Chappell RP, Chen GP, Bowles CQ, Chappelow CC (1992), J Dent Res 71: 598, Abstr No 662
243. Lam PWK, Piggott MR (1989) J Mater Sci 24: 4068
244. Lam PWK, Piggott MR (1989) J Mater Sci 24: 4427
245. Lam PWK, Piggott MR (1990) J Mater Sci 25: 1197
246. Piggott MR, Rosen S (1986) "Controlled shrinkage resin for fibre composites", 31st Inter SAMPE Symp, April 7–10, Las Vegas, NV, 541
247. Piggott MR, Woo MS (1986) Polymer Composites 7: 182
248. Woo M, Piggott MR (1987) J Compos Tech and Res 9: 101
249. Woo M, Piggott MR (1987) J Compos Tech and Res 9: 162
250. Woo M, Piggott MR (1988) J Compos Tech and Res 10: 16
251. Stansbury JW (1992) J Dent Res 71: 434
252. Craig RG, O'Brien WJ, Powers JM (1992), Dental Materials: Properties and Manipulations, 5th Edition. Mosby, St. Louis, MO
253. Powers JM, Hostetler RW, Dennison JB (1979) J Dent Res 58: 584
254. Watts DC, McAndrews R, Lloyd CH (1987) J Dent Res 66: 1576
255. Papadogiannis Y, Lakes RS, Petrou-Americanos A, Theothoridou-Pahini S (1993) Dent Mater 9: 118
256. Greener EH, Greener CS, Moser JB (1984) J Oral Rehab 11: 335
257. Draughn RA (1981) J Biomed Mater Res 15: 489
258. Silverstone LM (1982) in: Smith DC, Williams DF (eds) Biocompatibility of Dental Materials Volume 1, Characteristics of dental tissues and their response to dental materials, CRC Press, 39
259. Te Cate AR, Torneck CD (1982), in: Smith DC, Williams DF (eds) Biocompatibility of dental materials Volume 1, Characteristics of dental tissues and their response to dental materials, CRC Press, 75
260. Buonocore MG (1955) J Dent Res 34: 849
261. Eick JD, Johnson LN, Fromer JR, et al. (1972) J Dent Res 51: 780
262. Gwinnett AJ, Matsui A (1967) Arch Oral Biol 12: 1615
263. Brauer GM, Dulik DM, Hughes HN, et al. (1971) J Dent Res 60: 196
264. Mitchem JC, Turner LR (1974) J Am Dent Assoc 89: 1107
265. Hembree JH, Andrews JT (1976) J Am Dent Assoc 92: 414
266. Bowen RL (1985) US Patent 4,588,756
267. Nakabayashi N, Nakamura M, Yasuda N (1991) J Esthet Dent 3: 133
268. Suh BI (1991) J Esthet Dent 3: 139
269. Pashley DH, Ciucchi BM, Sano H, Horner JA (1993) Quintessence Int. 24: 618
270. Inokoshi S, Hosoda H, Harnirattisai C, Shimada Y (1993) Operative Dentistry 18: 8
271. Proceedings of the International Symposium on adhesives in dentistry (1992) operative dentristry 17 (Supplement 5).
272. Cox CF (1987) Operative Dentistry 12: 146
273. Newman SM, Porter EH, Szojka FP (1987) (Abstract # 1484) J Dent Res 66(special issue): 292

Tissue Regeneration Templates Based on Collagen-Glycosaminoglycan Copolymers

I. V. Yannas
Fibers and Polymers Laboratory, Massachusetts Institute of Technology, Bldg. 3-334, Cambridge, MA 02139-4307

Graft copolymers of type I collagen and a glycosaminoglycan have been shown capable of inducing regeneration of certain tissues in the adult mammal. These tissues are the dermis of skin, the sciatic nerve and the knee meniscus. Chemical composition alone does not suffice to describe the biological specificity of these regeneration templates. It is also necessary to specify the crosslink density and the architecture of their porous structure. Each tissue requires a different polymeric template. This article contains a review of the molecular structure of the constituent biopolymers, the procedure for synthesis and characterization of specific graft copolymer members of this family, and information on certain aspects of the mechanism by which these copolymers function as regeneration templates.

Advances in Polymer Science, Vol. 122
© Springer-Verlag Berlin-Heidelberg 1995

1 Introduction

The extracellular matrices (ECMs) of tissues are crosslinked networks comprising a large number of chemically distinct macromolecular species, mostly
proteins, glycosaminoglycans and abundant water. ECMs are hydrated macromolecular gels in which cells migrate and divide as well as synthesize or degrade
ECM components. Cells form intimate, highly specific connections with specific
ECM components outside their membrane. Such connections utilize macromolecules (e.g., integrins) which, while bound to structures within the cytoplasm,
extend through the cell membrane and form specific interactions with the
binding domains of ECM macromolecules (e.g., fibronectin, collagen).
Physicochemical modifications of the matrix immediately adjacent to a cell are
thereby recognized by the cytoplasm of the cell. The cell may modify its
functions appropriately as a result of such modifications in its environment.
Modifications of the ECM are especially critical during biological development
and during the healing of wounds in tissues. In these latter two cases cells
synthesize new tissues according to instructions which largely derive from the
nature and structure of ECM components.

The well-documented influence which the ECM exerts upon the cells adjacent to it suggests that synthetic analogs of ECMs could be used to study
cell-ECM interactions in model systems or to modify such interactions in order
to achieve a desired therapeutic objective. This article describes the synthesis
and characterization of simple ECM analogs as well as their use in the modification of the wound healing process. Such modification can be profound, changing
drastically the kinetics and mechanism of wound healing from repair to regeneration, thereby altering the choice of medical therapy in a significant way. The
ECM analogs which possess such biological activity have highly specific structures not only at the macromolecular but at the supermolecular level as well.
For this reason, the discussion below occasionally extends the scale by a factor
of 1000 from the description of morphological features common to synthetic
polymers (scale 1 to 100 nm) to the description of porous structures which
possess specificity of function at the scale of 1 to 100 μm.

The cell biology of wound healing is the subject of a comprehensive mono-
graph [1]. The biochemistry of collagen [2], fibronectin [3], proteoglycans [4]
and of other components of the ECM [5, 6] have been treated in some detail.
The synthesis and biological behavior of ECM analogs has been previously
reviewed [7, 8].

In this review we first summarize methods of synthesis of ECM analogs as
graft copolymers. The modification of supermolecular structure (crystallinity,
crosslink density, porosity) is then discussed. Finally, the relationships between
polymer structure and biologic activity are summarized.

2 Structure of Constituent Biopolymers

The composition and structure of extracellular matrix (ECM) varies from one tissue to the next. Typically, these matrices are highly hydrated macromolecular networks composed of various amounts of glycoproteins such as collagen, elastin, fibronectin, laminin and chondronectin; and glycosaminoglycans (GAG), including hyaluronic acid, chondroitin 6-sulfate, dermatan sulfate and heparan sulfate. Glycosaminoglycans usually occur as polysaccharide chains covalently attached to a protein core (proteoglycan). Macromolecular components of the ECM are synthesized in cells and are secreted in the extracellular space where further physicochemical modification, e.g., crystallization and covalent crosslinking of macromolecular chains, takes place [5, 6, 7].

The ECM analogs synthesized so far are graft copolymers of collagen and one each of several glycosaminoglycans (GAGs) and are, admittedly, very simple models of the chemically complex ECMs. The major constituent of these chemical analogs of ECM is collagen, the fibrous protein which accounts for about one-third of the total protein content of vertebrates. An unusual amino acid composition and a characteristic wide-angle X-ray diffraction pattern distinguish collagen clearly from other tissue components [2, 5]. Various levels of structural order in collagen are illustrated in Fig. 1. Collagen can be extracted from connective tissues such as cattle skin (hide) and tendon in relatively pure form and can be dispersed in aqueous acetic acid or other solvents in the form either of a solution of individual triple-helical macromolecules (tropocollagen) or a suspension of particles comprising naturally crosslinked aggregates of such macromolecules. The solid state can be recovered either by evaporation of the solvent or precipitation of the protein by use of a nonsolvent. Reconstituted collagen prepared thereby can be fashioned into membranes (films), tubes, fibers or tape [10]. Although collagen in the form of reconstituted membranes or sutures has been implanted surgically by several investigators at least since 1943 [11–15], the interaction between a casually reconstituted collagen and host tissue is passive and largely amounts to degradation of the triple-helical collagen molecules by collagenases secreted by cells adjacent to the implant. The glycosaminoglycans (GAGs) of the ECM comprise a family of aminopolysaccharides each of which possesses a chemical repeat unit consisting of a uronic acid and an N-acetylated sugar unit. The latter is sulfated except in the case of hyaluronic acid. For example, the repeat unit for chondroitin 6-sulfate, shown below, presents this GAG as an alternating copolymer of D-glucuronic acid and of an O-sulfate derivative of N-acetyl D-galactosamine [16]:

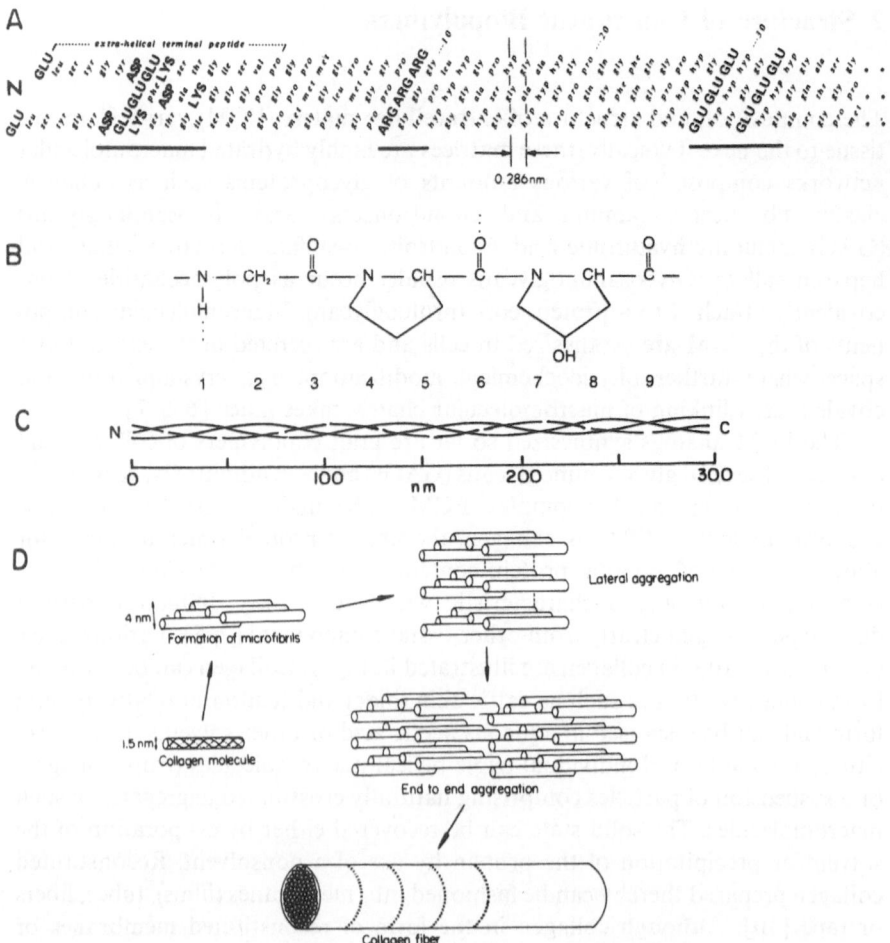

Fig. 1A–D. Collagen, like other proteins, is distinguished by several levels of structural order: A primary structure – the complete sequence of amino acids along each polypeptide chain. An example is the triple chain sequence of type I calf skin collagen at the *N*-end of the molecule. Roughly 5% of a complete molecule is shown. No attempt has been made to indicate the coiling of the chains. Amino acid residues participating in the triple helix are numbered, and the residue-to-residue spacing (0.286 nm) is shown as a constant within the triple helical domain, but not outside it. *Bold capitals* indicate charged residues which occur in groups (*underlined*) (Reprinted from [92], with permission); **B** secondary structure – the local configuration of a polypeptide chain. The triplet sequence Gly-Pro-Hyp illustrates elements of collagen triple-helix stabilization. The numbers identify peptide backbone atoms. The conformation is determined by trans peptide bonds (3–4, 6–7, and 9–1); fixed rotation angle of bond in proline ring (4–5); limited rotation of proline past the C=O group (bond 5–6); interchain hydrogen bonds (dots) involving the NH hydrogen at position 1 and the C̄O at position 6 in adjacent chains; and the hydroxy group of hydroxyproline, possibly through water-bridged hydrogen bonds (Reprinted from [93] with permission); **C** tertiary structure – the global configuration of polypeptide chains, representing the pattern according to which the secondary structures are packed together within the unit substructure. A schematic view of the type I collagen molecule, a triple helix 300 nm long (Reprinted from [94] with permission); **D** quaternary structure – the unit supermolecular structure. The most widely accepted unit is one involving five collagen molecules (microfibril). Several microfibrils aggregate end to end and also laterally to form a collagen fiber which exhibits a regular banding pattern in the electron microscope with a period of 65 nm (Reprinted from [95] with permission).

sodium chondroitin 6-sulfate

3 Nomenclature

Following nomenclature recommended by IUPAC [17], these polymers have been referred to as collagen-*graft*-chondroitin 6-sulfate copolymers, or, generically, as collagen-*graft*-glycosaminoglycan copolymers, or, briefly, as CG copolymers. Since the biological activity is dependent on their ultrastructure we will use the term copolymer matrix to denote the porous, hydrated, insoluble gel which results following physical processing of the copolymer. Implant refers to the sterile surgical device which is grafted at the site of the lesion.

The term regeneration template, or simply template, has been used to distinguish biologically active forms of CG polymers [8]. Templates are recognized by their ability to induce tissue regeneration in a well-defined animal lesion (wound) in which it has been previously demonstrated that regeneration does not occur spontaneously.

4 Synthesis of Graft Copolymers

Grafting of GAG chains onto collagen is conveniently carried out by first forming a coprecipitate of collagen and GAG and then treating this condensed state under conditions which favor formation of covalent bonds between the two macromolecules. Coprecipitation requires the presence of sulfate groups on the GAG and an acidic pH [16]. Hyaluronic acid, the only GAG which is nonsulfated, does not precipitate collagen out of solution [16]. Every one of the other members of this family of polysaccharides does; however, the precipitates formed in aqueous acetic acid readily dissolve when the pH is adjusted to neutrality [16]. The precipitate is an ionic complex probably formed by interaction between the anionic sulfate groups of the GAG and the amino groups in collagen which are positively charged at acidic pH [16, 18, 19].

Collagen-GAG coprecipitates can be made insoluble without use of a chemical crosslinking agent simply by drastic dehydration. This procedure is based on the discovery that removal of water below ca. 1 wt% insolubilizes collagen; gelatin, the totally amorphous state of collagen, also requires drastic dehydration in order to become insoluble [20, 21]. The nature of crosslinks formed can be inferred from results of earlier, independent studies using chemically modified gelatins. Gelatin which had been modified either by esterification of the carboxylic groups of aspartyl/glutamyl residues or by acetylation of the ε-amino groups of lysyl residues remained soluble in aqueous solvents after exposure of the solid protein to high temperature while unmodified gelatins lost their solubility [22]. Insolubilization of collagen and gelatin following severe dehydration has been, accordingly, interpreted as the result of drastic removal of the aqueous product of a condensation reaction which led to formation of interchain amide links [20]. The proposed mechanism is consistent with results, obtained by titration, showing that the number of free carboxylic groups and free amino groups in collagen were both significantly decreased following high temperature treatment [23, 24].

Removal of water to the extent necessary to achieve a density of crosslinks in excess of 10^{-5} mol crosslinks/g dry gelatin, corresponding to an average molecular weight between crosslinks, M_c, of about 70 kd, is achieved within hours by exposure to temperatures in excess of 105 °C under atmospheric pressure [25]. The possibility that the crosslinking achieved under these conditions is caused by a pyrolytic reaction can be ruled out on the basis of calorimetric data [26]. Furthermore, chromatographic data have shown that the amino acid composition of collagen remained intact after exposure to 105 °C for several days [27, 28]. In fact, it was observed that gelatin can be crosslinked by exposure to temperatures as low as 25 °C provided that a sufficiently high vacuum was present to achieve the drastic moisture removal which drives the reaction [20].

Exposure of highly hydrated collagen to temperatures in excess of ca. 37 °C is known to cause reversible melting of the triple helical structure [29]. The melting point increases with the collagen-diluent ratio from 37 °C, the helix-coil transition of the infinitely dilute solution, to ca. 120 °C for collagen swollen with as little as 20 wt% diluent [29] and up to ca. 210 °C, the melting point of anhydrous collagen [21]. Accordingly, it is possible to crosslink collagen using the drastic dehydration procedure described above without loss of the triple helical structure. It simply suffices to adjust the moisture content of collagen to a low enough level prior to exposure to the high temperature levels required for rapid dehydration. Conservation of the triple helix following prolonged exposure of solid, anhydrous collagen to 105 °C has been confirmed by wide-angle X-ray diffraction and infrared spectroscopy as well as measurement of the components of the optical activity tensor [21, 27].

The simple self-crosslinking treatment also crosslinks GAG chains to collagen [30]. The reaction kinetics are outlined in Fig. 2. The mechanism probably involves condensation of amino groups of collagen with carboxylic groups of glucuronic acid residues on the repeat unit of chondroitin 6-sulfate. Dehydra-

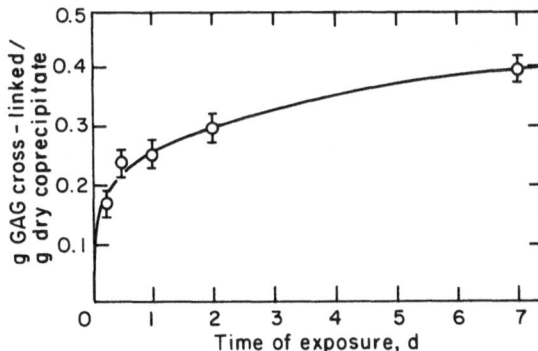

Fig. 2. Kinetics of cross-linking of chondroitin 6-sulfate, a glycosaminoglycan (GAG), to collagen following exposure to 105 °C under 6.7 Pa (50 mtorr). The mechanism of cross-linking is most probably interchain amide condensation involving ε-amino groups of lysyl residues on collagen chains with carboxylic groups on glucuronic acid residues in neighboring GAG chains (From [30] with permission).

tion treatment crosslinks as much as 40% of the GAG originally coprecipitated with collagen. Relative amounts of at least 10 wt% GAG (total dry copolymer basis) can be covalently bound to collagen and M_c can be varied in the range 2.5–25 kd [30]. Proof of formation of a macroscopic network is based on inability to extract GAG from collagen under conditions of temperature and ionic strength where control studies show that GAG separates readily and quantitatively from the coprecipitate [30]. Independent evidence of covalent cross-linking derives from the ability of the network to support an equilibrium tensile force under conditions where the network behaves as an ideal rubber and from the equilibrium swelling behavior (see below) [30].

Dialdehydes have been long known in the leather industry as effective tanning agents [31, 32] and in histological laboratories as useful fixatives [33]. Both of these applications are based on the reaction between the dialdehyde and the ε-amino group of lysyl residues in the protein [34–36] which induces formation of interchain crosslinks. The nature of the crosslink formed has been the subject of controversy, primarily due to the complex, apparently polymeric, character of the glutaraldehyde reagent. Considerable evidence supports the proposed anabilysine structure, derived from two lysyl side chains and two molecules of glutaraldehyde [36]:

$$HO_2C - CH(NH_2) - [CH_2]_4 - N$$

$$N^+ - [CH_2]_4 - CH(NH_2) - CO_2H$$

Evidence for other mechanisms has been presented [37]. By comparison with other aldehydes, glutaraldehyde has shown itself to be a particularly effective crosslinking agent [31, 32], as judged, for example, by its superior ability to increase the crosslink density or, conversely, decrease the average molecular weight between crosslinks, M_c [38]. A wide range of M_c values provides the experimenter with a series of collagens in which the enzymatic degradation rate can be studied over a wide range, thereby affording implants which effectively disappear from the tissue between a few days and several weeks following implantation. Such experimental flexibility is indispensable in efforts to establish, by scanning over the experimental range, whether a threshold degradation rate exists below which the implant exhibits biological activity. Scanning is particularly useful in studies which involve an "unknown" animal lesion, i.e., one which involves a different animal species or a wound on a different tissue site of a given species. Exposure of the collagen-GAG coprecipitate to aqueous glutaraldehyde in a neutral medium destabilizes the ionic macromolecular complex, and the yield of graft copolymer is accordingly very low [30]. Destabilization of the complex and correspondingly low yields of graft copolymer also occur when the ionic strength of the aqueous medium exceeds about 0.25. Up to about 3 wt% GAG (dry copolymer basis) can be incorporated at pH 3 and physiological ionic strength (0.15). By controlling the source of collagen, glutaraldehyde concentration and time of exposure, networks with M_c in the range 5 kd to 40 kd can be prepared [30].

Although the mechanism of the reaction between glutaraldehyde and collagen at neutral pH is understood in part, the reaction in acidic media has not been studied extensively. There is little doubt that covalent crosslinking is involved, since gelatin films prepared from collagen that has been previously exposed to glutaraldehyde support an equilibrium force almost indefinitely when stretched in 1M NaCl at 70 °C to various levels of extension [30, 38]; by contrast, gelatin which has been prepared from untreated collagen dissolves readily in the hot medium. Likewise, little is known about the mechanism by which GAG chains attach themselves tenaciously to collagen following treatment of the coprecipitate in glutaraldehyde. What is clear is that a gelatin-GAG complex prepared from a collagen-GAG coprecipitate which was treated in aqueous acidic glutaraldehyde successfully resists elution of the attached GAG under conditions (3h immersion in 1mol/l NaCl at 70 °C) where the untreated gelatin separates from the attached GAG in just seconds [30]. Additional evidence that glutaraldehyde-treated collagen-GAG coprecipitates have been converted to covalently crosslinked networks is based on their ability to support equilibrium tensile forces [30].

Current preparation procedures for collagen-GAG copolymers make use both of drastic dehydration and glutaraldehyde treatments for reasons that are related to the requirements for a biologically active implant. Dehydration of the highly porous solid which is produced by freeze-drying (see below) stiffens the coprecipitate by introducing crosslinks, thereby preventing pore collapse following prolonged exposure of the foam to ambient humidity. Pore collapse leads to irreversible loss of the high specific surface of the implant and results in complete loss of biological activity. In addition, both drastic dehydration and exposure to glutaraldehyde are, each in its own right and for reasons related to the protein crosslinking reaction which is promoted, efficient procedures for sterilization of the copolymer prior to implantation [30]. Finally, both of these treatments have been used to prepare a medical device, occasionally referred to as artificial skin or artificial dermis, which has been used to treat successfully over 100 massively burned patients without clinical evidence of toxicity [39, 40]. Dehydration crosslinking, which amounts to self-crosslinking as described above, obviously does not lend toxicity to these implants. Glutaraldehyde, on the other hand, is a toxic substance and devices treated with it require thorough rinsing and demonstration of the absence of glutaraldehyde in the rinse before further processing. Additional steps are occasionally followed, such as storage of the implant in media that react with residual reagent [38].

5 Modification of Crystallinity and Network Structure

Structural order in collagen, as in other proteins, occurs at several discrete levels of the structural hierarchy. We follow below the nomenclature proposed by Linderstrom-Lang and Kauzmann [41] to describe generally structural order in proteins, and we relate it to the structure of collagen [21] (Fig. 1). The primary structure of collagen is the complete sequence of amino acids along each of three polypeptide chains as well as the location of interchain crosslinks in relation to this sequence. The secondary structure is the local configuration of a polypeptide chain, i.e., the configuration of a 5- or 10-amino acid segment of the chain, that results from satisfaction of stereochemical angles and hydrogen-bonding potential of peptide residues. The tertiary structure refers to the global configuration of polypeptide chains; it represents the pattern according to which the secondary structures are packed together within the collagen molecule, and it constitutes the unit substructure that can exist as a stable entity in solution (the triple helical molecule). The fourth order or quaternary structure denotes the unit supermolecular structure, comprising several molecules packed in a specific lattice, which constitutes the basic element of the solid state. Higher levels of order, eventually leading to gross anatomical features which can be readily seen with the naked eye, have been proposed [42, 43].

Crystallinity in collagen can, according to Fig. 1, be detected at two discrete levels of structural order: the tertiary (triple helix) (Fig. 1C) and the quaternary (lattice of triple helices) (Fig. 1D). Biochemists have used optical rotation, circular dichroism and viscometry as reliable methods of detecting in solution the conversion of triple helical collagen to randomly coiled gelatin [44]; however, the use of optical methods to analyze the gelatin content of solid specimens, such as films cast from solution, has been complicated partly because of the optically anisotropic character of the solid state of this protein. Procedures for the quantitative analysis of collagen and gelatin in solid specimens which are imperfectly crystalline have been developed in order to study the substantial increase in enzymatic degradation rate which is observed when collagen is converted to gelatin [45, 46]. The presence of an intact tertiary structure (triple helix) can be detected in the solid state by mid- and far-infrared spectroscopy, using "helical marker" bands which provide a quantitative measure of the extent to which a given collagen preparation has been converted to gelatin [21, 47]. Measurements of optical rotation with solid specimens also yield a quantitative analysis of gelatin in solid specimens, provided that the tensorial character of optical activity is taken into consideration and corrections are also made to account for the frequent presence of birefringence in collagen specimens [48]. However, the detailed structure of banded collagen fibers has not yielded to X-ray crystallographic analysis detailed.

Electron microscopy has yielded the periodic banding pattern of collagen fibrils [49–52] and can also be used to provide a quantitative estimate of the length-average fraction of fibrils which possess discernible banding in a preparation of collagen fibrils [53, 54]. This morphological procedure can be supplemented with measurements of small-angle X-ray scattering (SAXS) [55]. Use of these procedures has led to the finding that the well-known loss of banding pattern which collagen fibrils undergo in aqueous acetic acid occurs relatively sharply at pH 4.25 ± 0.30 (Fig. 3) [53, 54]. Combined use of electron microscopy and IR spectroscopy showed that this transition amounts to disordering of the

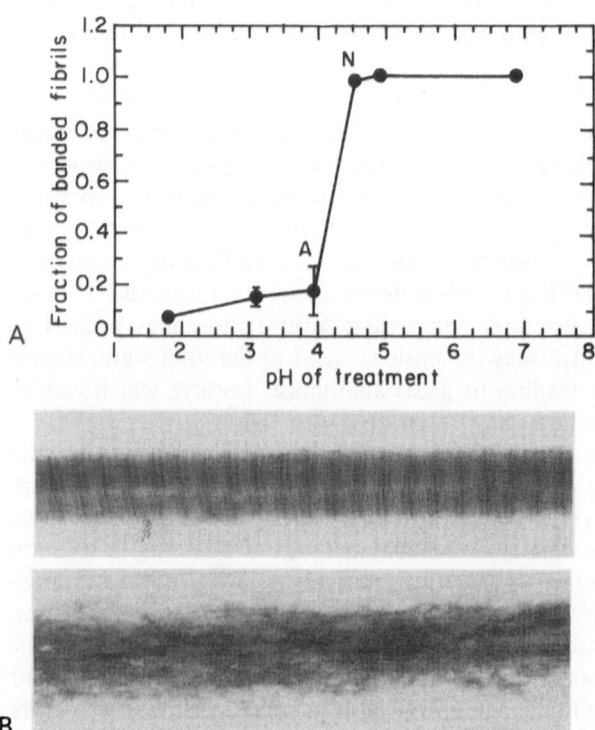

Fig. 3. A The banding pattern of collagen fibers from bovine hide persists down to pH 4.25 ± 0.30 in 0.05*M* acetic acid. The period is 65 nm. The fraction of banded fibrils was determined by transmission electron microscopy (graph). **B** Below pH 4.25 ± 0.30 the banding disappears from almost all fibers while the fiber diameter increases, indicative of swelling in the aqueous acetic acid diluent. Electron microscopy shows that a small fraction of banded fibrils, about 10% or less persists below the transition even after long exposure to the swelling medium (graph). The transformation abolishes the quaternary structure (packing order of helices, Fig. 1D) but leaves the tertiary structure (triple helix, Fig. 1C) intact. While banded collagen aggregates blood platelets, nonbanded collagen does not. Magnification for A and N: 75 000 X. (Reprinted from [53, 54] with permission)

lattice but not to loss of the triple helical structure [54]. Changes in pH can therefore be used to abolish selectively the quaternary structure while maintaining the tertiary structure intact. This experimental strategy has made it possible to show that the well-known phenomenon of blood platelet aggregation by collagen fibers is a specific property of the quaternary, rather than tertiary, structure [54]. Collagen has, therefore, been prepared in a thromboresistant form by selectively "melting out" the packing order of helices while preserving the helices intact [54]. Fig. 4 illustrates the banding pattern observed with collagen-GAG copolymer matrices. Notice that short segments of banded collagen fibrils occasionally interrupt long segments of nonbanded fibrils (Fig. 4, inset).

Network properties of crosslinked collagen-GAG copolymers can be analyzed structurally by methods which have been based on the theory of rubber elasticity. Under conditions where the theory holds it is possible to compute directly the crosslink density of the network from measurements of the equilibrium modulus [56]. Collagen specimens with high fraction of crystallinity, such as films cast at room temperature from a neutral buffer or naturally occurring tendon fibers, must be converted to gelatin before they show rubber-like elastic behavior [57]. Insoluble collagen which has been converted to gelatin by immersion in physiological saline between 65 and 80 °C for 1h, displays a resistance to uniaxial stretching (tensile modulus) which is directly proportional to the density of crosslinks c (mol crosslinks/g dry polymer) and inversely proportional to the number-average molecular weight between crosslinks, M_c (kd) [30, 38, 57]. Collagen-GAG copolymers show similar behavior provided that the collagen is maintained in a gelatinized condition during measurement [30, 38]. These relationships demonstrate that, unlike collagen, crosslinked and swollen gelatin is a network in which the chains are random coils [44]. Below about 65 °C recrystallization of gelatin occurs, leading to development of significant energetic interactions and these relationships fail [38]. In the range of applicability of the ideal rubber-like model the following constitutive relation [56, 58] describes the tensile behavior of appropriately gelatinized collagen [57] and collagen-GAG copolymers [30, 38]:

$$\sigma = (\rho RT/M_c)\,v_2^{1/3}(\alpha - \alpha^{-2}). \tag{1}$$

In Eq. (1), σ is the equilibrium stress (Nm^{-2}) supported by the swollen specimen; α is the stretched specimen length divided by the unstretched length (extension ratio); v_2 is the volume fraction of dry protein; and ρ is the density of dry protein. In the common case of tetrafunctional crosslinks, the concentration of network chains n (mol network chains/g polymer) is exactly one-half the concentration of crosslinks, so that n = 2c. The hypothesis that a specimen behaves as if it were an ideal rubber can be confirmed by observing a linear relation with zero intercept between σ and the strain function $(\alpha - 1/\alpha^2)$ and by establishing a direct proportionality between σ and the absolute temperature at constant value of the extension ratio, as stipulated by Eq. (1).

When a sample is smaller than about 1 cm it becomes impractical to use tensile measurements to study network structure. The swelling behavior of small

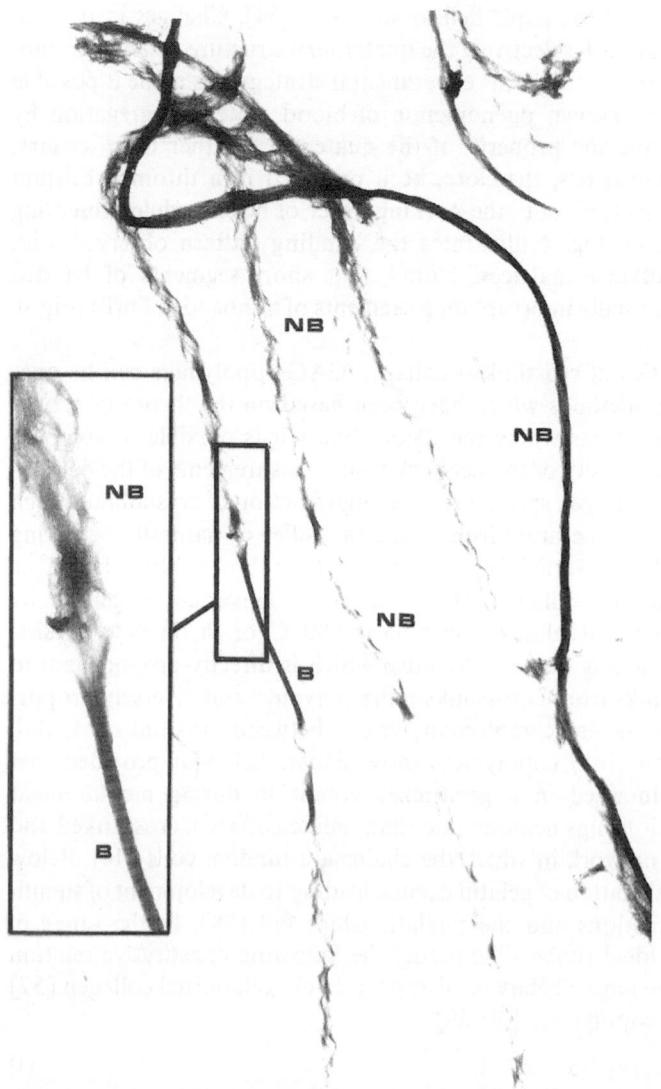

Fig. 4. Following exposure to pH below 4.25 ± 0.30 the banding pattern of type I bovine hide collagen practically disappears. Short lengths of banded collagen (B) do, however, persist next to very long lengths of nonbanded collagen (NB) which has tertiary but not quaternary structure. This preparation does not induce platelet aggregation provided that the fibers are prevented from recrystallizing to form banded structures when the pH is adjusted to neutral in order to perform the platelet assay. Stained with 0.5 wt% phosphotungstic acid. Banded collagen period, about 65 nm. Magnification: 15 000X. Inset mag.: 75 000X. (Reprinted from [53] with permission)

specimens can be used in such cases to calculate M_c. The method is based on the theory of Flory and Rehner [58, 59], who showed that the volume fraction of a swollen polymer v_2 depends on M_c through the following relationship:

$$\ln(1 - v_2) + v_2 + \chi v_2 + (\rho V_1/M_c)(v_2^{1/3} - v_2/2) = 0 \tag{2}$$

In Eq. (2) V_1 is the molar volume of the solvent, ρ is the density of the polymer, and χ is a constant characteristic of a specific polymer-solvent pair at a particular temperature. Although independent methods for estimating χ have been described [58], it is possible to obtain M_c from tensile measurements for a given polymer-solvent system and use it to compute a single value of χ for a large number of network variants that can be prepared from the same polymer or copolymer. Equation (2) holds under conditions where the network behaves as an ideal rubber. The use of Eq. (2) is illustrated in Fig. 5. Values of M_c computed by use of Eqs. (1) or (2) provide insight into the structure of the network. For example, since the average molecular weight of an amino acid residue is about 93 [60], an M_c value of 14 kd corresponds to a not very tightly crosslinked network with an average network chain of approximately 151 amino acids. Amino acid analysis of naturally occurring collagen from tendon yields about 27 mol lysyl residues per 1000 total mol residues [60]; on average, therefore, a lysyl residue occurs once every 37 residues. If we make the assumption, justified above, that glutaraldehyde attacks only lysyl residues we conclude that a M_c value of 14 kd implies that very roughly only 25% of the available ε-amino groups has reacted.

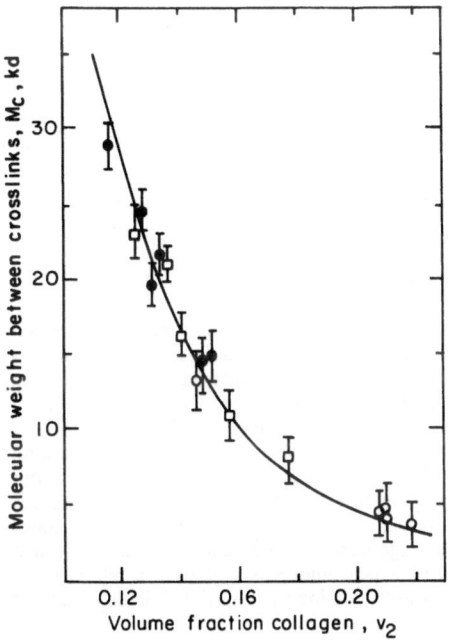

Fig. 5. Relation between the average molecular weight between cross-links, M_c, and the volume fraction of gelatinized collagen (gelatin), v_2. The swelling agent is 0.19 mol/l citric acid – phosphate buffer solution, pH 7.4 at 80 °C. Experimental data obtained after cross-linking with various aldehydes: *solid circles*, formaldehyde; *open circles*, glutaraldehyde; *squares*, glyoxal. The curve is predicted by Eq. (2) with $\chi = 0.52 \pm 0.04$ [38]

6 Construction of Porous Matrices

The porosity of a collagen-GAG copolymer is an indispensable component of its biological activity [61]. Pores are incorporated by first freezing a very dilute suspension of the collagen-GAG coprecipitate and then inducing sublimation of the ice crystals by exposing to vacuum at low temperatures [62]. The resulting

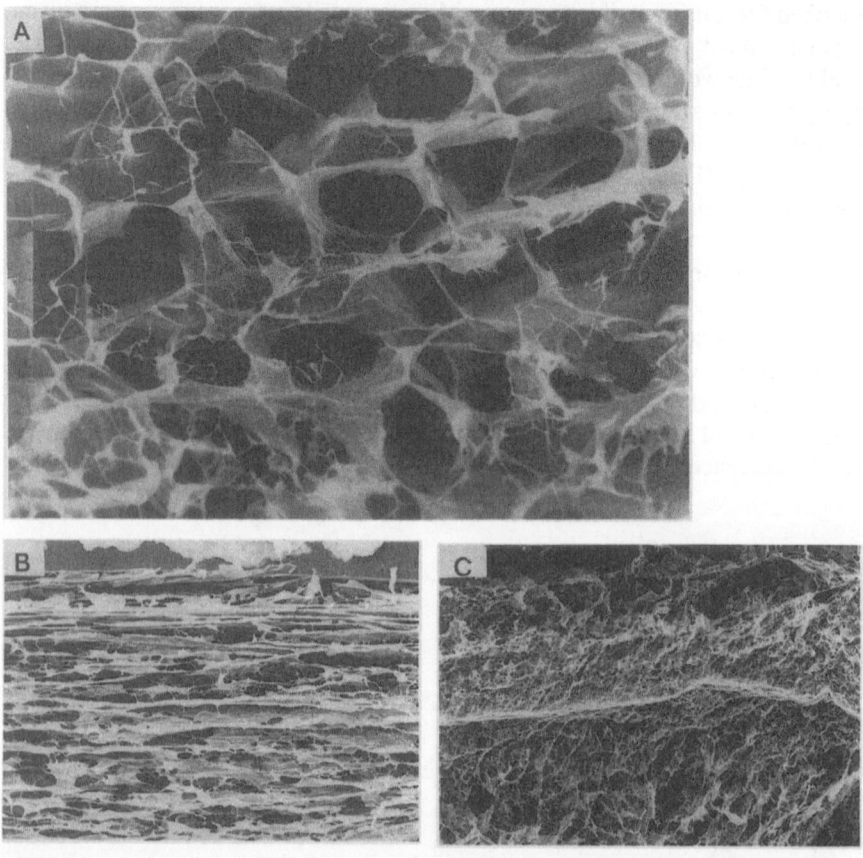

Fig. 6 A–C. Illustration of the variety of porous structures which can be obtained with collagen-GAG copolymers by adjusting the kinetics of crystallizaton of ice to the appropriate magnitude and direction. Pores form when the ice dendrites are eventually sublimed. Scanning electron microscopy: **A** The porous structure of collagen-GAG matrices is the negative replica of the ice crystal structure which is formed when the dilute suspension of collagen-GAG coprecipitate particles is quenched below 0 °C. The orientation of pore channel axes shown in this view is random and characterizes matrices which have induced skin regeneration. Average pore diameter, about 100 μm, Pore volume fraction, about 0.99; **B** Strong uniaxial orientation of pore channel axes is shown in this cross-section of a 1.5-mm diameter cylindrical matrix which was used to induce regeneration of rat sciatic nerve across a gap. Average pore diameter, about 100 μm; **C** Radial orientation of pore channel axes yields an implant which induced synthesis of a sciatic nerve regenerate that was much less functional than when the orientation was uniaxial. Cylinder diameter: 1.5-mm. Average pore diameter, about 50 μm

pore structure is, therefore, a negative replica of the network of ice crystals (dendrites). It follows that control of the conditions of ice crystal nucleation and growth can lead to a large variety of pore structures. In experimental implants the mean pore diameter has ranged between about 1 and 800 μm, volume fractions have ranged up to about 0.995, the specific surface has been varied between about 10^4 and 10^8 mm^2/g matrix and the orientation of axes of pore channels in cylindrical implants has ranged from strongly uniaxial to random to highly radial. Figure 6 illustrates the range of porous structures that can be made accessible by control of the kinetics of ice nucleation and crystal growth.

The biological activity of collagen-GAG copolymers depends generally on the structure of the porous matrix, i.e., on the volume fraction, specific surface, mean pore size and orientation of pores in the matrix. Determination of these properties is based on use of principles of stereology [63, 64], the discipline which relates the quantitative statistical properties of three-dimensional structures to those of their two-dimensional sections or projections. In reverse, stereological procedures allow reconstruction of certain aspects of three-dimensional objects from a quantitative analysis of planar images [64]. Semi- or fully automated apparatus for quantitative image analysis has greatly facilitated stereological measurements. A plane which goes through the two-phase structure may be sampled by random points (Fig. 7A), by a regular pattern of points (Fig. 7B), by a near-total sampling using a very dense array of points (Fig. 7C) or by arranging the sampling points to form a continuous line (Figure 7D). It can be shown [63, 64] that the volume fraction of pores, V_V, is equal to the fraction of total test points which fall inside pore regions, P_P; also equal to the total area fraction of pores A_A (Fig. 7C); and, finally, equal to the line fraction of pores, L_L, for a linear point array in the limit of infinitely close point spacing (Fig. 7D):

$$V_V = P_P = A_A = L_L \tag{3}$$

The sampling methods illustrated in Fig. 7 are the basis of all operations in stereology and are referred to as point counting (Fig. 7A, 7B), areal analysis

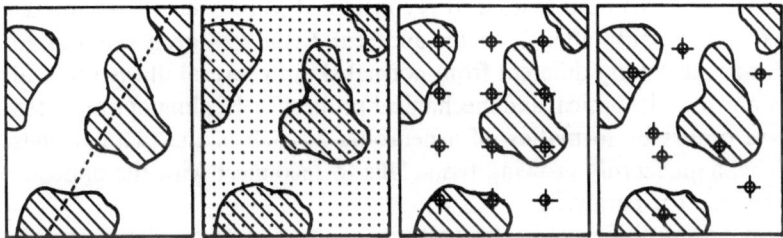

Fig. 7. Schematic representation of four procedures commonly used to sample a field in stereological analysis. These procedures have been used to study the porous structure of collagen-GAG matrices [74] and yield values for average pore diameter, pore volume fraction and other features. In this illustration, a phase A (*cross-hatched*) is embedded in a continuous phase B (*white background*). **A** Random point count; **B** systematic point count; **C** areal analysis; **D** lineal analysis. (Reprinted from [64] with permission).

(Fig. 7C) and lineal analysis (Fig. 7D). Clearly, randomness in the selection of a sampling procedure is of critical importance. Nearly complete characterization of the details of pore structure in collagen-GAG matrices have been obtained by systematic use of these procedures [62].

7 Physicochemical Correlates of Biological Specificity

"Biological activity" is often defined in a highly operational manner. Definitions of activity are couched in terms of a highly specific assay, involving cells or tissues, which typically incorporates a very narrowly defined hypothesis about the phenomenon of interest. Such an assay is itself defined strictly in operational terms, thereby encouraging reproducibility of the experimental results in different laboratories. Studies with collagen-GAG matrices have focused on regeneration of specific tissues which are well-known not to regenerate spontaneously. We seek, therefore, to construct assays to study the effect of certain structural features of collagen-GAG matrices on de novo synthesis of a given tissue. This problem is somewhat analogous to the search for experimental strategies to study the effect of a family of heterogeneous catalysts on the yield of a certain product in a relatively well-defined reactor space.

The activity of regeneration matrices has been studied in two relatively well established environments: the full-thickness, excised skin wound in the guinea pig and in man [65, 66]; and the 10-mm, as well as the 15-mm, transected gap in the sciatic nerve of the rat [67, 68]. Both of these "reactors" are wounds produced by a routine surgical procedure and in both cases the "kinetics" and the "mechanism" of the normal wound healing "reaction", i.e., healing in the absence of a biologically active graft, have been studied rather extensively. Normal healing of a full-thickness skin wound is characterized by contraction of the wound perimeter toward the wound interior which begins within several hours after the skin has been excised. It is considered to be over when contraction has ceased, roughly after 2 weeks, and scar tissue has been synthesized [65, 66]. Scar is distinctly different from normal skin in several ultrastructural features as well as in its optical and mechanical properties. Healing of transected sciatic nerve occurs by formation of a neuroma at each of the two cut ends [67, 68]. This haphazardly growing tissue fails to reconnect with the opposite side and the result is paralysis.

The most obvious difference between a regeneration matrix and a catalyst is that the former is consumed during the course of the reaction. Implanted collagen-GAG matrices are degraded by collagenases [69, 70], specific enzymes which attack the triple helical molecule at one specific locus. Two characteristic products result: the N-terminal three-quarter fragment and the C-terminal one-quarter fragment both of which are spontaneously denatured to gelatin at physiological temperatures [70]. The gelatinized fragments are then cleaved by several non-specific proteases. Collagenases are naturally present in healing wounds and are credited with the degradation of collagen fibers at the site of trauma. At about the same time that degradation of collagen and of other ECM components proceeds in the wound bed these components are being synthesized by cells in the wound bed. The combined process is referred to as remodeling [5–7]. The balance between the rates of the two processes has been frequently considered to be an important feature of wound healing [71–73].

The availability of chemical analogs of ECM which are degraded by collagenase over a relatively wide range of the time scale makes it possible to study quantitatively the effects of ECM transience on a model wound healing process. After the ECM analog has been brought in close physicochemical contact with the wound bed and after establishing that cells from the wound bed are migrating freely inside the pore structure of the analog one can seek an answer to the question: how does a deliberate variation in the degradation rate of the ECM affect the mechanism of wound healing? Although the system of interest is a model wound it is vitally important to establish a carefully designed in vitro experiment which can provide definitive tests of hypotheses about interactions between the matrix and specific components of the wound bed. Conclusive test results on isolated mechanistic steps frequently cannot be obtained in the complex in vivo model.

The degradation rate of collagen-GAG matrices in collagenase can be conveniently measured in vitro by at least two methods which yield complementary results. In the first, a suspension of the finely comminuted implant is incubated in a stirred, standardized bath of collagenase. Degradation produces oligopeptides which are determined by a photometric procedure [74, 75]. The second assay is a mechanochemical method in which small strips of matrix are stretched in a standardized bath of collagenase and the kinetics of degradation are monitored by measuring the force necessary to maintain the specimen at fixed extension [38, 45, 76]. The latter procedure was inspired by Tobolsky's use of chemical stress relaxation as a means of studying the kinetics and mechanism of chain scission in rubbers [77]. Whereas the first method measures the amount of solubilized protein the second monitors the amount of undigested protein which persists in the form of a stress-bearing network. Additional study is necessary to correlate the information obtained by each of these two, apparently complementary, procedures.

Certain interesting strategies for control of implant resistance to collagenolytic action have emerged from the use of in vitro assays. A systematic study of degradation rates of collagen specimens crosslinked with formaldehyde, glyoxal and glutaraldehyde in the M_c range ca. 5 to 25 kd showed an approximately 15-fold monotonous increase in the rate constant with increase in the mol. wt. between crosslinks [38, 45]. An approximately 10-fold increase in degradation rate with conversion to gelatin was also observed [45]. A particularly novel finding was a 5-fold increase in resistance to collagenolysis as the relative amount of chondroitin 6-sulfate grafted onto collagen increased up to about

8 wt%; no further increase in resistance was observed as the amount of GAG grafted onto collagen increased beyond that level [38, 73].

In vivo studies have confirmed the validity of these *in vitro* effects during the first 7–10 days following implantation; however, during the second week important and revealing deviations from *in vitro* predictions became evident. For example, an increase in M_c led to acceleration of the rate at which subcutaneously implanted matrices were degraded [61] during the 10 days following implantation, while grafting of collagen with chondroitin 6-sulfate decreased the rate of implant degradation during that period [73, 76]. Both of these early results were expected from *in vitro* studies. However, during the second week it was observed that implants in which the amount of grafted GAG exceeded about 2 wt% showed a net increase both in total dry weight and in relative collagen content while implants in which the GAG content was lower or was zero continued to lose weight monotonously and maintained a constant relative collagen content during this period [61, 73, 76]. Histological studies have shown that net increases in total dry weight followed by increases in relative collagen content are associated with synthesis of new connective tissue in very close proximity to the degrading matrix [61, 73]. These findings showed that the simple in vitro assays for degradation rate of matrices had important predictive power which, however, was apparently limited to events occurring during the early phases of wound healing (1–3 weeks).

The effect of physicochemical manipulations of the matrix on the kinetics and mechanism of wound healing has turned out to be remarkable. One of the most obvious events which follows a total surgical removal (excision) of skin in rodents and in man is contraction of the wound perimeter [65, 66]. In the rodent model, where the contraction is especially vigorous, this process stops when two apposed wound edges have come in approximate contact and scar tissue has been synthesized in the narrow space separating the two edges. In the human, contraction stops at an earlier stage with a greater amount of scar tissue separating the wound edges. It is now clear that skin wound contraction can be delayed so that it starts about 12 days after the wound is generated, rather than within 1–2 days after trauma, provided that the collagen-GAG matrices possess a narrowly defined constellation of properties: the resistance to degradation is higher than a critical level (Fig. 8), corresponding to loss of about 50% of dry implant weight within 10 days; and the average pore diameter is between 20 and 120 μm (Fig. 9), corresponding to a specific surface in the approximate range 10^6–10^7 mm^2/g dry matrix [78]. Furthermore, preliminary studies have shown that substantial delay in onset of wound contraction occurs when collagen banding in implants was maintained at near-zero levels (Fig. 4) while the triple helical structure remained intact [54, 79] and when the pore volume fraction was higher than about 0.95 [79].

When collagen-GAG matrices which can significantly delay the onset of wound contraction are seeded with a minimum density of easily separable skin cells (basal cells) from the same animal, and the cell-seeded matrices are then grafted, wound healing is affected in an even more profound way: In this case,

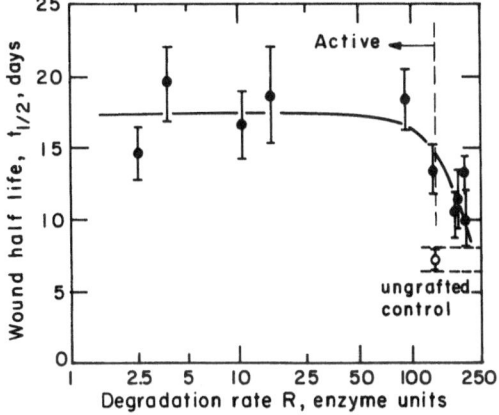

Fig. 8. Variation of the skin wound half-life with degradation rate R (in collagenase) of collagen-GAG matrix. The half-life is the time required for a wound to contract to 50% of the original area. The degradation is in empirical units, which are defined in terms of an in vitro assay. A somewhat arbitrary *broken vertical line* is drawn near R = 140 enzyme units. This line shows the level of degradation rate above which the half-life of matrices rapidly drops to the level of the ungrafted wound. The horizontal scale is logarithmic [79]

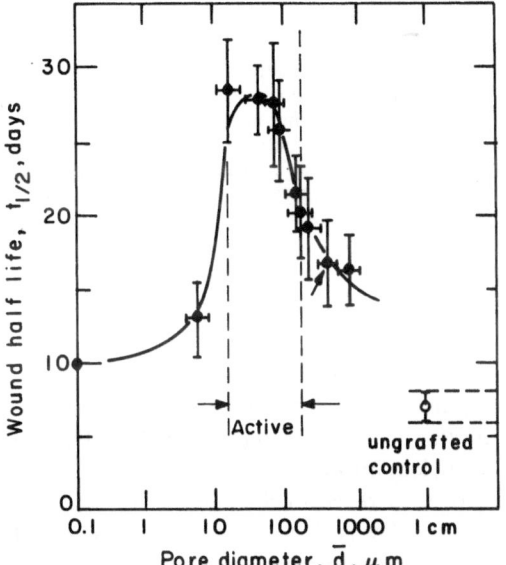

Fig. 9. Variation of half-life of skin wound with the average pore diameter of collagen-GAG matrices used as grafts for full-thickness skin wounds in the guinea pig. The *vertical broken lines* at about 20 and 120 µm mark the limits of matrix activity. Outside these approximate limits the wound half-life rapidly drops to the level of the ungrafted wound. The horizontal scale is logarithmic [79]

not only is the onset of wound contraction delayed by approximately 10 days but, a few days after it starts, contraction is arrested midway before complete closure; its direction is then reversed and the wound perimeter expands over several days until it slows down and finally stops expanding (Fig. 10). Several weeks after grafting the wound perimeter encloses an area about 65% of the original wound area and new skin, complete with a dermis and an epidermis, fills this area [79, 80]. If the matrix has not been previously seeded with skin cells contraction is still delayed significantly but the wound perimeter eventually closes with formation of scar. The new skin synthesized with the help of

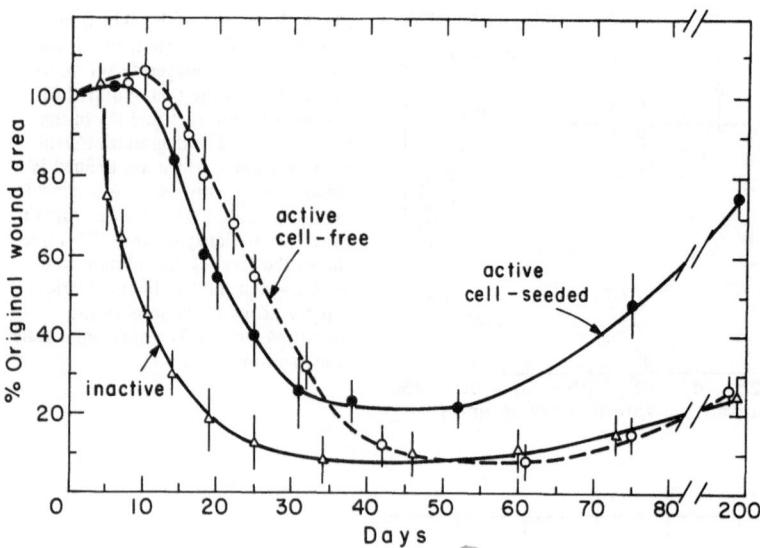

Fig. 10. The kinetics of skin wound contraction for two types of matrix and an ungrafted control. Change in the original wound area with time observed when full-thickness guinea pig skin wounds were grafted with inactive collagen-GAG matrices (Δ), active cell-free collagen-GAG matrices (○), and active cell-seeded collagen-GAG matrices (●). Active cell-free matrices significantly delayed the onset of wound contraction but did not eventually lead to arrest of contraction and skin regeneration. Active cell-seeded matrices delayed the onset of contraction, eventually arrested it, and induced skin regeneration within a woundbed which was expanding in area [79]

cell-seeded matrices is hairless but there is strong evidence from light scattering studies [81] as well as histological studies [79–80] that it is distinct from scar and is similar, though not identical, to normal skin. Morphological and physical (light scattering) evidence which distinguishes new skin from scar is illustrated in Fig. 11. The skin wound contraction kinetics are summarized in Fig. 12. The kinetic data separate collagen-GAG matrices into three classes, namely inactive matrices, active matrices which are not seeded with cells, and active matrices which are seeded with an adequate number of cells [82]. This is the first report of successful regeneration of the dermis in the adult mammal [78–80]. Other collagen-GAG matrices have induced regeneration of almost intact rat sciatic nerve to occur across a gap of 15 mm without the need for cell seeding [83].

Although the epidermis regenerates readily over an intact or partly intact dermal bed, it is well-known that *de novo* synthesis of the dermis does not occur spontaneously [65, 66]. Likewise, even though the rat sciatic nerve is regenerated spontaneously across a 5-mm gap or occasionally across a 10-mm gap (provided that the cut ends of the nerve are inserted in a saline-filled rubber tube), no such regeneration is spontaneously observed across a 15-mm gap [67, 68]. The appropriate use of collagen-GAG matrices leads to synthesis both of skin, complete with dermis and epidermis [79, 80, 84] and of new sciatic nerve [83]. It is this ability to induce *de novo* synthesis of nearly physiological tissue

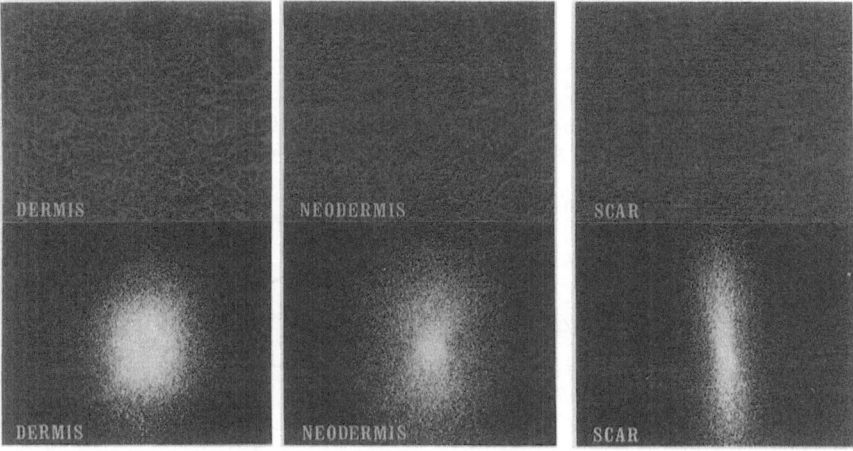

Fig. 11. Comparison of histological data obtained by light microscopy of tissue sections (*top row*) with small-angle patterns obtained by scattering a visible laser beam from the same tissue sections (*bottom row*). Tissue sections were stained with hematoxylin and eosin, but the nature of the histological stain does not appear to affect the patterns. (*Left*) View of normal guinea pig dermis shows a relatively random arrangement of collagen fibers and an elliptical scattering pattern. (*Middle*) Regenerated dermis shows a less random orientation of collagen fibers and a scattering pattern which is clearly more anisometric than is true for intact dermis. (*Right*) Dermal scar shows a highly oriented array of collagen fibers and a scattering pattern which is indicative of very high orientation of fibers and significantly different from the pattern for neodermis [8]

under conditions where such synthesis does not occur spontaneously that characterizes regeneration matrices from the very large array of collagen or collagen-GAG matrices which are biologically inactive in this respect.

The clinical applications of ECM analogs appear to be unprecedented. Patients who have suffered massive burns and have lost a large fraction of their skin require immediate replacement of the lost skin area in order to regain control of their dehydration rate as well as to stem large-scale infection. The best treatment currently is the skin autograft, i.e., the patient's own skin, which can be obtained only after subjecting the patient to a serious operation. Clinical studies have shown that an ECM analog, virtually identical to the one which induces synthesis of both a dermis and an epidermis in the guinea pig except that no cells have been previously seeded into it, induces synthesis of new physiologic dermis within about 15 days in patients with massive burns [39]. The new dermis forms a suitable bed on which to graft a thin layer of the patient's epidermis, which can be obtained without a serious operation, and the open wound is thereby closed permanently. A recent randomized clinical trial involved 106 patients, all massively burned [40]. This 11-hospital trial has confirmed the early finding [39] and has shown that treatment with the ECM analog gives results which are clinically equivalent to treatment with the autograft [40]. The inability of patients to synthesize dermis spontaneously makes this clinical finding especially interesting. It is very likely that other

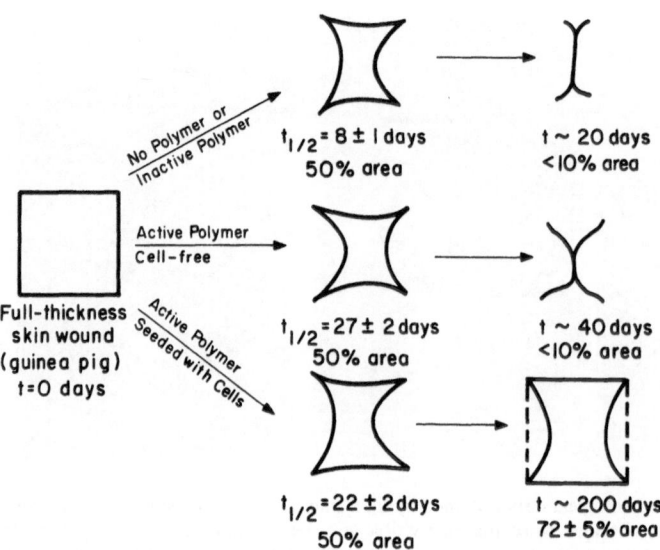

Fig. 12. The kinetics of contraction of full-thickness guinea pig skin wounds separate collagen-GAG matrices into three classes. The wound half-life $t_{1/2}$ is the number of days necessary to reduce the original wound area to 50%. An inactive matrix does not delay wound contraction significantly relative to the ungrafted control and eventually allows formation of a "linear scar." An active, cell-free matrix delays wound contraction by about 20 days but eventually allows full contraction to occur. An active matrix, which has been seeded with a minimal number of skin cells, delays contraction significantly, later arrests it , and eventually induces synthesis of a new dermis and epidermis within an expanding wound perimeter

tissues which do not regenerate spontaneously during wound healing can be induced to do so if appropriately synthesized ECM analogs are used to treat them.

The necessary attributes of a regeneration matrix can be most simply understood in terms of a highly specific cell-matrix interaction which diverts decisively the mechanism of wound healing away from contraction and scar synthesis to arrest of contraction and tissue regeneration in the wound bed [79]. The quantitative relations between matrix structure and wound contraction kinetics, summarized in Figs. 8–10, can be used to deduce the physicochemical requirements: what seems to be necessary is the presence of a collagen-GAG surface which is a) sufficiently extensive (upper limit of pore diameter; Fig. 9); b) sufficiently non-diffusible over a period of about 10 days (upper limit of degradation rate; Fig. 8); and c) endowed with pores which are sufficiently large to allow ready access to cells migrating into the wound bed (lower limit of pore diameter; Fig. 9) [79]. Additional criteria which are currently being studied include an apparent requirement for nearly zero crystallinity (banding) (Fig. 4) and a pore volume fraction in excess of 0.95. Although skin regeneration requires that the matrix be seeded with a critical density of cells, the physicochemically defined structure suffices, by itself, to delay wound contraction (Figs. 10, 12) and probably sets the biological stage for regeneration to occur. In studies of nerve regeneration it has also been observed that a matrix which has not been seeded with cells can, in fact, induce synthesis of new, functional nerve tissue over large gaps [83].

It is fairly obvious that the unusual biological activity of certain collagen-GAG matrices is due to specific cell-matrix interactions which take place when these matrices are in contact with the skin wound bed or with the cut end of the nerve. The molecular character of these interactions is currently under study.

8 Light Scattering Studies of Regenerated Tissue

Physiologic dermis and scar tissue differ not so much in chemical composition as in the morphology of the collagen fibers constituting each type of tissue. The most common method for differentiating between dermis and scar involves the expertise of a histopathologist who judges the nature of the tissue by viewing an appropriately stained tissue section in the optical microscope. Although very widespread and usually quite reliable, this method loses its value in cases where the tissue synthesized by the animal possesses a morphology which is intermediate in character between that of physiologic tissue and scar. This in fact is the case in studies with ECM analogs where the tissue synthesized by an "active" analog, i.e., by a regeneration template, is very similar to physiologic tissue and quite dissimilar to scar, while use of an 'inactive' analog leads to synthesis of scar.

A light scattering technique [81] which can be used to analyze the degree of orientation and diameter of collagen fibers in histological sections of connective tissue has been used to study quantitatively the fidelity with which the dermis was regenerated following use of an "active" cell-seeded collagen-GAG copolymer [79]. The diameter of collagen fiber bundles in connective tissue is roughly equal to between ten and fifty times the wavelength of visible light [85]. For this reason, collagen fiber bundles scatter light at small angles, roughly between 1° and 6° [86]. Light scattered from histological sections of dermis produces a scattering pattern which contains information on the orientation and size of the constituent fibers. Kronick and Sacks [87] have independently developed a light scattering procedure to study connective tissues.

The problem of calculating the degree of orientation and size from a scattering pattern has been widely addressed in the scientific literature [86, 88]. In the study of regenerated tissues a two-dimensional analog of the Hermans orientation index [89, 90], previously used to specify the orientation of a polymer from X-ray scattering data, was adapted [81, 91].

The degree of orientation [91] was defined by an orientation index, S, which has an upper limit indicative of perfect orientation along a given axis and a lower limit corresponding to an ideally random orientation. A particularly useful definition for S is:

$$S = 2\langle \cos^2 a \rangle - 1 \tag{4}$$

where a is the angle between an individual fiber and the mean axis of the fibers, and $\langle \cos^2 a \rangle$ denotes the square cosine of a averaged over all the fibers in the sample. Since $\langle \cos^2 a \rangle$ equals 1/2 for a random arrangement of fibers and 1 for a perfectly aligned arrangement, S will vary from 0 to 1, respectively.

The final expression for S used to calculate the orientation index from experimental data is (see [81, 91] for derivation):

$$S = 2 \frac{\int_0^{\pi/2} I(\beta) \cos^2 \beta \, d\beta}{\int_0^{\pi/2} I(\beta) \, d\beta} \tag{5}$$

Data obtained by this procedure with guinea pig tissues [73] show S values of 0.25 ± 0.09, 0.45 ± 0.11 and 0.84 ± 0.05 for normal dermis, regenerated dermis and scar, respectively. This data shows that although the regenerated tissue is clearly not scar it is also not identical in orientation to intact dermis. The quantitative procedure [81, 91] can be readily employed as a definitive measurement in an iterative procedure which could lead to modifications of collagen-GAG copolymer. Such modifications would quite probably lead to an ECM analog of high biological activity, conceivably yielding regenerated tissues which resemble intact dermis more closely.

9 Conclusion

Although the extracellular matrices are exceedingly complex macromolecular networks it is possible to synthesize relatively simple analogs of ECM which possess substantial biological activity. One of these analogs, a type I collagen-graft-chondroitin 6-sulfate copolymer with narrowly defined porosity and degradation rate, has been shown capable of partially regenerating the dermis in animals and humans. Another analog with the same chemical composition but quite different porosity and degradation rate has induced regeneration of the sciatic nerve in animals. The knee meniscus in dogs has also been regenerated by use of an ECM analog [96]. It is highly likely that much more sophisticated structures which resemble naturally occurring ECMs more faithfully could revolutionize the treatment of tissue and organ loss.

This manuscript is based in part on the author's article in [9].

10 References

1. Clark RAF, Henson PM (eds) (1988) The molecular and cellular biology of wound repair. Plenum, New York
2. Nimni ME (ed) (1988) Collagen, Vol I, Biochemistry, CRC Press, Boca Raton
3. Hynes RO (1990) Fibronectins. Springer, Berlin Heidelberg New York
4. Silbey JE (1987) Advances in the biochemistry of proteoglycans. In: Uitto J, Perejda AJ (eds) Connective tissue diseases. Marcel Dekker, New York
5. Piez KA, Reddi AH (eds) (1984) Extracellular matrix biochemistry. Elsevier, New York
6. Hay ED (ed) (1981) Cell biology of extracellular matrix. Plenum Press, New York

7. Trelstad RL (ed) (1984) The Role of Extracellular Matrix in Development, Alan R. Liss, New York
8. Yannas IV (1988) In: Encyclop Polymer Sci Eng, Second Ed, vol 13, p 317
9. Yannas IV (1990) Angewandte Chemie 29: 20
10. Chvapil M, Kronenthal RL, van Winkle W Jr (1973) In: Int Rev Conn Tissue Res Vol 6, Academic Press, New York, p 1
11. Schmitt FO (1985) Ann Rev Biophys Chem 14: 1
12. Grillo HC and Gross J (1962) J Surg Res 2: 69.
13. Abbenhaus JI, MacMahon RA, Rosenbrantz JG, Paton BC (1965) Surg Forum 16: 477
14. Chvapil M, Holusa R (1968) J Biomed Mater Res 2: 245
15. Rubin AL, Stenzel KH (1969) In: Biomaterials, Stark L, Agarwal G (eds) Plenum, New York, p 157
16. Mathews MB (1975) Connective tissue. Springer, Berlin Heidelberg New York
17. Ring W, Mita I, Jenkins AD, Bikales NM (1985) Pure Appl Chem 57: 1427
18. Toole BP, Lowther DA (1968) Biochem J 109: 857
19. Podrazky V, Steven FS, Jackson DS, Weiss JB, Leibovich SJ (1971) Biochim Biophys Acta 229: 690
20. Yannas IV, Tobolsky AV (1967) Nature 215: 509
21. Yannas IV (1972) Rev Macromol Chem C7: 49
22. Bello J, Riese-Bello H (1958) Sci Indust Photograph 29: 361
23. Silver FH (1977) GAG inhibition of collagen-platelet interaction. Ph.D. Thesis, Massachusetts Institute of Technology, Cambridge, MA
24. Silver FH, Yannas IV, Salzman EW (1979) J Biomed Mater Res 13: 701
25. Yannas IV, Tobolsky AV (1966) J Macromol, Chem 1: 723
26. Yannas IV, Tobolsky AV (1968) Eur Polym J 4: 257
27. Sung N-H (1972) Structure and properties of collagen and gelatin in the hydrated and anhydrous solid state. ScD Thesis, Massachusetts Institute of Technology, Cambridge, MA
28. Bowes JH, Taylor JE (1971) J Amer Leather Chem Assoc 66: 96
29. Flory PJ, Garrett RR (1958) J Amer Chem Soc 80: 4836
30. Yannas IV, Burke JF, Gordon PL, Huang C, Rubenstein RH (1980) J Biomed Mater Res 14: 107
31. Cater CW (1963) J Soc Leather Trades Chem 47: 259
32. Bowes JH, Cater CW (1964) J Appl Chem 14: 296
33. Hopwood D (1977) In: Bancroft JD, Stevens A (eds) Theory and practice of histological techniques. chap 2, Churchill Livingstone, Edinburgh
34. Richards FM, Knowles JR (1968) J Mol Biol 37: 231
35. Hardy PM, Nicholls AC, Rydon HN (1976) J.C.S. Perkin I, p 958
36. Hardy PM, Hughes, GJ, Rydon HN (1976) J.C.S. Perkin I, p 2282
37. Nimni ME, Cheung DT, Strates B, Kodama M, Sheikh K (1988) In: Nimni ME (ed) Collagen, Vol. III, Biotechnology, Chap 1, CRC Press, Boca Raton
38. Huang C (1974) Plysicochemical studies of collagen and collagen-mucopolysaccharide composite materials (model materials for skin). Sc.D. Thesis, Massachusetts Institute of Technology, Cambridge MA
39. Burke JF, Yannas IV, Quinby WC Jr., Bondoc CC, Jung WK (1981) Ann Surg 194: 413; Yannas IV, Burke JF, Warpehoski M, Stasikelis P, Skzabut EM, Orgill D, Giard DJ (1981) Trans Am Soc Artif Iutern Organs 27: 19–22
40. Heimbach D, Luterman A, Burke J, Cram A, Herndon D, Hunt J, Jordan M, McManus W, Solem L, Warden G, Zawacki B (1988) Ann Surg 208: 313
41. Kauzmann W (1959) Adv Protein Chem 14: 1
42. Baer E, Cassidy JJ, Hiltner A (1988) in Collagen, Vol II Biochemistry and biomechanics. Chap 9, CRC Press, Boca Raton
43. Yannas IV, Huang C, (1972) J Polym Sci A2(10): 577
44. Veis A (1964) The macromolecular chemistry of gelatin. Academic Press, New York
45. Huang C, Yannas IV (1977) J Biomed Mater Res Symp No 8 p 137
46. von Hippel PH, Harrington WF (1959) Biochim Biophys Acta 36: 427
47. Gordon PL, Huang C, Lord RC, Yannas IV (1974) Macromolecules 7: 954
48. Yannas IV, Sung N-H, Huang C (1976) J Phys Chem 72: 2935
49. Hall CE, Jakus MA, Schmitt FO (1942) J Amer Chem Soc 64: 1234
50. Wolpers C (1943) Klin Wschr 22: 624
51. Highberger JH, Gross J, Schmitt FO, (1950) J Amer Chem Soc 72: 3321
52. Highberger JH, Gross J, Schmitt FO (1951) Proc Natl Acad Sci USA 37: 286

53. Forbes MJ (1980) Cross-flow filtration, Transmission electron micrographic analysis and blood compatibility testing of collagen composite materials for use as vascular prostheses. M.S. Thesis, Massachusetts Institute of Technology, Cambridge, MA
54. Sylvester M, Yannas IV, Salzman EW (1989) Thromb Res 55: 135
55. Bear RS (1942) J Amer Chem Soc 64: 727
56. Treloar LRG (1975) The physics of rubber elasticity. 3rd Edition, Clarendon Press, Oxford
57. Wiederhorn NM, Reardon GV (1952) J Polym Sci 9: 4, 314
58. Flory PJ (1953) Principles of Polymer Chemistry. Cornell University Press, Ithaca
59. Flory PJ, Rehner J (1943) J Chem Phys 11: 512
60. Eastoe JE (1967) In: Ramachandran GN (ed) Treatise on Collagen, Vol 1, Chemistry of Collagen. chap 1, Academic Press, London
61. Yannas IV (1981) In: Dineen P (ed) The surgical wound. Chap 15, Lea & Febiger, Philadelphia
62. Dagalakis N, Flink J, Stasikelis P, Burke JF, Yannas IV (1980) J Biomed Mater Res 14: 511
63. Underwood EE (1969) Quantitative Stereology. Addison-Wesley, Reading
64. Fischmeister HF, (1973) In: Proc Int Symp RILEM/IUPAC Prague Final Report Part II, p C-439
65. Billingham RE, Medawar PB (1955) J Anat 89: 114
66. Billingham RE, Medawar PB (1951) J Exp Biol 28: 385
67. Lundborg G (1987) Acta Orthop Scand 58: 145
68. Lundborg G, Dahlin LB, Danielsen N, Gelberman RH, Longo FM, Powell HL, Varon S (1982) Exp Neurol 76: 361
69. Gross J, Lapiere CM (1962) Proc Natl Acad Sci USA 48: 1014
70. Woolley DE (1984) In: Piez KA, Reddi AH (eds) Extracellular matrix biochemistry chap 4, Elsevier, New York
71. Mainardi CL (1987) In: Uitto J, Perejda AJ (eds) Connective tissue disease. Chap 23, Marcel Dekker, New York
72. Yannas IV, Burke JF (1980) J Biomed Mater Res 14: 65
73. Yannas IV (1988) In: Collagen, Vol. III, Biotechnology Nimni ME, (ed) Chap 4, CRC Press, Boca Raton
74. Mandl I, Maclennan JD, Howes EL (1953) J Clin Invest 32: 1323
75. Yannas IV, Lee E, Bentz MD (1988) In: Applied bioactive polymeric materials. Gebelein CG (ed) Plenum, New York, p 313
76. Yannas IV, Burke JF, Huang C, Gordon PL (1975) Polym Prepr Am Chem Soc Div Polym Chem 16: 209
77. Tobolsky AV (1960) Properties and structure of polymers. John Wiley, New York
78. Yannas IV, Orgill DP, Skrabut EM, Burke JF (1984) In: Polymeric materials and artificial organs. Gebelein CG (ed) ACS Symp Series, No 256, Chap 13, Amer Chem Soc., Washington
79. Yannas IV, Lee E, Orgill DP, Skrabut EM, Murphy GF (1989) Proc Natl Acad Sci USA 86: 933
80. Yannas IV, Burke JF, Orgill DP, Skrabut EM (1982) Science 215: 174
81. Ferdman AG (1987) The measurement of collagen fiber orientation in tissue by small-angle light scattering. Ph.D. Thesis, Massachusetts Institute of Technology, Cambridge, MA
82. Yannas IV (1989) In: Cutaneous Development, Aging and Repair. Abatangelo G, Davidson JM (eds) Fidia Research Series, Vol. 18 Liviana Press, Padova, p 131
83. Yannas IV, Orgill DP, Silver J, Norregaard TV, Zervas NT, Schoene WC (1987) In: Advances in Biomedical Polymers. Gebelein CG (ed) Plenum, New York, p 1
84. Murphy GF, Orgill DP, Yannas IV (1990) Lab Invest 63: 305
85. Kessel RG, Kardon RH (1979) Tissues and organs: a text-atlas of scanning electron microscopy, WH Freeman, San Francisco
86. Van de Hulst HC (1957) Light Scattering by Small Particles, John Wiley and Sons, New York
87. Kronick P, Sacks MS (1991) Connective Tissue Research 27: 1
88. Guinier A, Fournet G (1955) Small-angle scattering of X-rays, John Wiley, New York
89. Hermans PH (1948) Contributions to the physics of cellulose fibers, Elsevier, Holland
90. Alexander LE (ed) (1969) X-ray Diffraction Methods in Polymer Science, John Wiley & Sons, New York
91. Ferdman AG, Yannas IV (1993) J Invest Dermatol 100(5): 710
92. Chapman JA, Hulmes DJS (1984) In: Ruggeri A, Motta PM (eds): Ultrastructure of the Connective Tissue Matrix, Martinus Nijhoff, Boston 1984, Chapter 1, Fig 1
93. Piez KA in Ref 5, Chapter 1, Fig 1.6
94. Piez KA in Ref 5, Chapter 1, Fig 1.22
95. Nimni ME, Harkness RD in Ref 2, chap 1, Fig 10
96. Stone KR, Rodkey WG, Webber RJ, McKinney L, Steadman JR (1990) Clin Orthop 252:

Biodegradable Polymer Scaffolds to Regenerate Organs

R. C. Thomson[1], M. C. Wake[1], M. J. Yaszemski[2], and A. G. Mikos[1,*]

[1] Cox Laboratory for Biomedical Engineering, Department of Chemical Engineering and Institute of Biosciences and Bioengineering, Rice University, P.O. Box 1892, Houston, TX 77251-1892, USA

[2] Department of Orthopaedic Surgery, Wilford Hall Medical Center, Lackland AFB, TX 78236, USA

The problem of donor scarcity precludes the widespread utilization of whole organ transplantation as a therapy to treat many diseases for which there is often no alternative treatment. Cell transplantation using biodegradable polymer scaffolds offers the possibility to create completely natural new tissue and replace organ function. Tissue inducing biodegradable polymers can also be utilized to regenerate certain tissues and without the need for in vitro cell culture. Biocompatible, biodegradable polymers play an important role in organ regeneration as temporary substrates to transplanted cells which allow cell attachment, growth, and retention of differentiated function. Novel processing techniques have been developed to manufacture reproducibly scaffolds with high porosities for cell seeding and large surface areas for cell attachment. These scaffolds have been used to demonstrate the feasibility of regenerating several organs.

---

* To whom correspondence should be addressed

1 Introduction

As technology progresses and human longevity increases, the medical profession is faced with an increasing number of organ failures. Not only do the patients and their families suffer, but the health care costs for treating patients suffering from tissue loss or organ failure are sizable. Complete organ failure has traditionally been treated by transplanting a healthy organ from a donor. Although organ transplantation has saved many lives, the harsh reality remains that the need for donor organs far outweighs the supply. In 1990, approximately 30 000 deaths resulted from chronic liver disease or cirrhosis but fewer than 3000 liver donors are available annually [1]. Finding a suitable donor is difficult and, once found, it is a costly coordinated effort to harvest and transport the required organ.

Scientists and physicians have sought a feasible alternative to organ transplantation. Materials such as metals, ceramics, and polymers have been used extensively to replace the mechanical function of hard tissues such as cartilage and bone [2]. Devices constructed primarily of polymers, such as artificial hearts and membrane oxygenators, have also been used to partially replace organ function, at least on a temporary basis [3, 4]. Tissue engineering may provide an alternative to existing therapies, which would allow the restoration of organ function by creating completely natural tissue, with the required mechanical or metabolic features, in vivo. Though still in its experimental stages, the use of tissue engineering techniques to regenerate organs has the potential to affect the quality and length of life of millions of people.

2 Tissue Engineering

Tissue engineering involves the creation of natural tissue with the ability to restore missing organ function. This may be achieved either by transplanting cells seeded into a porous material (Fig. 1a) or, in some cases, by relying on ingrowth of tissue and cells into such a material. The process by which regeneration is effected by ingrowth from surrounding tissue is known as tissue induction. Figure 1c shows how this process can be facilitated or enhanced by the release of chemotactic agents from the scaffold which attract cells to the area of regeneration. In either case, the porous material acts as a scaffold, which provides the cells with a substrate and organizes cell growth and tissue formation [5]. Since many cell types are anchorage-dependent (their function is dependent upon specific cell-substrate interactions), their direct transplantation without a scaffold results in cell death or loss of function [6].

Transplanted cells may be autogeneic (from the same genotype), allogeneic (from the same species but with a different genotype), or xenogeneic (from

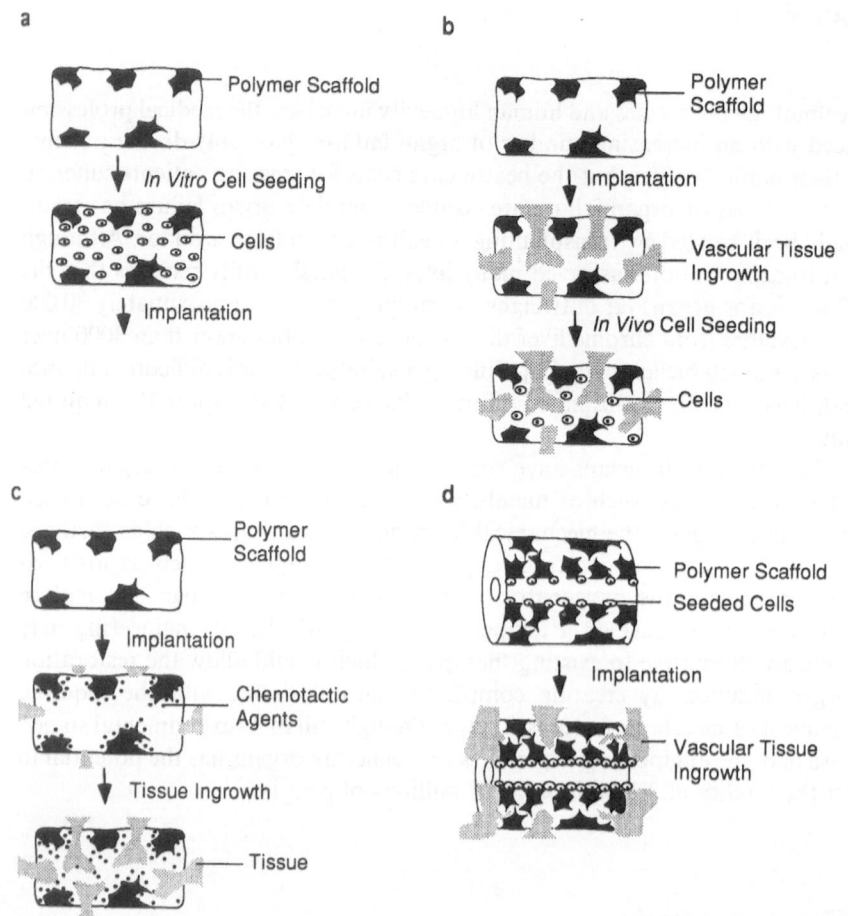

Fig. 1a–d. Schematic diagrams of various organ regeneration techniques: **a** cell transplantation into porous scaffolds to regenerate tissues such as cartilage; **b** prevascularization of scaffolds and subsequent cell seeding to regenerate metabolic organs such as liver; **c** tissue induction via release of chemotactic agents which attract the desired cells into a porous scaffold; **d** cell seeding into scaffolds with an annular space and subsequent tissue ingrowth to regenerate tubular tissues

a different species). Autogeneic cell transplantation is the most preferable choice since it avoids many of the problems associated with immune rejection of foreign tissue. Autogeneic cells may be used provided the cell type in question is capable of proliferation. In this case, a small number of harvested cells can be expanded in vitro until a cell mass sufficient to replace organ function is achieved. The cells may then be seeded into a scaffold and re-implanted into the host. If the failing organ function is a result of a genetic disease, autologous cell transplantation cannot be used directly. Advances in gene therapy may allow genetic abnormalities to be corrected in vitro in a small number of cells which may then be expanded and re-implanted [7]. However, gene therapy is still in its infancy and there are many problems to overcome before this type of treatment becomes a practical reality.

Allogeneic cell transplantation may provide a more readily attainable solution. In this instance, cells from a close relative of the patient would be expanded in vitro, seeded into a scaffold, and transplanted into the patient. Although rejection of the graft by the host's immune system is a risk, it may be reduced by administering immuno-suppressant drugs. If the cell type required to restore organ function is incapable of proliferation, xenogeneic cells may be used to achieve a sufficient cell mass for functional replacement. The successful use of xenogeneic cells hinges on the practicality of preventing graft rejection. There are several possible solutions to this problem, including immuno-suppression, alteration of cell surface proteins to reduce rejection potential, and encapsulation of cells in a polymer membrane. Extensive research has been performed with encapsulated xenogeneic pancreatic islets to treat insulin deficiencies in diabetics [8]. In this case, the polymer acts both as a cell substrate and as a permanent protective barrier to the immune system. Research has so far utilized mainly autologous cells. Such cells need only a temporary substrate, until they secrete their own in the form of extra-cellular matrix (ECM), and do not require a barrier to the immune system. As a result, there has been a great deal of research involving cell culture on biodegradable materials and development of techniques to manufacture degradable scaffolds.

Tissue engineering using cell transplantation has been studied to restore the function of tissues such as cartilage [9, 10], bone [11], skin [12, 13], nerve [14, 15], kidney [16], and liver [17]. This method of restoring organ function offers a number of advantages over whole organ transplantation. The two most important of these are: overcoming the problem of donor scarcity; and the reduced risk of rejection due to the use of autologous cells [2]. Cell transplantation for tissue and organ regeneration is an interdisciplinary field requiring working knowledge not only of biology but also of fundamental engineering, polymer science, medical principles, and surgical techniques. Increased knowledge and improved understanding of the basic premises involved in engineering tissues is the first step in enabling scientists to apply cell transplantation successfully to a variety of clinical disorders.

3 Design of a Scaffold for Cell Transplantation

There are many engineering and biological principles which must be taken into account when designing scaffolds for cell transplantation. Some are peculiar to the specific application while others are essential requirements which must be observed for all organs. The primary role of a scaffold is to provide a temporary substrate to which transplanted cells can adhere. Many cell types are anchorage dependent and require the presence of a substrate to retain their ability to proliferate and perform their specific or differentiated function. Cell adhesion is also an important factor in determining the rate of cell proliferation (if the cell type in question is capable of division). Strong cell adhesion to a substrate leading to cell spreading usually favors proliferation [18]. Rounded cell morphology, associated with moderate cell-substrate interaction, is generally indicative of a cell performing its differentiated function. A scaffold must therefore act as a substrate to promote cell adhesion and maintenance of differentiated function without hindering proliferation. The effectiveness of a given material in achieving these goals is dependent mainly on its surface chemistry, which determines the interaction between cell and substrate [19]. However, the function of many organs, e.g. liver, is dependent not only on cell morphology but also on the three-dimensional spatial relationship of cells and ECM [6]. A suitable scaffold for organ regeneration should therefore act as a template to organize and direct the growth of cells and the formation of ECM. It has been shown that porous scaffolds with well defined interconnected pore networks are capable of performing such an organizational role [17].

Restoration of organ function requires a large number of cells which must be accommodated in the scaffold. Highly porous scaffolds are desirable in this respect since they provide a large void volume into which transplanted cells may be seeded. Since many cell types are anchorage dependent, it is also important to provide a sufficient surface area for cell attachment. The volume of a cylindrical pore of fixed length varies as the square of its radius while the surface area is directly proportional to the radius. A high surface area per unit volume therefore favors the use of small diameter pores. One physical consideration limiting the minimum pore size which may be employed is the diameter of the cell. In order to seed cells into the scaffold, the pore size must be greater than the diameter of a cell in suspension (typically around 10 μm). Pores less than this diameter are therefore redundant since cells cannot enter them and hence cannot make use of the additional surface area created. A sufficiently large surface area may normally be attained using scaffolds of high porosity with small diameter pores ($10-10^3$ μm).

Any material proposed for implantation, whether for cell transplantation or some other application, must be biocompatible; i.e. it must not provoke an adverse response from the host's immune system. If this goal is not met the implant may be rejected. To this end it is important that the material be easily sterilized either by exposure to high temperatures, ethylene oxide vapor, or gamma radiation. A suitable material must therefore remain unaffected by one of these three techniques. However, biocompatibility is not simply a question of sterility. The chemistry, structure, and physical form of a material are all important factors which determine its biocompatibility. Although our understanding of the human immune system is advancing rapidly, it is not yet possible to predict the immune response to a new material. This can only be determined by in vivo experiments.

It is normally desirable to utilize biodegradable materials for organ regeneration. With such materials it is then also essential that the degradation products are biocompatible and may be removed from the body either through the respiratory system, via metabolic pathways, or through the renal system. Potential problems with long term biocompatibility are therefore avoided. The major advantage of utilizing biodegradable materials for organ regeneration is that they act only as temporary substrates which are eventually removed from the body leaving only natural tissue. Because tissues regenerate at different rates, it is highly desirable to utilize materials with degradation rates which may be tailored to suit the needs of a particular application. In many cases the material should be designed to degrade completely once sufficient ECM has been secreted to provide the cells with a natural substrate and an artificial scaffold is no longer required. However, in the case of load bearing tissues such as cartilage and bone, the scaffold must serve a further function; it must provide temporary mechanical support sufficient to withstand in vivo stresses. In such cases the material must be designed with a degradation rate such that the strength of the scaffold is retained until the regenerating tissue can assume this structural role. The shape of a hard tissue such as bone is often critical to its function. A suitable material in this case must be easily formed into a porous scaffold with any desired three-dimensional geometry.

Many highly metabolic organs such as liver and pancreas have high nutritional demands. The transplantation of sufficient cell mass for functional replacement of such organs may be limited by the diffusion of nutrients. Although an interconnected pore-structure enhances the diffusion rates to and from the center of the scaffold, transport of nutrients and waste products may not be sufficient until angiogenesis (the formation of new blood vessels) occurs. The creation of a vascularized bed may ensure the survival and function of seeded cells which have access to the vascular system for nutrition, gas exchange, and elimination of waste products. Vascularization of a scaffold may be effected by relying on capillary ingrowth into the pores from surrounding tissue. It may also be possible to control the degree and rate of vascularization by incorporating angiogenic or anti-angiogenic factors in the degrading scaffold material [20]. In this way it is possible to control the release of these factors to achieve the desired effect. Advances in controlled drug delivery systems which utilize biodegradable polymers have demonstrated the feasibility of such release mechanisms [3, 21]. The controlled release of bioactive molecules from degradable scaffolds need not be limited to angiogenic factors. Delivery of growth and differentiation factors could provide a powerful tool to aid the regeneration of organs.

4 Material Options for a Scaffold

The selection of a scaffold material is both a critical and difficult choice. There are many biocompatible materials available; metals, ceramics, and polymers.

Due to their wide range of physical properties, these materials have been utilized in numerous biomedical applications, from hip prostheses to artificial hearts. However, a scaffold material for organ regeneration must possess the following key characteristics: (1) surface properties promoting cell adhesion, proliferation, and differentiation; (2) biodegradable with a controllable degradation rate; (3) biocompatible with degradation products that can be excreted via physiologic metabolic pathways; (4) high surface area/volume ratio; (5) easily processed into three-dimensional shapes of irregular geometry; (6) mechanical properties sufficient to withstand any in vivo stresses. The criterion of biodegradability excludes the use of all metals and most ceramics as scaffold materials. Although biodegradable ceramic materials, such as tri-calcium phosphate and sea coral, have been used with some success mainly in orthopedic applications, they have two major limitations. First, they are difficult to process into porous materials with complex shapes, and second, it is currently not possible to control their rate of degradation. Polymers, on the other hand, are easily formed into any shape. There are many biocompatible polymers available and their usefulness as biomaterials is apparent by their numerous biomedical applications [4]. Both natural and synthetic degradable polymers have been utilized as scaffolds for organ regeneration. However, synthetic polymers offer greater versatility due to the ease with which their properties may be tailored to the needs of a specific application.

4.1 Natural Polymers

In the biomedical field there are several natural polymers which are used currently and many others which are under development [4]. The major advantage of natural over synthetic polymers is that they often have a highly organized structure, at both molecular and macroscopic levels, which imparts favorable characteristics to the polymer as a biomaterial, such as strength, or the ability to induce tissue ingrowth. Polypeptides with certain amino acid sequences can impart information important to cell adhesion and function, and therefore can be utilized as substrates for cell attachment and transplantation, or guided tissue ingrowth. However, one problem with these polymers is their antigenicity which, when the material is implanted into a host, stimulates an immune response leading to possible immune rejection. In addition, because the degradation of naturally occurring polymers almost always relies on enzymatic processes, there will inevitably be some patient to patient variation in the degradation rate depending on the activity of the specific degradative enzyme in each individual.

Natural polymers were the first to be used as scaffold materials for organ regeneration. Collagen fibers, bonded by glycosaminoglycans to form a matrix, have been used as a scaffold material for the regeneration of damaged skin in order to prevent fibrous scar formation during the healing process [13]. This is a special case of organ regeneration which relies on the migration of the desired cells into the scaffold from tissue adjacent to the implant. It is known as tissue induction and is discussed in detail later. Collagen/glycosaminoglycan substrates have been widely employed in this method of tissue reconstruction and have been successful in their application to cell seeded dressings for the treatment of severe burns in humans. A similar methodology using the same scaffold material has been used in animal studies for nerve regeneration.

4.2 Synthetic Polymers

Synthetic biodegradable polymers offer several advantages over natural polymers. Due to the fact that they are chemically synthesized, it is possible to control (with varying degrees of accuracy depending on, among other factors, the type of polymerization reaction) their molecular weight and molecular weight distribution. These characteristics have a profound effect on the physical properties of the polymer, such as strength and degradation rate. The degradation of synthetic polymers is, in general, brought about by simple hydrolysis, although in some cases enzymatic processes assist in the degradation mechanism [22]. Simple hydrolysis is desirable because the degradation rate does not vary from person to person. In addition, degradation rate data obtained in animal studies can be expected to be reproducible in humans. In the case of biodegradable copolymers, the degradation rate may be controlled by varying the ratio of the monomer units in the polymer [23]. In some instances, it is possible to copolymerize a non-degradable polymer monomer unit with a small proportion of a degradable polymer monomer unit [24]. This technique can be used to make an otherwise hydrolytically stable polymer degradable. An additional advantage of synthetic polymers is that they tend to be more easily processed into a finished product. This is particularly important in organ regeneration since it requires the use of a porous, degradable polymer scaffold with a large surface area and a defined pore morphology.

The homopolymers poly(L-lactic acid) (PLLA) and poly(glycolic acid) (PGA) as well as poly(DL-lactic-co-glycolic acid) (PLGA) copolymers are poly(α-hydroxy esters) (Fig. 2). They are biocompatible, biodegradable and are among the few biodegradable polymers with Food and Drug Administration (FDA) approval for human clinical use. PLLA, PLGA copolymers, and PGA are produced by a catalyzed ring-opening polymerization of lactide and/or glycolide [25]. Their degradation in vivo is brought about by simple, non-enzymatic hydrolysis of the polyester bond. Hydrolysis continues to break the polyester bonds at random sites along the polymer chains until, eventually, lactic acid and/or glycolic acid is produced. These acids occur naturally in the human body, and upon formation they are processed through normal metabolic pathways [22]. Elimination from the body is ultimately achieved through the respiratory system as carbon dioxide. PGA is a highly crystalline, hydrophilic poly(α-hydroxy ester) and was one of the first synthetic degradable polymers to find application as a biomaterial [26]. Its degradation rate is dependent on the crystallinity of the

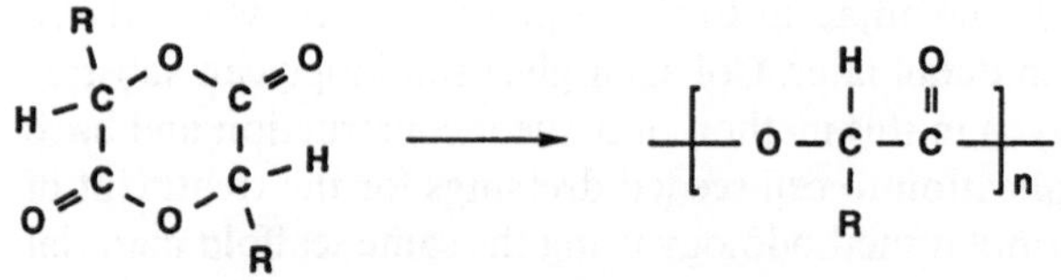

Fig. 2. Formation of poly(glycolic acid) and poly(L-lactic acid) by ring-opening polymerization of glycolide (R = H) and L-lactide (R = CH$_3$) respectively

polymer. One of the first applications of PGA was as a biodegradable suture material. PGA has also been used as an artificial scaffold in cell transplantation and organ regeneration. In all its applications, PGA, and its degradation products, have proven to be non-toxic and biocompatible, and hence this polymer has gained FDA approval for a number of applications.

PLLA is less crystalline and more hydrophobic than PGA, and therefore degrades at a slower rate [22]. The major degradation mechanism is simple hydrolysis, but there is a small but nonetheless significant contribution from in vivo enzymatic degradation [22]. The lower degradation rate of PLLA, as compared to PGA, is due to the presence of the methyl group, which gives PLLA its hydrophobic character and sterically hinders ester bond cleavage. PLLA is one of the strongest known biodegradable polymers and has therefore found many applications in areas, such as orthopedics, where structural strength is an important criterion [27]. The mechanical strength of PLLA allows its use as a scaffold material for both soft tissues such as liver, and hard, load bearing tissues such as cartilage and bone. A porous form of PLLA has been used as a scaffold for both chondrocyte and hepatocyte transplantation in the regeneration of cartilage [10] and liver [6] respectively. The surface chemistry and the pore morphology of this material are such that they provide an environment for cell attachment, growth, and differentiation.

PLGA copolymers, with glycolic acid contents between 25% and 70%, are amorphous [22] and are therefore not as strong as partially crystalline PLLA. However, by altering the copolymer ratio of lactic to glycolic acid, the degradation rate can be tailored to the requirements of the specific application. This attribute is especially useful in tissue engineering since the time the scaffold is required as a substrate for cells varies from one tissue to the next. PLGA copolymers have been utilized in intestine and liver regeneration. PLGA copolymers have also been employed in controlled drug delivery and therefore offer the possibility of delivering bioactive molecules in a controlled manner as the scaffold degrades.

PLLA, PLGA copolymers, and PGA have proven to be biocompatible materials and are FDA approved for several applications. However, one drawback to their use as scaffold materials for organ regeneration is the acidity caused by the release of lactic and glycolic acid, which at high concentrations becomes toxic to surrounding tissues. Initially, the amount of acid released during the course of degradation is very small but increases rapidly as the polymer is broken down to low molecular weight oligomers. However, a sudden rise in the acid concentration in vivo can render the local environment acidic and induce an inflammatory reaction or even tissue necrosis. Consequently, for tissue engineering applications, a constant and modulated release of acid at levels which can be accommodated by the adjacent tissues is desired. The feasibility of modulated lactic acid release has been demonstrated in vitro using blends of both low and high molecular weight PLLA [28]. Initially, as degradation of these blends proceeds, the low molecular weight PLLA chains form lactic acid while the high molecular weight PLLA form smaller polymer chains without significant lactic acid production. By the time the large chains begin to produce significant quantities of lactic acid, the low molecular weight PLLA has already completely degraded. Thus, the production of lactic acid is spread over the entire degradation period.

The range of physical properties and degradation rates which may be achieved using PLLA, PLGA copolymers, and PGA makes these synthetic degradable polymers extremely versatile as scaffold materials. In addition, their FDA approval for other biomedical applications and their proven ability to act as suitable substrates for several cell types has made these polymers attractive choices as scaffold materials to engineer many tissues. The development of processing techniques to prepare porous scaffolds with inter-connected pore networks using these materials has therefore become an important area of research.

5 Processing Methods to Prepare Scaffolds

The technique used to manufacture scaffolds for organ regeneration is dependent on the properties of the polymer and its intended application. Foaming agents are frequently used in industry to form porous polymers for packaging materials such as polystyrene. This technique, however, can only be used during fast polymerization or cross-linking reactions since it involves the transient formation of a liquid foam which must solidify quickly before it collapses. PLLA, PLGA copolymers, and PGA are linear, uncrosslinked polymers which polymerize slowly. They can therefore not be formed into porous materials using conventional techniques which utilize foaming agents. These polymers are commercially available in the form of small solid chunks or in some cases as long fibers. Scaffold manufacturing techniques were therefore developed using the commercially available starting materials. The techniques which were developed involved either heating the polymers or dissolving them in a suitable organic solvent. The viscous behavior of the polymers above their glass transition or melting temperatures and their solubility in various organic solvents were two of the more important characteristics to consider during process development.

The major requirement of any proposed scaffold processing technique is that it should utilize only biocompatible materials and should in no way affect the biocompatibility of the polymer. It should also allow the manufacture of scaffolds with controlled porosity and pore size since these are important factors in organ regeneration. When the shape of a regenerated tissue is important to its function the processing technique must allow the preparation of scaffolds with irregular three-dimensional geometries. A further design consideration is the incorporation of bioactive molecules into the polymer matrix. Inactivation of bioactive molecules by exposure to high temperatures or a harsh chemical environment, is the major obstacle to successful drug incorporation and delivery from a degradable scaffold. However, if localized controlled delivery of bioactive molecules to the site of growing tissue can be achieved it would provide a powerful tool to aid in organ regeneration.

5.1 Fiber Bonding

Fibers provide a large surface area to volume ratio and are therefore desirable as scaffold materials. One of the first biomedical uses of PGA was as a degradable suture material and it is therefore commercially available in the form of long fibers. PGA fibers in the form of tassels and felts were therefore utilized as scaffolds in organ regeneration feasibility studies [29]. However, these scaffolds lacked the structural stability necessary for in vivo use. A fiber bonding technique was therefore developed to prepare interconnecting fiber networks with different shapes for use as scaffolds in organ regeneration [30].

PLLA was dissolved in methylene chloride and the resulting polymer solution was cast over a non-woven mesh of PGA fibers in a glass container. Methylene chloride is a non-solvent for PGA. The solvent was allowed to evaporate and residual amounts were removed by vacuum-drying. A composite material was thus produced consisting of non-bonded PGA fibers embedded in a PLLA matrix. The PLLA-PGA composite was then heated to a temperature above the melting point of PGA for a given time period. During heating the PGA fibers join at their cross-points as melting commences but the two polymers do not, due to their immiscibility in the melt state. The composite was quenched to prevent any further melting of the PGA fibers during cooling. After heat treatment, the PLLA matrix of a PLLA-PGA composite membrane was selectively dissolved in methylene chloride and the resulting bonded PGA fibers were vacuum-dried. Using this technique, the fibers are physically joined without any surface or bulk modification and retain their initial diameter. The PLLA matrix is required to prevent collapse of the PGA mesh and to confine the melted PGA to a fiber-like shape. The heating time is also of critical importance since prolonged exposure to the elevated temperature results in the gradual transformation of the PGA fibers into spherical domains.

Bonded PGA fiber structures with high porosities and area/volume ratios were produced using this manufacturing method and utilizing only biocompatible materials (Fig. 3). In this respect, the structures produced were highly suitable as scaffolds for organ regeneration. In addition, they demonstrated structural integrity which was lacking in the PLGA tassels and felts used in organ regeneration feasibility studies. However, the stipulations concerning the choice of solvent, immiscibility of the two polymers, and their relative melting temperatures, restricts the general application of this technique to other polymers. In addition, the technique does not lend itself to easy and independent control of porosity and pore size.

An alternative method of fiber bonding has recently been developed which involves coating a non-bonded mesh of PGA fibers with solutions of PLLA or PLGA [31]. The mesh is attached to a rotating Teflon cylinder and is sprayed with an atomized polymer solution. As spraying continues, a coat of PLLA (or PLGA) builds up on the PGA fibers and forms bonds at their cross-points. Longer spraying times result in thickening of the PGA fibers due to PLLA deposition. Although a bonded PGA mesh is formed by this process, the surface of the scaffold presented to transplanted cell populations is a thin coat of PLLA or PLGA. Cell attachment, growth, and function will therefore be determined by the coating rather than the PGA mesh. This may be advantageous if the mechanical properties of PGA fibers are required but the surface properties of PLLA or PLGA are more desirable for the transplanted cell population. This method of fiber bonding does not address the problem of creating scaffolds with complex three-dimensional shapes but it has proven successful for producing hollow tubes which have been proposed for use in intestine regeneration.

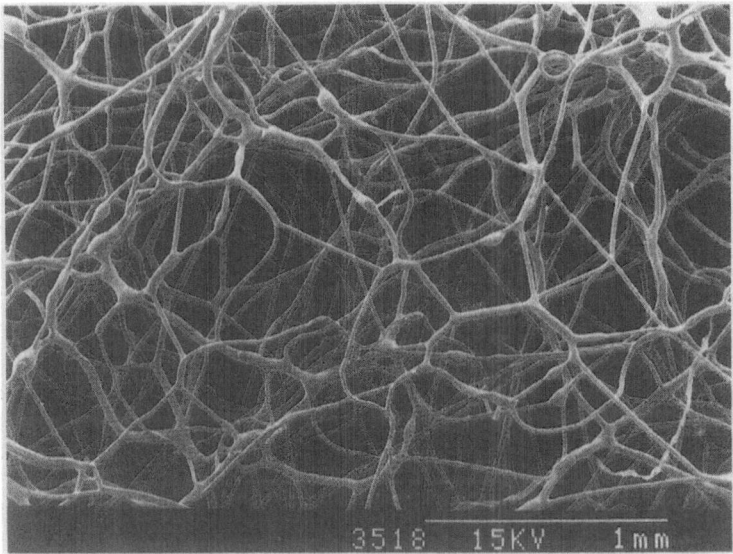

Fig. 3. Scanning electron photomicrographs of bonded PGA fiber meshes prepared by a fiber bonding method [30]. Note the formation of inter-fiber bonds at the fiber cross-points. (Reproduced with permission from [30])

5.2 Solvent-Casting and Particulate-Leaching

In order to overcome some of the drawbacks associated with the fiber bonding
preparation, a solvent-casting and particulate-leaching technique was developed
[32]. With appropriate thermal treatment porous constructs of synthetic biode-

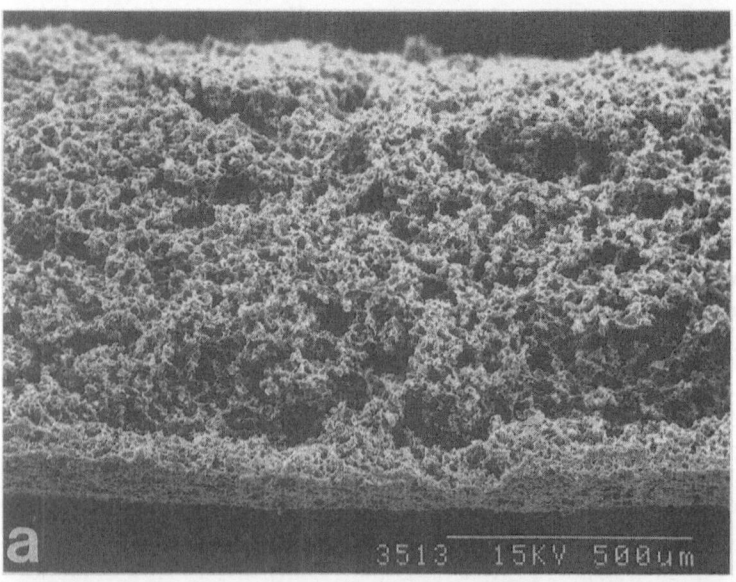

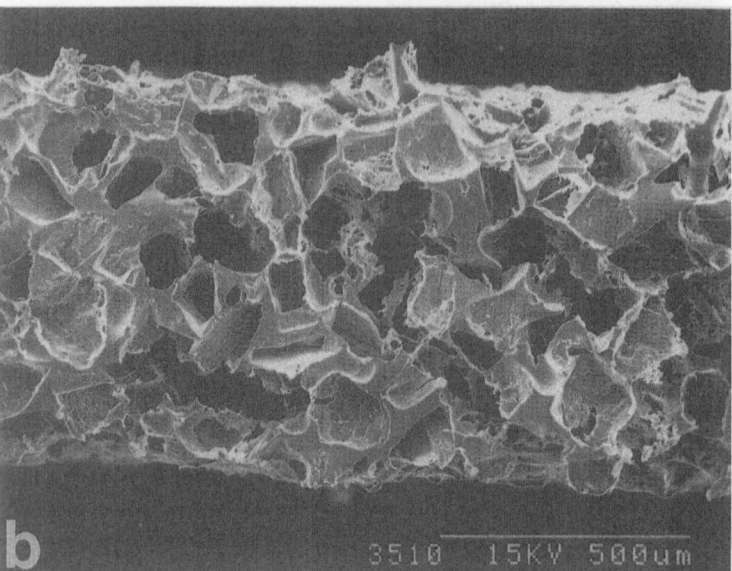

Fig. 4a, b. Scanning electron photomicrographs of amorphous poly(L-lactic acid) foams: **a** 92%
porosity and 30 μm median pore diameter; **b** and 91% porosity and 94 μm median pore diameter.
Prepared by a solvent-casting and particulate-leaching method [32] using 90 wt% sieved sodium
chloride particles of size range between 0–53 μm and 106–150 μm, respectively

gradable polymers were prepared with specific porosity, surface/volume ratio, pore size, and crystallinity for different applications. The technique was validated for PLLA and PLGA scaffolds but can be applied to any other polymer which is soluble in a solvent such as chloroform or methylene chloride.

Sieved salt particles were dispersed in a PLLA/chloroform solution and cast into a glass container. The salt particles utilized were insoluble in chloroform. The solvent was allowed to evaporate and residual amounts were removed by vacuum-drying. The resulting PLLA/salt composite membranes were heated at a temperature above the PLLA melting temperature to ensure complete melting of the polymer crystallites formed during the previous step. The melted PLLA membranes with dispersed salt particles were either annealed by cooling at a slow, controlled rate to produce semi-crystalline membranes with specific crystallinity or quenched to produce amorphous membranes. Finally, the membranes were immersed in distilled-deionized water to leach out the salt and the resulting salt-free PLLA membranes were then dried.

Highly porous membranes with an inter-connected pore structure were produced using this solvent-casting and particulate-leaching technique (Fig. 4a, b). The porosity of porous PLLA membranes could be controlled by varying the amount of salt used to construct the composite material (Fig. 5a).

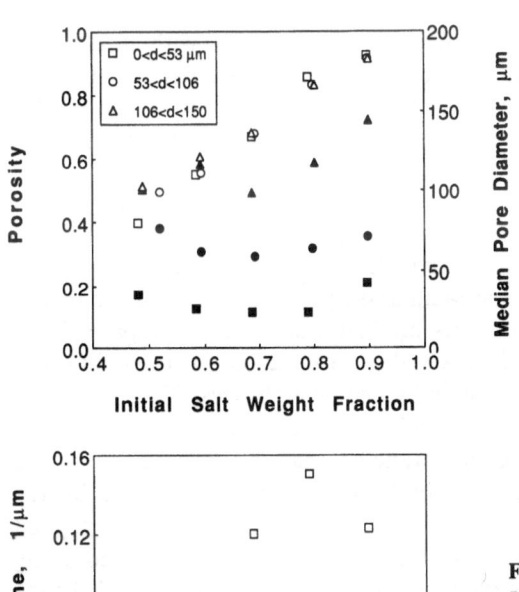

Fig. 5. a Porosity (*open symbols*) and median pore diameter (*filled symbols*) of semi-crystalline foams. **b** Surface/volume ratio of semi-crystalline PLLA foams. Prepared by a solvent casting and particulate-leaching method [32] using sodium tartrate particles, as a function of the initial salt weight fraction and salt particle size

The pore size of the membrane could also be controlled independently of the porosity by altering the size of the salt particles (Fig. 5a). Membranes with high surface area/volume ratios were produced and the ratio was dependent on both salt weight fraction and particle size (Fig. 5b). In addition, the crystallinity of PLLA membranes can be tailored to that desired for each application. These characteristics are all desirable properties of a scaffold for organ regeneration. The major disadvantage of this technique is that it can only be used to produce thin wafers or membranes (up to 2 mm in thickness). A three-dimensional scaffold cannot be directly constructed. This problem may be circumvented however, by membrane lamination.

5.3 Membrane Lamination

The construction of scaffolds with three-dimensional anatomical shapes is necessary for the regeneration of hard tissues such as cartilage and bone, whose function is partially dependent on geometry. Membrane lamination offers a means of constructing highly-porous biomaterials with anatomical shapes [33]. However, the lamination procedure is useful only if it preserves the uniform porous structure of the original membranes. Also, the boundary between two layers must be indistinguishable from the bulk of the device. The methodology to process biodegradable polymer scaffolds with anatomical shapes involves the construction of a contour plot of the particular three-dimensional shape. The shapes of the contours are cut from highly porous biodegradable membranes prepared using the solvent-casting and particulate-leaching technique described above. A small quantity of chloroform is then coated onto the contacting surfaces of adjacent membranes and a bond is formed. The desired three-dimensional shape is constructed layer by layer.

The porous structure of the PLLA and PLGA membranes used to validate the technique was unaffected by the lamination procedure, and no boundary between the layers was observed. This method was used to prepare three-dimensional anatomically shaped PLLA and PLGA scaffolds whose bulk properties were identical to those of the individual membranes. This technique cannot be applied to the lamination of PGA since it is only soluble in highly toxic solvents. A similar method of membrane lamination was used to produce tubular stents [34]. Flat porous scaffolds of PLGA were manufactured using the solvent-casting and particulate-leaching method. These were then wrapped around a Teflon cylinder and the overlapping edges were laminated using chloroform. The Teflon tube was then removed to leave an open-ended tube which was then capped by laminating end pieces. These constructs were proposed as scaffolds for the regeneration of tubular tissues such as intestine.

5.4 Melt Molding

Melt molding is an alternative method of constructing three-dimensional scaffolds which has many advantages over membrane lamination. PLGA scaffolds were produced by leaching PLGA/gelatin microsphere composites formed using a molding technique [35]. A fine PLGA powder was mixed with previously sieved gelatin microspheres and poured into a Teflon mold which was heated above the glass transition temperature of the polymer. The PLGA/gelatin microsphere composite was subsequently removed from the mold and placed in distilled-deionized water. The gelatin, being soluble in water, was leached out leaving a porous PLGA scaffold with a geometry identical to the shape of the mold (Fig. 6).

Using this method it is possible to construct PLGA scaffolds of any shape simply by changing the mold geometry. The porosity can be controlled by varying the amount of gelatin used to construct the composite material. The pore size of the scaffold can also be altered, independently of the porosity, by using microspheres of different diameters. In addition, because this technique does not utilize organic solvents and is carried out at relatively low temperatures, it has potential for the incorporation and controlled delivery of bioactive molecules. These molecules may be incorporated either into the polymer or into the gelatin microspheres. Gelatin has been used in the past as a vehicle for controlled delivery [4]. This scaffold manufacturing technique can also be applied to PLLA and PGA. However, higher temperatures are required because these polymers are semi-crystalline and must therefore be heated above their melting temperatures. The use of higher temperatures to mold these polymers may exclude the possibility of drug incorporation. It also entails the use of

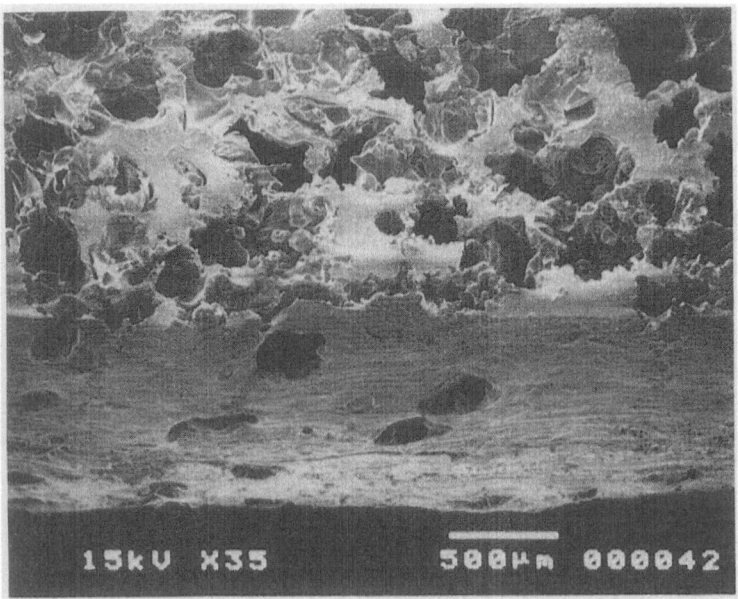

Fig. 6. Scanning electron photomicrograph of a PLGA 50:50 scaffold in the shape of a half-cylinder prepared by a melt molding method [35] using 35 wt% gelatin microspheres with diameters in the size range 300–500 μm. (Reproduced with permission from [35])

a leachable component other than gelatin since this material becomes insoluble in water after exposure to the required elevated temperatures. This manufacturing method satisfies many of the design criteria previously discussed and offers an extremely versatile means of scaffold preparation. Although gelatin microspheres were utilized to validate the feasibility of this technique, alternative leachable components such as salt or other polymer microspheres may be used.

6 Cell Seeding into Scaffolds

Seeding cells into a scaffold for the purpose of organ regeneration is one of the critical stages in cell transplantation. The seeding procedure must be atraumatic for the cells and should result in their even distribution throughout the scaffold. Since many of the polymers used as scaffold materials are hydrophobic, normal plating methods involving the application of a cell suspension to the surface of a porous scaffold are ineffective. Liquid culture medium is prevented from entering the pores by surface tension forces. One means of overcoming this problem is to pre-wet the polymer scaffold with a liquid which is both miscible with the culture medium and able to penetrate into the scaffold pores [36]. When highly-porous polymer scaffolds of PLLA, PLGA 85:15, and PLGA 50:50 were placed in ethanol, the pores were filled with liquid. (The ratios 85:15 and 50:50 designate the copolymer ratio of lactic acid:glycolic acid.) Subsequent immersion of the pre-wet polymer scaffolds in water resulted in ethanol dilution and eventual replacement as water diffused into the pores. Cells may then be easily seeded into the scaffold by plating cell suspensions onto the scaffold surface. This pre-wetting technique was used successfully in recent studies to seed PLLA scaffolds with chondrocytes [10]. Cell seeding into pre-wet polymer scaffolds was also achieved by injecting a cell suspension through a catheter positioned at the center of a scaffold [37]. Attainment of uniform cell distribution using this technique requires a careful choice of scaffold pore size and injection rate. In addition, the injection apparatus must be designed to prevent exposure of cells to high shear rates, which could result in cell damage.

Porous, biodegradable polymer scaffolds seeded with cells have been used to determine the feasibility of regenerating a number of organs. The goals and specific problems associated with regenerating several different organs are outlined below.

6.1 Cartilage

Current therapies for cartilage reconstruction or replacement include implantation of artificial prosthetic devices [2] and, occasionally, transplantation of osteo-chondral allograft to the injury site. In the latter case, the transplanted cartilage has its subchondral bone still attached, so bone to bone healing occurs at the transplantation site. Problems associated with artificial implants include increased risk of infection, failure of the implant at the host-prosthesis interface, and inability to remodel in response to changes to the local stress environment. Creation of completely natural cartilage at the site of injury would provide a means to circumvent such problems. Initial cartilage regeneration studies utilized biodegradable scaffolds of PGA fiber-based felt or short, unbraided sections of Vicryl sutures (PLGA 90:10 coated with calcium stearate and PLGA 70:30) [10, 38]. These materials were used to assess the feasibility of culturing chondrocytes on degradable polymer scaffolds. In vitro studies demonstrated that chondrocytes adhered readily to these scaffolds in multiple layers. Chondrocyte proliferation and maintenance of differentiated function were also observed during the six week culture period as polymer degradation proceeded. Differentiated function was indicated by the formation of a basophilic matrix within the scaffold and also by the presence of sulphated glycosaminoglycans (S-GAG), which serve as cross-links in normal cartilage. In vivo experiments showed that new cartilage could be formed by implanting these chondrocyte seeded degradable polymer scaffolds into mice. After seven weeks in culture the presence of type II collagen, indicative of cartilage formation, was observed. Gross examination of the implants revealed polymer degradation and long term implants appeared similar to normal human fetal cartilage with little evidence of polymer. It was also noted that the newly formed tissue maintained the approximate shape and dimensions of the scaffold. The important role played by the scaffold was demonstrated by the failure of subcutaneously injected chondrocytes to produce cartilage.

In vitro and in vivo culture of chondrocytes has also been carried out on PLLA scaffolds produced by the solvent-casting and particulate-leaching technique described earlier [10, 39]. The in vitro studies showed that chondrocyte proliferation and S-GAG production was much higher on PGA than PLLA scaffolds. Nevertheless, both PGA and PLLA scaffolds seeded with chondrocytes were successful in regenerating cartilage in vivo. SEM micrographs of PLLA scaffolds, produced by the solvent-casting particulate-leaching technique, after one month of chondrocyte culture show the formation of individual collagen fibers and networks (Fig. 7a, b).

6.2 Bone

Bone is a unique organ with respect to its ability to regenerate following injury or to remodel following changes in local stresses. However, if a severe injury occurs, the bone does not heal correctly, a fibrous nonunion results, and mechanical function is not restored. In such instances surgical intervention is required to aid the healing process. Existing therapies provide orthopedic surgeons with a number of practical alternatives to restore the structural

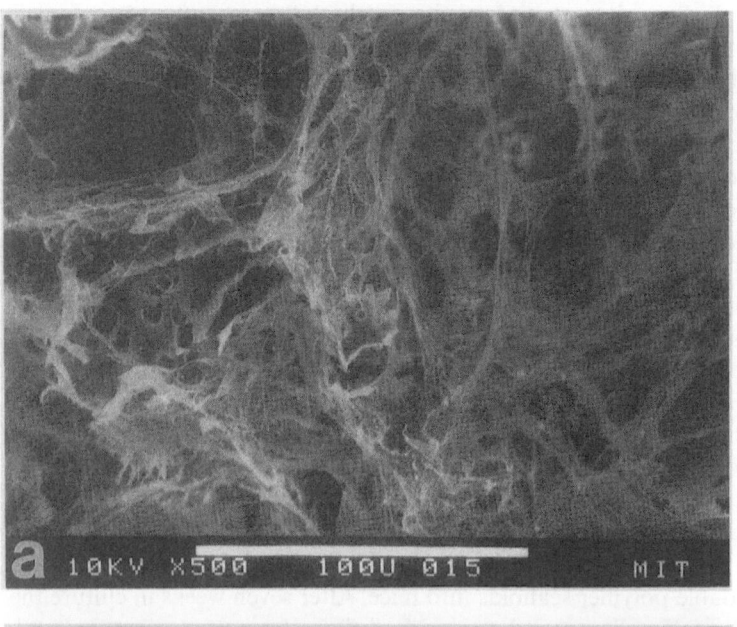

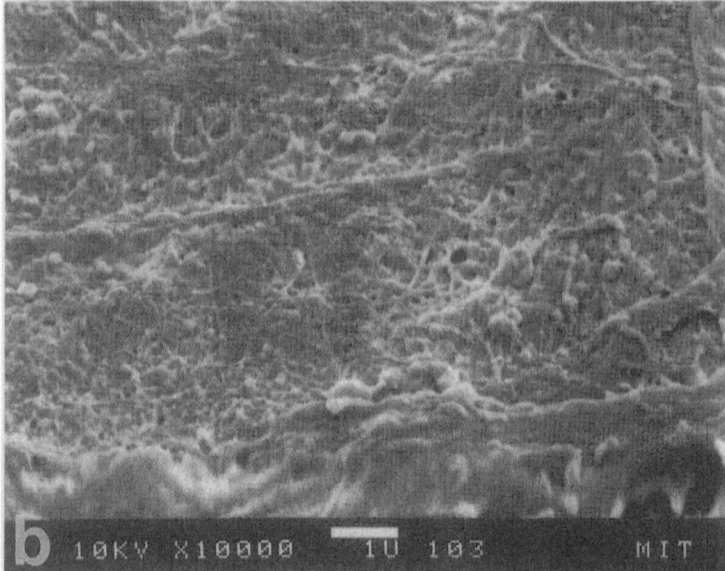

Fig. 7a, b. Scanning electron photomicrographs of PLLA foams of pore size 250–500 μm: **a** cross-section at low magnification after 30 days of chondrocyte culture; **b** and on the surface at high magnification after 28 days of chondrocyte culture (Reproduced with permission from [39])

function of bone; however, each has it drawbacks and limitations [2]. Due to the self-regenerative capacity of bone, there has been extensive research into the development of biomaterials which allow tissue induction from surrounding bone (see Sect. 8). The use of cell transplantation into degradable polymer scaffolds has recently been proposed as alternative therapy [40]. PLLA, PLGA copolymers, and PGA have been shown to be suitable biodegradable substrates for osteoblasts [40]. Osteoblasts cultured on films of these polymers were found to attach, proliferate and perform their differentiated function. Two-dimensional culture of osteoblast-like cells on polyphosphazenes has also been shown to be feasible [41]. Three-dimensional osteoblast culture has been carried out in vitro on a nonwoven PGA mesh [11]. Osteoblasts readily adhered to the polymer fibers, proliferated, and maintained their differentiated function. In vivo studies showed that cartilage formed during the first 6 weeks and that this was gradually replaced by bone over the following 14 weeks. Although the PGA meshes were implanted into non-load bearing sites, this study demonstrated the feasibility of this method of bone regeneration. Restoring the mechanical function of load-bearing bone using autologous cell transplantation will require the development of degradable scaffolds with improved mechanical properties [35]. These scaffolds must support in vivo loads until newly formed bone can assume this structural role.

6.3 Urothelium

Natural tissues, such as bowel, or synthetic material such as polytetrafluoroethylene, can be used for urothelial replacement in surgical reconstructive procedures. However, their success is limited by their inability to replace both structure and function. The feasibility of uroepithelial cell transplantation was therefore studied as a means of regenerating urothelium [42]. Nonwoven PGA meshes were seeded with rabbit uroepithelial cells and cultured in vitro for 1 to 4 days before being implanted into athymic mice. During the in vitro culture stage, cells were observed to attach to the PGA substrate. Vascular ingrowth was observed 5 days post-implantation and by day 30 multiple epithelial layers, attached to the PGA fibers, were present in some of the implants. In addition, the presence of a protein prominently expressed in rabbit bladder epithelium suggested that differentiated function was maintained even as polymer degradation occurred. These results demonstrated the feasibility of urothelial replacement using biodegradable polymers seeded with uroepithelial cells. Regeneration of tubular tissues using this technique will require the use of specially constructed scaffolds with an annular space, the inner-surface of which may be seeded with epithelial or endothelial cells. Figure 1d shows a schematic representation of tubular tissue replacement using cell transplantation.

6.4 Intestine

Following major resection of small bowel, malabsorption and malnutrition can occur. Replacement with allograft small intestine is a promising but experimental solution and its widespread use would be hindered by the problem of donor scarcity. The feasibility of intestine regeneration was first demonstrated using Vicryl fibers seeded with fetal rat intestinal cells [6]. Further studies were performed with nonwoven PGA fibers seeded with intestinal epithelial cells (enterocytes) obtained from adult rats [43]. In vitro, enterocytes attached to the PGA fibers 15 to 60 min after seeding. PGA fibers seeded with enterocytes were rolled around silastic stents to form tubular scaffolds which were then implanted into rats. Seven days after implantation, vascular ingrowth into the scaffolds had occurred and randomly oriented enterocytes were observed. In one specimen, enterocytes formed a stratified epithelium around the PGA scaffold. Intestinal regeneration is still in its early stages but its development could provide surgeons with an exciting alternative to existing techniques.

6.5 Nerve

One novel approach to nerve regeneration utilized hydrogels to entrap cells prior to implantation [15]. Astrocytes and undifferentiated neurones were entrapped within a porous non-degradable hydrogel matrix based on N-(2-hydroxypropyl) methacrylamide and methylene bisacrylamide copolymers containing collagen. This was achieved by polymerizing the crosslinked hydrogel around the cells. In vitro culture of the entrapped cells demonstrated viability, normal differentiation, and maturation. The ultimate goal of this technique is to be able to replace damaged brain tissue and to promote axonal regeneration. Although this is a long way from the current achievements, this technique has enormous potential as a means of neural tissue reconstruction using biodegradable matrices for cell entrapment.

6.6 Liver

Recent advances in medicine and surgery have seen an increase in the utilization of liver transplantation as a therapy for end-stage liver disease. The widespread use of this therapy, however, is limited by the shortage of suitable donors. The feasibility of liver regeneration using biodegradable polymer scaffolds was assessed by implanting hepatocytes grown on Vicryl tassels into rats [6]. Hepatocyte attachment to the polymer substrate and subsequent survival was observed at day seven post-implantation. In vitro studies with a PLGA 85:15 and a blend of this polymer with PLLA showed that hepatocytes attached to films of these materials [44]. Cell survival, and retention of differentiated function over the 5 day culture period was also demonstrated. In contrast, negligible cell attachment to PLLA films was observed. In vivo studies using nonwoven PGA mesh as a substrate showed that hepatocytes attached and survived on this biodegradable polymer substrate [29]. Although there was an initial loss of viable cells, clusters of hepatocytes remained after one week at which point vascular ingrowth was observed. The presence of albumin six months post-implantation indicated the retention of hepato-specific function over this period. It was also noted that more cells were evident at the periphery of the scaffold than toward its center. This suggests that cell survival was limited by the diffusion of essential nutrients from the vascular supply. Prevascularization of scaffolds to create a vascular bed prior to cell seeding was therefore postulated as a means of improving nutrient supply. Non-degradable poly(vinyl alcohol) (PVA) sponges were used to assess the feasibility of this approach [17]. These sponges were implanted for several days prior to hepatocyte seeding. The pre-seeding period resulted in the generation of a vascular bed with sufficient potential for nutrient transport to enable hepatocyte survival. Histological sections of the implant 4 days after hepatocyte injection showed the presence of vascular tissue and organized clusters of viable hepatocytes in proximity to the polymer.

7 Prevascularization of Scaffolds

The supply of essential nutrients to cells with high metabolic demands is an important consideration when attempting to culture large cell numbers within porous scaffolds. The presence of capillary networks in natural metabolic organs results in a short diffusion distance between the nutrient source in the blood stream and the cells. Liver regeneration studies showed that the survival and proliferation of transplanted hepatocytes was limited by the diffusion of necessary nutrients. Although the scaffolds became vascularized post-implantation, the rate of ingrowth was insufficient to prevent cell death at the scaffold center. This problem gave rise to the notion of prevascularizing the scaffold prior to the introduction of cells (Fig. 1b). To assess the feasibility of this approach, non-degradable poly(vinyl alcohol) scaffolds were implanted for several days prior to cell seeding [17]. The pre-seeding period resulted in the generation of a vascular bed with sufficient potential for nutrient transport to enable hepatocyte survival.

   The formation of vascular beds using biodegradable polymers has also been studied [45, 46]. Scaffolds of PLLA and PLGA copolymers manufactured using the solvent-casting and particulate-leaching technique were used in these experiments. The rate of vascular tissue ingrowth was found to increase as the porosity and/or the pore size of the implanted devices increased (Fig. 8). In addition, the time required for vascular tissue to fill the device was dependent on the polymer crystallinity and was shorter for amorphous polymers. Although prevascularization of scaffolds provides a means of improving nutrient transport, the in-

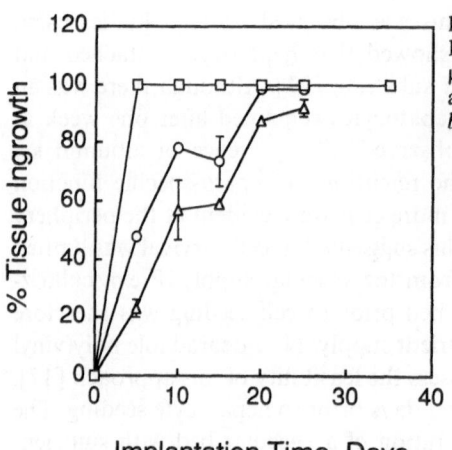

Fig. 8. Normalized tissue growth in amorphous PLLA devices of different pore sizes (~ 500 μm (□); 179 μm (○); and 91 μm (△)) as a function of implantation time [45]. The *error bars* designate means ± range of two sections

Implantation Time, Days

growth of vascular tissue limits the void space available for cell seeding. This is currently one of the major problems associated with scaffold prevascularization.

8 Tissue Induction

Tissue induction is the process by which ingrowth of tissue adjacent to a bio-material is effected. Vascularization of porous scaffolds is one example of tissue induction; however, the same ideas can be used to regenerate certain tissues. The requirements of a tissue inducing biomaterial are similar to those of a cell transplantation scaffold. In the case of tissue induction, however, challenges associated with cell seeding are traded for those involved in achieving migration of cells into the material. The material must be placed in contact with the tissue to be regenerated and must induce the migration and proliferation of the required cell types. It should also act as a substrate to maintain differentiated cell function and organize the growth of cells and formation of tissue. This approach has been employed to regenerate several tissues.

8.1 Skin

Although severely damaged skin is capable of healing, it does so by forming fibrous scar tissue. Collagen fibers, bonded by glycosaminoglycans to form a matrix, have been used as a tissue induction scaffold for the regeneration of damaged skin without scar formation [13]. In this application, the wound is covered with a microporous form of the matrix which provides many potential sites for cell attachment. The outer surface of the scaffold is covered with a thin silicone rubber sheet to control moisture loss. Over a period of several days, epithelial cells migrate into the scaffold from adjacent tissue and secrete ECM which forms the basis of a new, completely natural layer of skin. As the scaffold degrades, its supportive role is superseded by ECM produced by the epithelial cells until only natural skin remains. Collagen/glycosaminoglycan substrates have been widely employed in this method of skin reconstruction and have been successful in their application to cell seeded dressings for the treatment of severe burns in humans.

8.2 Nerve

A biomaterial construct, similar to that described for skin, was used to regenerate nerve [47]. Surgically produced gaps (15 mm in length) in sciatic nerve were bridged with a silicone tube filled with the same porous collagen matrix used in skin regeneration. Six weeks post-implantation extensive vascularization and axon formation was observed indicating new sciatic nerve formation. Controls, bridged only with silicone tubes, showed no evidence of nerve regeneration. The collagen scaffold is therefore essential to the induction of nerve.

Nonwoven PGA fibers in the form of a U-shaped duct with a flat cover were used to assess the effectiveness of such a device to repair median nerve [48]. At 6 and 12 months post-implantation, repair of transected nerves was observed and the PGA scaffold was completely resorbed with no deleterious effect to surrounding tissue. The degree of nerve repair, assessed both histologically and by electrophysiological function, could not be distinguished from that achieved using a standard microsurgical suture technique. The advantage of using tubular constructs as scaffolds, as opposed to suture techniques, is that nerve repair may be achieved even if some nerve tissue is lost.

8.3 Esophagus

Replacement of esophagus following its surgical resection may be achieved using autologous tissue grafts or synthetic biomaterials. However, the first of these alternatives involves multiple surgical procedures and the second often results in infection or leakage at the anastomosis between esophagus and biomaterial. The possibility of esophageal regeneration using a tissue inducing stent was therefore studied [49]. The stent consisted of a silicone tube whose outer surface was coated with a porous collagen sponge. When such stents were used to replace sections of excised esophagus in dogs, the collagen sponge was replaced by neoesophageal tubes after two weeks. Epithelial development, emanating from both biomaterial-esophagus anastomoses, was also evident at this time. Maturation of the new epithelium occurred over the following 2–3 weeks. This study showed the promise of tissue inducing biomaterials to regenerate esophagus; however, prevention of stenosis following removal of the silicone tube must be overcome before functional replacement may be achieved.

8.4 Anterior Cruciate Ligament

Braided and plied crosslinked collagen fibers were studied as anterior cruciate ligament (ACL) substitutes [50]. When these constructs were implanted into goats to replace host ACL, connective tissue ingrowth into the collagen constructs was observed. Only a small amount of collagen resorbed during the time course of the experiment; however, there was a rapid decline in the breaking strength of the construct. Although the decline of implant mechanical properties was too rapid for this application, this study demonstrated that connective tissue could be induced to grow into a degradable material.

8.5 Bone

Bone tissue has a rather unique capability among human tissues in response to injury or other changes that alter its mechanical requirements: it can continuously remodel itself to meet those mechanical needs best [51, 52]. For example, when a bone fractures, the body creates new bone to connect the broken fragments together, and then remodels the new bone to optimize its mechanical function in the particular region of the skeleton that the fracture has occurred. Human skeletal tissue has several functions, but, when a bone is fractured by trauma, or removed surgically due to disease or tumor formation, we only seek solutions so that its mechanical function can be carried out [53, 54].

The body usually accomplishes any necessary bone regeneration, and intervention is required when this system fails to work. When a bone fractures, the healing process begins immediately [55, 56]. The broken blood vessels in the vicinity of the fracture fill the space in and around the fractured bone ends with a fracture hematoma. New blood vessels grow into the region of repair, and bone forming cells migrate into the area. The cells (osteoblasts) secrete a nonmineralized substance called osteoid, which will become mineralized in an orderly fashion and become bone. The bone thus formed is called woven (or fibrous) bone. It is isotropic, and still needs to be remodelled by the sequential action of removal and redeposition along local stress lines. Osteoclasts are the multinucleated cells that accomplish the removal, and the osteoblasts redeposit bone where it is needed. This remodelling occurs continuously in the body, and results in the anisotropic structure of bone that is optimized for each area of the skeleton.

When this fracture repair system fails or is not expected to work, then an attempt must be made to induce bone regeneration for skeletal repair or reconstruction. The surgeon endeavors to make local skeletal conditions mimic those that the body would expect to encounter in a situation that requires fracture repair. In a very real sense, he or she tries to "trick" the body into initiating the fracture repair system. Current surgical options to accomplish this

task include transplantation of autograft or allograft bone to the site of reconstruction [57]. Autograft bone is bone from a different location in the same
person. Allograft bone is donated from deceased members of the same species as
the recipient. These are the most common methods of reconstruction. The
transplantation of bone to a skeletal region needing repair brings one of the
requirements for bone regeneration: a scaffold upon which new bone can grow.
The region of bone regeneration also needs cellular elements (osteoblasts), and
the bioactive molecules that cause the osteoblasts to synthesize and secrete
osteoid, the bone matrix.

The various tissue engineering strategies to regenerate bone supply the
region of skeletal repair with one or more of the factors necessary for bone
formation [58, 59]. A factor is considered osteoconductive if it will enable bone
to occur in a place and at a time that it otherwise would occur, given the proper
local environment. The various scaffold materials are osteoconductive, and
include autograft bone, allograft bone, tricalcium phosphate, hydroxyapatite,
polymer foams, and ceramics. A substance is considered osteoinductive if its
presence causes bone to occur at a time and in a place where it otherwise would
not occur. Factors that consist of a scaffold plus cells or bioactive molecules, or
both, are osteoinductive. There is the theoretical possibility that autograft bone
carries live osteoblast cells to the recipient site, and is thus osteoinductive. Three
dimensional polymer scaffolds seeded with cultured osteoblasts are osteoinductive. Polymeric biomaterials that contain and deliver bioactive molecules such
as bone morphogenic protein (BMP) or transforming growth factor beta (TGF-
β) are also examples of osteoinductive materials. These materials will ideally
impart adequate mechanical properties to the region of skeletal repair, encour-
age new bone to occur there, and then degrade to allow the regenerated skeletal
region to remodel along the lines of the local stress field. The materials should be
designed to degrade into nontoxic molecules that the body can excrete via
normal physiologic pathways.

Poly(methyl methacrylate) is a non-degradable bone cement which is widely
used to fill bone defects or provide fixation of prostheses to bone. However, it is
a biologically inert material and acts as a permanent barrier preventing fracture
healing or integration with host bone. A biodegradable bone cement would have
many potential applications. For a biodegradable bone cement to be useful it
must [60]: (1) initially be easily formed into the required shape and subsequently
harden in situ; (2) possess mechanical properties, once hardened, sufficient to
withstand in vivo stresses; (3) be osteoinductive and replaced by bone as it
degrades; (4) maintain its mechanical properties during degradation until newly
formed bone can provide adequate structural support. The development of
a biodegradable bone cement based on either natural or synthetic [60, 61]
polymers holds promise as a means of regenerating bone.

This discussion has introduced the problem of bone regeneration clinical
situations where the body's fracture repair system either has failed or has not
been induced to action. The overall strategy to address this problem includes

providing adequate initial mechanical properties to the region of skeletal repair, inducing the body to form bone there, and maintaining the mechanical properties while the newly formed bone gradually remodels and provides the mechanical strength to the reconstructed region.

9 Conclusions

Tissue regeneration may be achieved either by cell transplantation or by tissue induction. In either case, biodegradable scaffolds act as temporary substrates to which cells can adhere, proliferate, and retain their differentiated function. PLLA, PLGA copolymers, and PGA are biocompatible, biodegradable polymers which have proven to be suitable substrates for many cell types. Novel manufacturing techniques have been developed to process these polymers into porous scaffolds with large void volumes for cell seeding and sufficient surface area for cell attachment. The feasibility of organ regeneration using biodegradable polymer scaffolds as temporary cell substrates has been demonstrated for a number of different organs, including cartilage, liver, and nerve. There are many future challenges in this field which must be addressed before the goal of functional organ replacement can be achieved. Metabolic organ regeneration will require the creation of a vascular bed, which can provide an adequate nutrient supply to a large cell mass, while still leaving sufficient void space for cell seeding. Reconstruction of load bearing tissues will require the development of porous degradable scaffolds with sufficient mechanical strength to withstand in vivo loads. In the future, advances in polymer technology, controlled drug delivery, and cell biology may combine to make organ regeneration using biodegradable polymer scaffolds a practical alternative to whole organ transplantation.

Acknowledgements. This work was supported by grants from the National Science Foundation (BCS-9213197), the Orthopaedic Research and Education Foundation (93-017), and a T.N. Law Fund for Biotechnology Research. M.C. Wake would like to acknowledge the support of a National Science Foundation Graduate Fellowship.

10 References

1. U.S. Department of Health and Human Services, P.H.S., Center for Disease Control (1991) Monthly Vital Statistics Report, 39: 1–28
2. Ishaug SL, Thomson RC, Mikos AG, Langer R (in press). In: Meyers RA (Ed) Encyclopedia of Molecular Biology and Biotechnology, VCH, New York

3. Peppas NA, Langer R (1994) Science, 263: 1715–1720
4. Thomson RC, Ishaug SL, Mikos AG, Langer R (in press). In: Meyers RA (Ed) Encyclopedia of Molecular Biology: Fundamentals and Applications, VCH Publishers, New York
5. Langer R, Vacanti JP (1993) Science, 260: 920–926
6. Vacanti JP, Morse MA, Saltzman WM, Domb AJ, Perez–Atayde A, Langer R (1988) J Pediatr Surg, 23: 3–9
7. Wilson JM, Birinyi LK, Salomon RN, Libby P, Callow AD, Mulligan RC (1989) Science, 244: 1344–1346
8. Mikos AG, Papadaki MG, Kouvroukoglou S, Ishaug SL, Thomson RC (1994) Biotechnol Bioeng, 43: 673–677
9. Freed LE, Vunjak-Novakovic G, Marquis JC, Langer, R (1994) Biotechnol Bioeng, 43: 597–604
10. Freed LE, Marquis JC, Nohria A, Emmanual J, Mikos AG, Langer R (1993) J Biomed Mater Res, 27: 11–23
11. Vacanti CA, Kim W, Upton J, Vacanti MP, Mooney D, Schloo B, Vacanti JP (1993) Transplant Proc, 25: 1019–1021
12. Wilkins LM, Watson SR, Prosky SJ, Meunier SF, Parenteau NL (1994) Biotechnol Bioeng, 43: 747–756
13. Yannas IV (1988) in: Nimni ME (Ed) Collagen III, CRC Press, Boca Raton, Florida, pp 87–115
14. Bellamkonda R, Aebischer P (1994) Biotechnol Bioeng, 43: 543–554
15. Woerly S (1993) Biomaterials, 14: 1056–1058
16. Cieslinski DA, Humes HD (1994) Biotechnol Bioeng, 43: 678–681
17. Uyama S, Kaufmann PM, Tadeka T, Vacanti JP (1993) Transplantation, 55: 932–935
18. Mooney D, Hansen L, Vacanti J, Langer R, Farmer S, Ingber D (1992) J Cell Physiol, 151: 497–505
19. Healy KE, Lom B, Hockberger PE (1994) Biotechnol Bioeng, 43: 792–800
20. Folkman J, Klagsbrun M (1987) Science, 235: 442–447
21. Langer R (1990) Science, 249: 1527–1533
22. Gilding DK (1981) in: DF Williams (Ed) Biocompatibility of Clinical Implant Materials, CRC Press, Boca Raton, Florida, pp 209–232
23. Pulapura S, Kohn J (1992) J Biomater Appl, 6: 216–250
24. Sawhney AS, Pathak CP, Hubell JA (1993) Macromolecules, 26: 581–587
25. Vert M, Christel P, Chabot F, Leray J (1984) in: Hastings GW, Ducheyne P (Ed) Macromolecular Biomaterials, CRC Press, Boca Raton, Florida, pp 119–142
26. Benicewicz BC, Hopper PK (1991) J Bioact Compat Polym, 6: 64–94
27. Engelberg I, Kohn J (1990) Biomaterials, 12: 292–304
28. von Recum HA, Cleek RL, Eskin SG, Mikos AG (in press) Biomaterials
29. Cima LG, Vacanti JP, Vacanti C, Ingber D, Mooney D, Langer R (1991) J Biomech Eng, 113: 143–151
30. Mikos AG, Bao Y, Cima LG, Ingber DE, Vacanti JP, Langer R (1993) J Biomed Mater Res, 27: 183–189
31. Mooney DJ, Mazzoni CL, Organ GM, Puelacher WC, Vacanti JP, Langer R (1994) in: Mikos AG, Murphy RM, Bernstein H, Peppas NA (Ed) Biomaterials for Drug and Cell Delivery, MRS Symposium Proceedings, Vol 331, Materials Research Society, Pittsburgh, Pennsylvania, pp 47–52
32. Mikos AG, Thorsen AJ, Czerwonka LA, Bao Y, Langer, R, Winslow DN, Vacanti JP (1994) Polymer, 35: 1068–1077
33. Mikos AG, Sarakinos G, Leite SM, Vacanti JP, Langer R (1993) Biomaterials, 14: 323–330
34. Mooney DJ, Organ G, Vacanti JP, Langer R (1994) Cell Transplantation, 3: 203–210
35. Thomson RC, Yaszemski MJ, Powers JM, Mikos AG (in press) J Biomater Sci Polym Ed
36. Mikos AG, Lyman MD, Freed LE, Langer R (1994) Biomaterials, 15: 55–58
37. Wald HL, Sarakinos G, Lyman MD, Mikos AG, Vacanti JP, Langer R (1993) Biomaterials, 14: 270–278
38. Vacanti CA, Langer R, Schloo B, Vacanti JP (1991) Plast Reconstr Surg, 88: 753–759
39. Wald HL (1989) Chondrocyte Culture on Biodegradable Polymer Substrates. M.S. Thesis Massachusetts Institute of Technology
40. Ishaug SL, Yaszemski MJ, Bizios R, Mikos AG (1994) J Biomed Mater Res, 28: 1445–1453
41. Laurencin CT, Norman ME, Elgendy HM, El–Amin SF, Allcock HR, Pucher SR, Ambrosio, AA (1993) J Biomed Mater Res, 27: 963–973
42. Atala A, Vacanti JP, Peters CA, Mandell J, Retik AB, Freeman MR (1992) J Urol, 148: 658–662

43. Organ GM, Mooney DJ, Hansen LK, Schloo B, Vacanti JP (1992) Transplant Proc, 24: 3009–3011
44. Cima LG, Ingber DE, Vacanti JP, Langer R (1991) Biotechnol Bioeng, 38: 145–158
45. Wake CM, Patrick CW, Mikos AG (1994) Cell Transplantation, 3: 339–343
46. Mikos AG, Sarakinos G, Lyman MD, Ingber DE, Vacanti JP, Langer R (1993) Biotechnol Bioeng, 42: 716–723
47. Yannas IV, Lee E, Orgill DP, Skrabut EM, Murphy GF (1989) Proc Natl Acad Sci USA, 86: 933–937
48. Tountas CP, Bergman RA, Lewis TW, Stone HE, Pyrek JD, Mendenhall HV (1993) J Appl Biomater, 4: 261–268
49. Natsume T, Ike O, Okada T, Takimoto N, Shimizu Y, Ikada Y (1993) J Biomed Mater Res, 27: 867–875
50. Chvapil M, Speer DP, Holubec H, Chvapil TA, King DH (1993) J Biomed Mater Res, 27: 313–325
51. Cowin SC, Moss-Salentijn L, Moss ML (1991) J Biomech Eng, 113: 191–197
52. Rubin CT, Hausman MR (1988) Rheum Dis Clin N Amer, 14: 503–517
53. Gibson LJ (1985) J Biomech, 18: 317–328
54. Carter DR, Hayes WC (1976) Science, 194: 1174–1176
55. Sandberg MM, Aro HT, Vuorio EI (1993) Clin Orthop Related Res, 289: 292–312
56. Yaszemski MJ, Payne RG, Hayes WC, Langer RS, Mikos AG (submitted) Biomaterials
57. Gross TP, Cox QGN, Jinnah RH (1993) Orthopedics, 16: 895–900
58. Einhorn TA, Majeska RJ (1991) Clin Orthop Related Res, 262: 286–297
59. Covey DC, Albright JA (1989) Orthop Rev, 18: 857–863
60. Gerhart TN, Miller RL, Kleshinski SJ, Hayes WC (1988) J Biomed Mater Res, 22: 1071–1082
61. Yaszemski MJ, Payne RG, Hayes WC, Langer RS, Aufdemorte TB, Mikos AG (in press) Tissue Engineering

Author Index Volumes 101-122

Subject Index

Springer-Verlag
and the Environment

We at Springer-Verlag firmly believe that an international science publisher has a special obligation to the environment, and our corporate policies consistently reflect this conviction.

We also expect our business partners – paper mills, printers, packaging manufacturers, etc. – to commit themselves to using environmentally friendly materials and production processes.

The paper in this book is made from low- or no-chlorine pulp and is acid free, in conformance with international standards for paper permanency.